Marine Engines Performance and Emissions

Marine Engines Performance and Emissions

Editor

María Isabel Lamas Galdo

MDPI • Basel • Beijing • Wuhan • Barcelona • Belgrade • Manchester • Tokyo • Cluj • Tianjin

Editor
María Isabel Lamas Galdo
Universidade da Coruña,
A Coruna
Spain

Editorial Office
MDPI
St. Alban-Anlage 66
4052 Basel, Switzerland

This is a reprint of articles from the Special Issue published online in the open access journal *Journal of Marine Science and Engineering* (ISSN 2077-1312) (available at: https://www.mdpi.com/journal/jmse/special_issues/marine_engines).

For citation purposes, cite each article independently as indicated on the article page online and as indicated below:

LastName, A.A.; LastName, B.B.; LastName, C.C. Article Title. *Journal Name* **Year**, *Volume Number*, Page Range.

ISBN 978-3-0365-0964-8 (Hbk)
ISBN 978-3-0365-0965-5 (PDF)

Contents

About the Editor

María Isabel Lamas Galdo has a Ph.D. in Industrial Engineering, which she obtained in 2013. She has worked as an associate professor at the Higher Polytechnic University College of University of Coruña, Spain, since 2008. Moreover, she collaborated with the Norplan Engineering S.L. Her main interest is on computational fluid dynamics (CFD), specially applied to thermal systems, marine engines, and pollution reduction. She has written several books. She is the author of dozens of publications in ISI journals and she has participated in several national and international conferences and research projects. She also has experience in engineering projects.

Preface to "Marine Engines Performance and Emissions"

Marine engines are key components in most ships. Nowadays, diesel engines power most of the ships in the world. These engines are efficient in comparison with other thermal machines, but emit harmful species such as nitrogen oxides, soot, carbon dioxide, sulfur oxides, carbon monoxide, etc. Several international, national, and regional policies have been developed to limit pollutants from engines. Due to these increasingly restrictive regulations, several pollution reduction methods have been developed in recent years. Besides, the performance of marine engines is vital for the efficiency, environment, and safety. According to this, innovative solutions are being increasingly developed in the recent years. This book contains a collection of peer-review scientific papers about developments in the research of marine engines performance and emissions. Innovative researchers are provided. These papers were developed by 39 authors from The Netherlands, Finland, Sweden, Korea, Slovakia, China, Chile, Ukraine, Poland, and Spain. Most papers treat about emissions from marine engines, but the performance is also treated.

María Isabel Lamas Galdo
Editor

Journal of
Marine Science and Engineering

Article

Reduction of the Gaseous Emissions in the Marine Diesel Engine Using Biodiesel Mixtures

Michal Puškár [1,*], Melichar Kopas [1], Dušan Sabadka [1], Marek Kliment [1] and Marieta Šoltésová [2]

[1]   Faculty of Mechanical Engineering, Technical University of Košice, Letná 9, 042 00 Košice, Slovakia; melichar.kopas@tuke.sk (M.K.); dusan.sabadka@tuke.sk (D.S.); marek.kliment@tuke.sk (M.K.)

[2]   Faculty of Mining, Ecology, Process Control and Geotechnology, TU Košice, Park Komenského 19, 040 01 Košice, Slovakia; marietta.soltesova@tuke.sk

*   Correspondence: michal.puskar@tuke.sk

Received: 17 April 2020; Accepted: 2 May 2020; Published: 8 May 2020

Abstract: Taking into consideration the quality of air, it is necessary to ensure a continued reduction of the gaseous emissions that are produced by the maritime transport. The most effective solution of this serious worldwide problem is application of a suitable fuel mixture, which contains a bio-component, i.e. the biofuel. The presented scientific study is focused on influence of the biofuels on production of the gaseous emissions in the case of a diesel auxiliary engine, which is used in the ship transport. There were created various fuel mixtures with different content of the bio-component in order to investigate their emission characteristics. The individual experimental measurements were performed at the different engine loading levels and using a variable engine speed spectrum. The obtained results demonstrated a significant influence of the fuel mixtures on the whole combustion process, on the heat release process, on the pressure time behaviour as well as on the engine emission characteristics.

Keywords: reduction; gaseous emissions; biodiesel mixtures

1. Introduction

It is a well-known fact that a large amount of the international transport of various goods and materials is carried out by the sea. However, it is also necessary to emphasize such reality that the sea transport is characterized by a high fuel consumption as well as by production of a large volume of the harmful gaseous emissions. There is often neglected a relevant fact that the ship transport causes more serious and more dangerous air pollution than the standard motor vehicles. With regard to these negative circumstances, the International Maritime Organization (IMO) determined the new internationally valid regulations in order to reduce the increasing amount of the most dangerous air pollutants, namely the sulphur oxides and the nitrogen oxides.

The diesel engines installed in the trans-continental ships are burning the heavy fuel oils (HFO) both in the main driving engines and in the auxiliary diesel engines. This kind of fuel produces a large amount of the harmful gaseous emissions. In the view of the above mentioned facts, it is necessary to look for other solutions, such as biodiesel. Biodiesel is a renewable natural fuel source with only a minimum content of sulphur and aromatic hydrocarbons, as well as it is characterised by a high value of the cetane number and biodegradability [1–3]. At the same time, it does not require constructional modification of the diesel engine itself.

Various studies were elaborated in order to investigate the real environmental impacts of the gaseous emissions caused by biodiesel fuel. The results gained from the performed studies show a positive fact that the values of sulfur, CO emissions, unburned hydrocarbons, and particulate matter were significantly reduced in the exhaust gases compared to the standard diesel fuels. Biodiesel is soluble in normal diesel oil and for that reason it can be combined with diesel oil in any proportion. All the studies concerning application of biodiesel fuels were focused only on diesel engines working in motorcars [4–10]. However, it is also necessary to investigate gaseous emissions which are emitted from the auxiliary diesel engines installed in ships as diesel-powered electric generators. Another considerable problem is a lack of the scientific research studies describing the relation between the NO_2 and NO emissions that are caused by combustion of the fuel mixture "biodiesel–diesel oil" [11–14].

The main task of the performed research was to analyse the influence of various mixtures of biodiesel and diesel fuels on the combustion process and on the NO_X emissions in the case of diesel combustion engines that are installed in the ships.

2. Experimental Engine and Measuring Specifications

2.1. Experimental Engine and Diesel Mixtures

The testing experimental engine, which was applied during realisation of the experiments, is an auxiliary six-cylinder diesel engine (Figure 1). The technical data of this engine are presented in Table 1. The individual diesel fuel mixtures were created by mixing of the ULSDF (Ultra Low Sulphur Diesel Fuel) with the biodiesel using various mixing ratios. The tested engine was connected to a test stand [15–17].

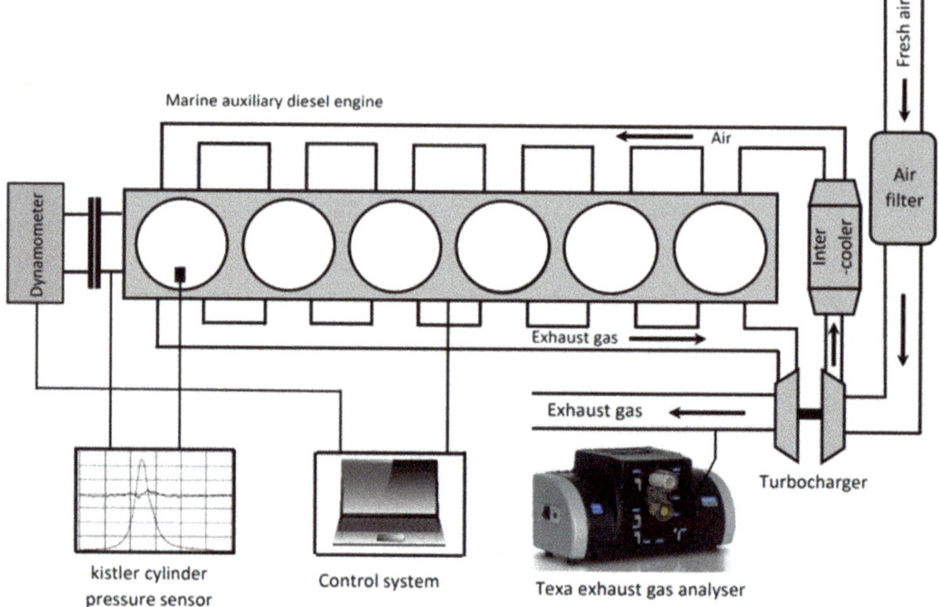

Figure 1. Experimental engine.

Table 1. Technical data of the engine.

Model	6TWGM
Type	6 cylinders, direct injection
Compression ratio	17.0:1
Bore × stroke	100×127 mm
Displacement	5.99 L
Rated output	125.5 kw
Engine speed	1500 rpm
Air intake way	Turbocharged, air to water cooled

The ULSDF, which was used during testing, contained less than 10 ppm of sulfur. Parameters of the applied biodiesel are in accordance with the European standard EN 14214 (Table 2).

Table 2. The ULSDF and biodiesel specifications.

Specification	Biodiesel	ULSDF
Density, 15 [kg/m^3]	860–900	810–850
Viscosity (40 °C) [mm^2/s]	3.5–5.0	2.0–4.0
Cetane number	Min.51	Min.48
Sulphur content [mg/kg]	<10	<10
Heat of evavoration [kJ/kg]	250–290	282–338
Flash point [°C]	101	82
Carbon content	81.5	97.1

The tested fuels were obtained by mixing of the ULSDF with the biodiesel, applying various mixing ratio values, namely, 0%, 30%, 70%, and 100%, using the following marking: BU (0:100), BU (30:70), BU (70:30), and BU (100:0), respectively, whereby the letter B means biodiesel and the letter U means the ULSDF. There were also tested diesel engine fuels with other ratios between the biodiesel and the ULSDF, but with regard to the main goal of this study, the above-mentioned fuel mixing ratios were the most suitable. The first type of the tested fuel was the pure ULSDF and the last tested fuel was the pure diesel oil. The fuel consumption of the tested engine was measured for each one of the fuels. The operational conditions during the measuring process are summarized in Table 3.

Table 3. Fuel consumption (FC).

Engine Speed	Engine Loading	ULSDF FC [kg/h]	B30:U70 FC [kg/h]	B70:U30 FC [kg/h]	Biodiesel FC [kg/h]
	267 Nm (30%)	11.50	11.98	12.47	13.25
1000 rpm	534 Nm (60%)	18.93	19.72	20.99	21.91
	801 Nm (90%)	26.85	27.98	28.83	29.71
	321 Nm (30%)	14.52	14.84	15.84	16.59
1500 rpm	642 Nm (60%)	24.68	25.03	26.64	27.12
	963 Nm (90%)	31.72	32.98	34.70	35.48

2.2. Conditions of Measuring Process

The pressure inside the cylinder was measured using the Kistler measuring equipment, which contains a pressure converter mounted in the first cylinder of the given engine [18–21].

The pressure sensor was connected with the amplifier as well as with the sensor of crankshaft angular displacement. This arrangement enabled measuring the pressure values in the cylinder within the 0.5-degree range of the crankshaft angular displacement. The emissions of NO_X were monitored by means of a special measuring system, which was intended for recording of the emissions [22,23].

2.3. Methodology of Measuring Process

The tested engine was operated at two stabilized engine speeds: 1000 rpm or 1500 rpm and at different engine loading levels (Table 3). The cooling water temperature was maintained between 75 °C and 85 °C. The lubricating oil temperature was kept between 90 °C and 100 °C, depending on the operational conditions. The values of engine fuel consumption, exhaust gas temperature and NO_X emissions were continuously monitored in three-minute intervals, whereby the measured results were averaged.

3. Results and Discussion

3.1. Influence of Various Fuel Mixtures on Output Characteristics

The average values of the measured data for various fuel mixtures are given on Figures 2–7. More than 100 measuring cycles were performed in order to reduce in this way an influence of random changes during the experiment [24–26]. It is visible from Figures 4 and 7 that the value of pressure inside the cylinder decreased with higher engine operational loading in the case of increasing proportion of biodiesel in the experimental fuel. The curves, which describe the intensity of heat generation, are various in the operational regime utilizing the ULSDF and the biodiesel fuel mixture. The pressure value inside the cylinder grew more rapidly for the fuel mixture biodiesel–ULSDF compared to the clean ULSDF. The cetane number that belonged to the biodiesel and also the value of flame speed were higher in comparison with ULSDF. The speed of heat release was reduced with the increasing amount of biodiesel in the tested fuel mixtures and the ignition delay decreased if the volume of the biodiesel in the tested fuel was higher.

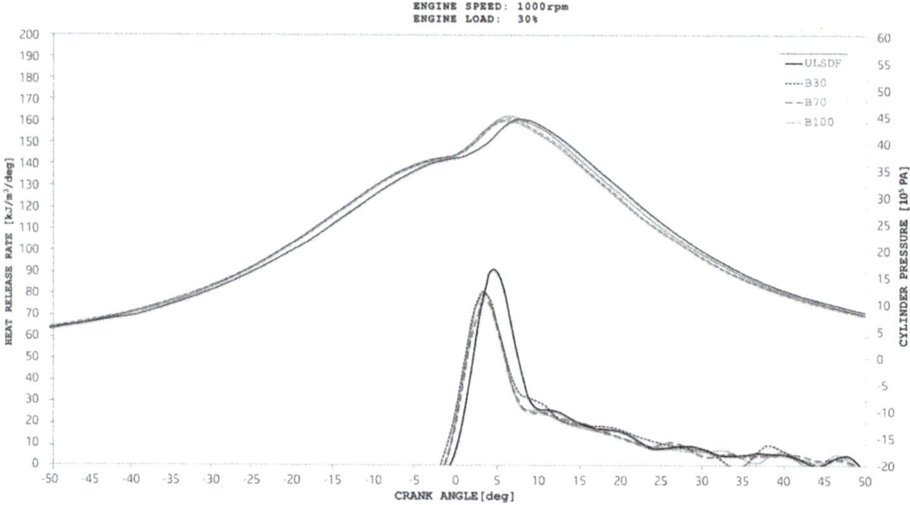

Figure 2. The dependence of fuels on the heat release rate and cylinder pressure (1000 rpm/30%).

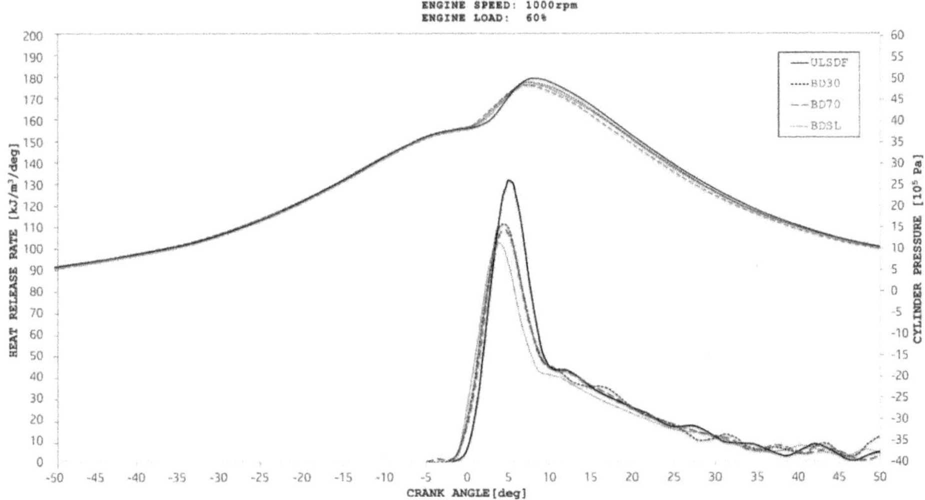

Figure 3. The dependence of fuels on the heat release rate and cylinder pressure (1000 rpm/60%).

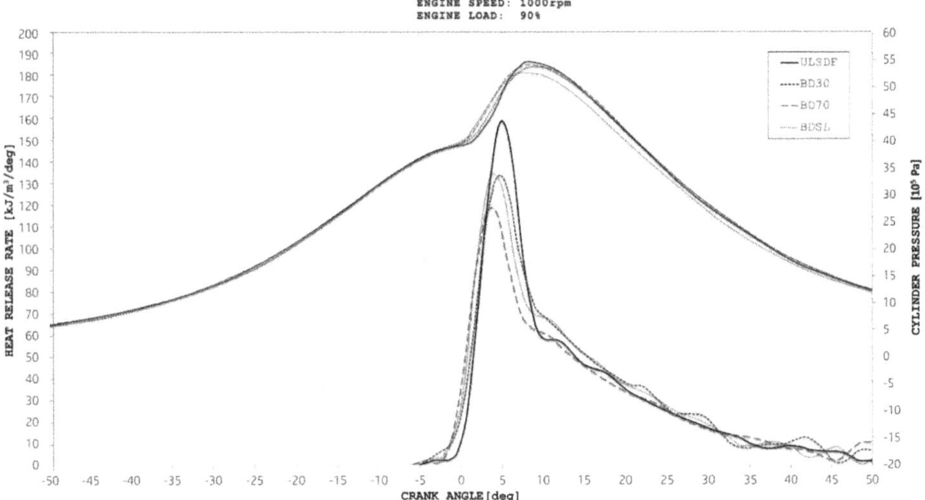

Figure 4. The dependence of fuels on the heat release rate and cylinder pressure (1000 rpm/90%).

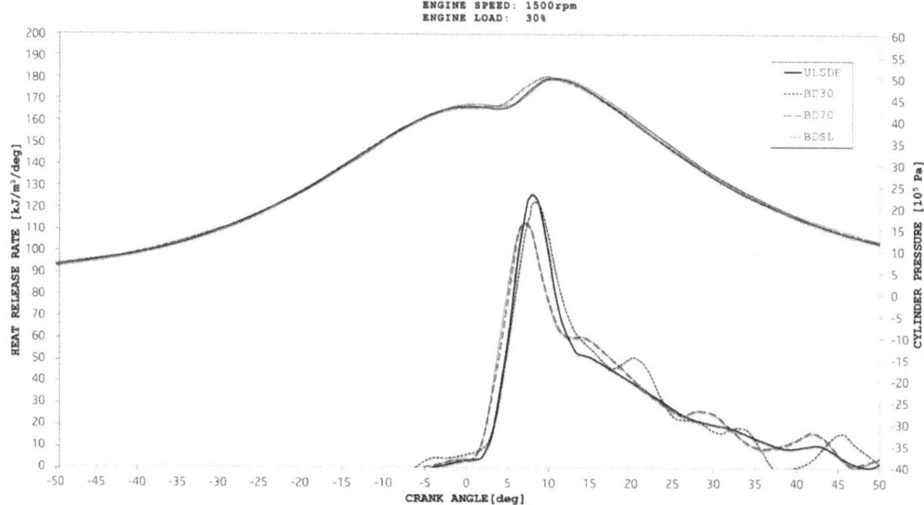

Figure 5. The dependence of fuels on the heat release rate and cylinder pressure (1500 rpm/30%).

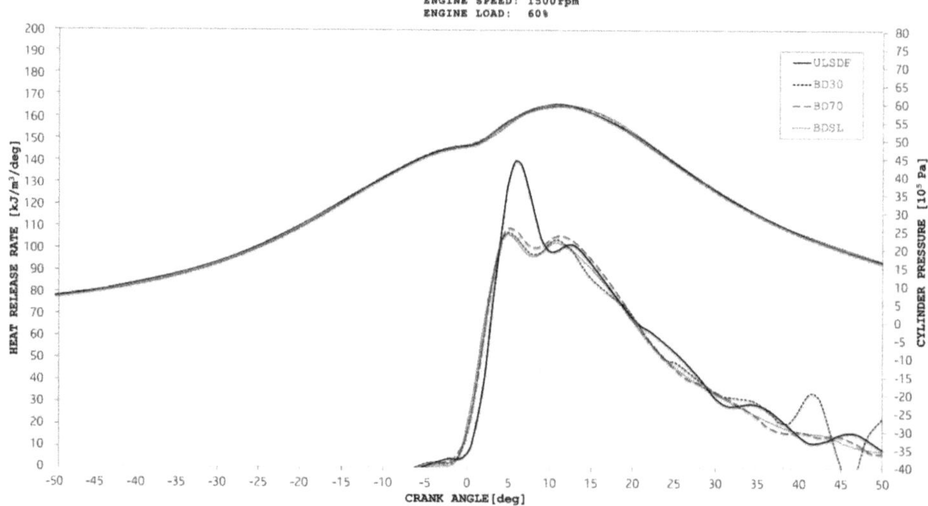

Figure 6. The dependence of fuels on the heat release rate and cylinder pressure (1500 rpm/60%).

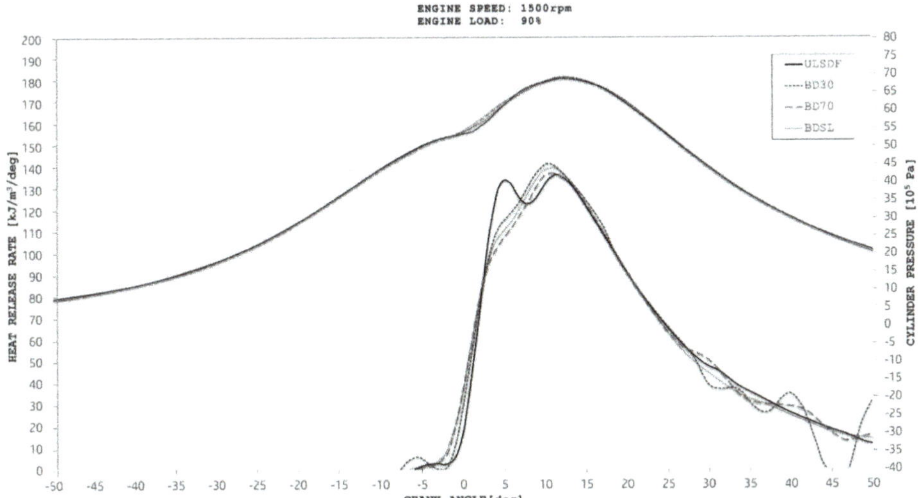

Figure 7. The dependence of fuels on the heat release rate and cylinder pressure (1500 rpm/90%).

3.2. Influence of Fuel Mixture Compositions on NO_X Emissions

The NO_X emissions, which belong among the most polluting substances, consist of the following components: the nitric oxide (NO), the nitrogen dioxide (NO_2) and the dinitrogen monoxide (N_2O). The most relevant are the NO emissions, because their amount is more than 60 percent. The portion of the NO_2 emissions is less than 40 percent and volume of the N_2O is negligible. It is evident, from the Figures 8 and 9, that a higher amount of biodiesel in the fuel mixture significantly affects formation of the NO_X pollutants. Higher content of biodiesel in the tested fuel mixtures minimized the NO_X emissions. It is possible to mention a fact as an illustrative example that operation of the tested engine in the ULSDF fuel mode at low engine speed and low operational loading generated approximately 145 ppm of the NO_X gaseous emissions, whereby in the case of pure biodiesel it was only 100 ppm (Figure 8).

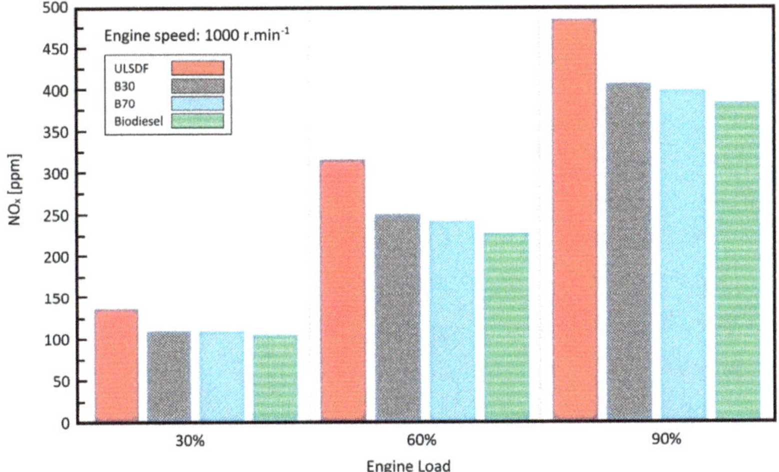

Figure 8. The dependence of testing fuels on the NO_X emissions (1000 rpm).

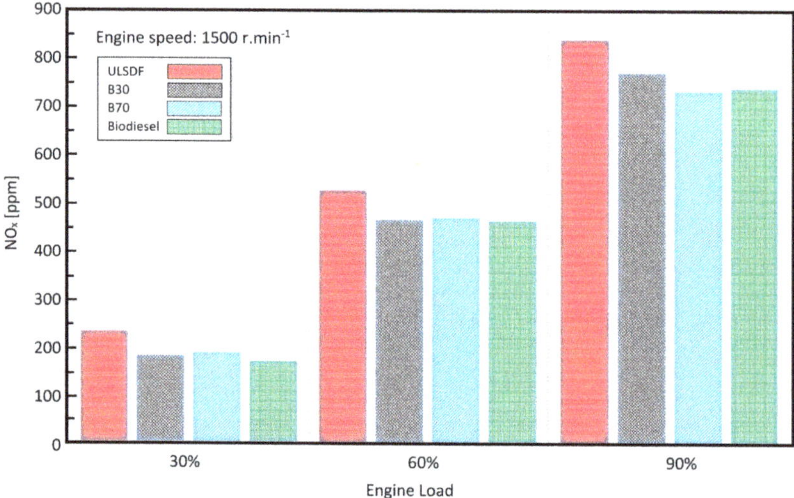

Figure 9. The dependence of testing fuels on the NOX emissions (1500 rpm).

Higher levels of the engine operational loading and engine speed caused an increased amount of the NO_X emissions in the case of all testing fuels. (i.e. up to 850 ppm for ULSDF and 750 ppm for biodiesel) (Figure 9). At the same time, the exhaust gas temperature increased with higher engine speed values and with higher engine operational loading. This fact had a major impact on growth of the NO_X emissions.

3.3. Influence of Fuel Mixture Compositions on NO Emissions and NO₂ Emissions

The Figures 10 and 11 illustrate the influence of various fuel mixtures on the NO emissions and NO_2 emissions. It is interesting that as the engine load increased, the volume of the NO emissions was higher. At the same time, however, these emissions were minimized with increasing amount of biodiesel in the tested fuels. Higher engine loading and higher portions of biodiesel in the tested fuel mixtures minimized the gaseous NO_2 emissions.

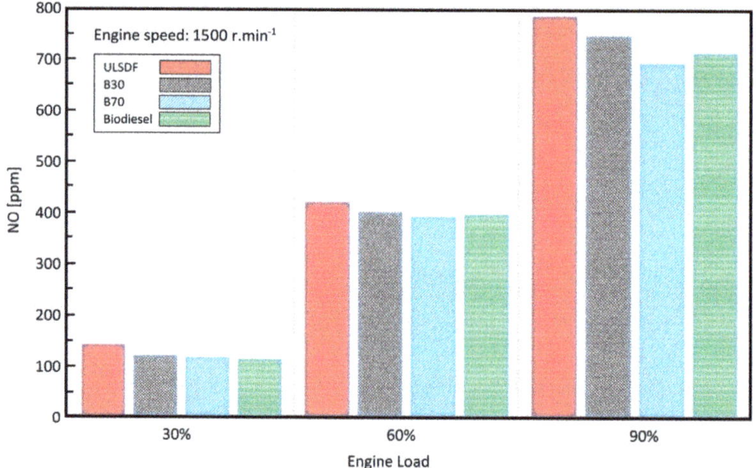

Figure 10. The dependence of testing fuels on the NO emissions (1500 rpm).

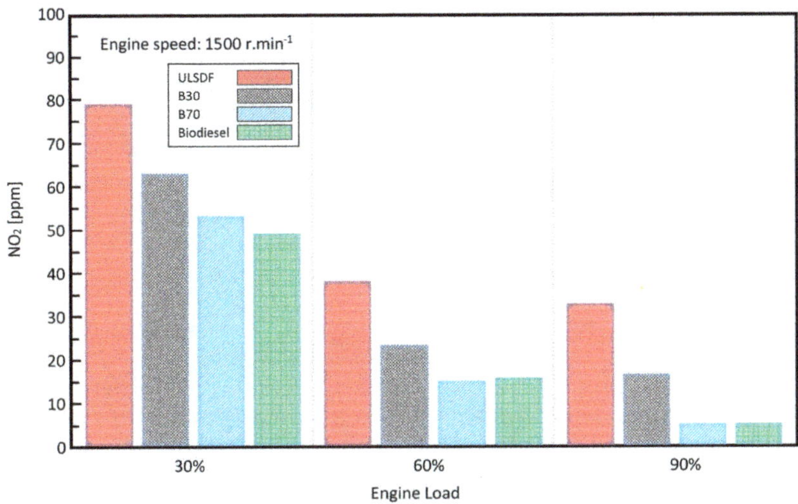

Figure 11. The dependence of testing fuels on the NO_2 emissions (1500 rpm).

4. Conclusions

The main task of this scientific-research work was to fill gaps within the current knowledge base concerning influence of bio-component fuels on production of the gaseous emissions. All the tests were performed using a 6-liter auxiliary diesel engine, which is usually installed in the marine ships. Various testing fuels were created by mixing of the biodiesel with the ULSDF. The applied mixing ratios were as follows: 0%, 30%, 70%, 100%, whereby the fuel mixtures were marked in the following order: B0: U100, B30: U70, B70: U30, B100: U0. Thus, the first tested fuel was the pure ULSDF (B0: U100) and the last tested fuel was the pure biodiesel (B100: U0). The individual experiments were performed using the engine speed values 1000 rpm or 1500 rpm and at the engine loading levels from interval between 30% and 90%. The obtained results can be summarized as follows:

- As the proportion of biodiesel in the experimental fuel increases, so the pressure of gas in the cylinder increases and intensity of heat release is significantly lower in this case.
- The ignition delay is shortened with increasing volume of biodiesel in the experimental fuel and at the same time the combustion process is faster. The ignition delay was on average about 2° longer.
- The NO_X emissions are higher if the engine speed and the engine loading levels increase. This fact is valid for every experimental fuel. On the other hand, a higher volume of biodiesel in the experimental fuel means a significant reduction of the NO_X emissions.
- The emissions of NO are higher if the engine loading level increases, but these emissions are decreasing with higher amount of biodiesel in the experimental fuel. The NO_2 emissions are significantly minimized if the engine loading level is increased and if amount of biodiesel in the experimental fuel is higher.

Author Contributions: Conceptualization, M.P. and M.K. (Melichar Kopas); methodology, M.P. and M.K. (Melichar Kopas); software, M.P. and M.K. (Melichar Kopas); validation, M.P. and D.S.; formal analysis, , M.P. and D.S.; investigation, , M.P. and M.K. (Marek Kliment); resources, M.P. and M.K. (Marek Kliment); writing—original draft preparation, M.P. and M.Š.; writing—review and editing, M.P. and M.Š. All authors have read and agreed to the published version of the manuscript.

Funding: This work was supported by the Slovak Research and Development Agency under the contract No. APVV-16-0259. The article was written in the framework of Grant Projects: APVV-16-0259 "Research and development of combustion technology based on controlled homogenous charge compression ignition in order to reduce nitrogen oxide emissions of motor vehicles", VEGA 1/0473/17 "Research and development of technology for homogeneous charge self-ignition using compression in order to increase engine efficiency and to reduce vehicle emissions", and KEGA 006TUKE-4/2020 "Implementation of Knowledge from Research Focused on Reduction of Motor Vehicle Emissions into the Educational Process".

Conflicts of Interest: The authors declare no conflict of interest.

References

1. Sinay, J.; Tompos, A.; Puskar, M.; Petkova, V. Multiparametric Diagnostics of Gas Turbine Engines. *Int. J. Marit. Eng.* **2014**, *156*, 149–156.
2. Balland, O.; Erikstad, S.O.; Fagerholt, K. Concurrent design of vessel machinery system and air emission controls to meet future air emissions regulations. *Ocean Eng.* **2014**, *84*, 283–292. [CrossRef]
3. Puškár, M.; Bigoš, P.; Kelemen, M.; Markulik, Š.; Puškárová, P. Method for accurate measurement of output ignition curves for combustion engines. *Measurement* **2013**, *46*, 1379–1384. [CrossRef]
4. Puškár, M.; Bigoš, P. Output Performance Increase of Two-stroke Combustion Engine with Detonation Combustion Optimization. *Strojarstvo: Časopis za teoriju i praksu u strojarstvu* **2010**, *52*, 577–587.
5. Puškár, M.; Bigoš, P. Method for accurate measurements of detonations in motorbike high speed racing engine. *Measurement* **2012**, *45*, 529–534. [CrossRef]
6. Puškár, M.; Bigoš, P.; Puškárová, P. Accurate measurements of output characteristics and detonations of motorbike high-speed racing engine and their optimization at actual atmospheric conditions and combusted mixture composition. *Measurement* **2012**, *45*, 1067–1076. [CrossRef]
7. Puškár, M.; Bigoš, P.; Balážiková, M.; Peťková, V. The measurement method solving the problems of engine output characteristics caused by change in atmospheric conditions on the principle of the theory of optimal temperature range of exhaust system. *Measurement* **2012**, in press.
8. Puškár, M.; Bigoš, P.; Kelemen, M.; Tonhajzer, R.; Šima, M. Measuring method for feedback provision during development of fuel map in hexadecimal format for high-speed racing engines. *Measurement* **2014**, *50*, 203–212. [CrossRef]
9. Puškár, M.; Bigoš, P. Measuring of acoustic wave influences generated at various configurations of racing engine inlet and exhaust system on brake mean effective pressure. *Measurement* **2013**, *46*, 3389–3400. [CrossRef]
10. Lamas, M.I.; Rodriguez, C.G.; Rodriguez, J.D.; Telmo, J. Internal modifications to reduce pollutant emissions from marine engines. A numerical approach. *Int. J. Naval Archit. Mar. Eng.* **2013**, *5*, 493–501. [CrossRef]
11. Puškár, M.; Brestovič, T.; Jasminská, N. Numerical simulation and experimental analysis of acoustic wave influences on brake mean effective pressure in thrust-ejector inlet pipe of combustion engine. *Int. J. Veh. Des.* **2015**, *67*, 63–76. [CrossRef]
12. Czech, P. Application of probabilistic neural network and vibration signals for gasket under diesel engine head damage. *Sci. J. Sil. Univ. Technol. Ser. Transp.* **2013**, *78*, 39–45.
13. Toman, R.; Polóni, M.; Chríbik, A. Preliminary study on combustion and overall parameters of syngas fuel mixtures for spark ignition combustion engine. *Acta Polytech.* **2017**, *57*, 38–48. [CrossRef]
14. Chríbik, A.; Polóni, M.; Lach, J.; Ragan, B. The effect of adding hydrogen on the performance and the cyclic variability of a spark ignition engine powered by natural gas. *Acta Polytech.* **2014**, *54*, 10–14. [CrossRef]
15. Brezinová, J.; Guzanová, A. Friction conditions during the wear of injection mould functional parts in contact with polymer composites. *J. Reinf. Plast. Compos.* **2009**, *28*, 1–15.
16. Lamas, M.I.; Rodríguez, C.G. Numerical model to study the combustion process and emissions in the Wärtsilä 6L 46 four-stroke marine engine. *Polish Marit. Res.* **2013**, *20*, 61–66. [CrossRef]
17. Kučera, P.; Píštěk, V.; Prokop, A.; Řehák, K. Measurement of the powertrain torque. In Proceedings of the Engineering Mechanics, Svratka, Czech Republic, 14–17 May 2018; pp. 449–452, ISBN 978-80-86246-88-8.
18. Rybár, R.; Beer, M.; Cehlár, M. Thermal power measurement of the novel evacuated tube solar collector and conventional solar collector during simultaneous operation. *Measurement* **2016**, *88*, 153–164. [CrossRef]
19. Lamas, M.I.; Rodriguez, C.G.; Telmo, J.; Rodriguez, J.D. Numerical analysis emissions from marine engines using alternative fuels. *Polish Marit. Res.* **2015**, *22*, 48–52. [CrossRef]

20. Kučera, P.; Píštěk, V. Testing of the Mechatronic Robotic System of the Differential Lock Control on a Truck. *Int. J. Adv. Robot. Syst.* **2017**, *14*, 1–7. [CrossRef]
21. Czech, P. Diagnosis of Industrial Gearboxes Condition by Vibration and Time-Frequency, Scale-Frequency, Frequency-Frequency Analysis. *Metalurgija* **2012**, *51*, 521–524.
22. Czech, P. Identification of Leakages in the Inlet System of an Internal Combustion Engine with the Use of Wigner-Ville Transform and RBF Neural Networks. In Proceedings of the 12th International Conference on Transport Systems Telematics Location, Katowice Ustron, Poland, 10–13 October 2012; Jerzy, M., Ed.; Telematics in the Transport Environment. Book Series: Communications in Computer and Information Science. 2012; Volume 329, pp. 414–422.
23. Madej, H.; Czech, P. Discrete Wavelet Transform and Probabilistic Neural Network in IC Engine Fault Diagnosis. *Eksploatacja I Niezawodnosc-Maint. Reliab.* **2010**, *4*, 47–54.
24. Lamas Galdo, M.I.; Castro-Santos, L.; Rodriguez Vidal, C.G. Numerical analysis of NOx reduction using ammonia injection and comparison with water injection. *J. Mar. Sci. Eng.* **2020**, *8*, 109. [CrossRef]
25. KUČERA, P.; PÍŠTĚK, V. Prototyping a System for Truck Differential Lock Control. *Sensors* **2019**, *19*, 3619. [CrossRef]
26. PÍŠTĚK, V.; KLIMEŠ, L.; MAUDER, T.; KUČERA, P. Optimal design of structure in rheological models: An automotive application to dampers with high viscosity silicone fluids. *J. Vibroeng.* **2017**, *19*, 4459–4470. [CrossRef]

Journal of
Marine Science and Engineering

Editorial

Marine Engines Performance and Emissions

María Isabel Lamas Galdo

Escola Politécnica Superior, Universidade da Coruña, 15001-15011 A Coruna, Spain; isabel.lamas.galdo@udc.es

Citation: Galdo, M.I.L. Marine Engines Performance and Emissions. *J. Mar. Sci. Eng.* **2021**, *9*, 280. https://doi.org/10.3390/jmse9030280

Received: 22 February 2021
Accepted: 2 March 2021
Published: 5 March 2021

Publisher's Note: MDPI stays neutral with regard to jurisdictional claims in published maps and institutional affiliations.

Marine engines are key components in most ships. Nowadays, diesel engines power most of the ships in the world. These engines are efficient in comparison with other thermal machines, but emit harmful species such as nitrogen oxides, soot, carbon dioxide, sulfur oxides, carbon monoxide, etc. Several international, national, and regional policies have been developed to limit pollutants from engines. In the marine field, the European Commission and the Environmental Protection Agency limit emissions in the European Union and the United States, respectively. On an international level, the International Maritime Organization (IMO) regulates pollution and other aspects. In 1973, the IMO adopted Marpol 73/78, the International Convention for the Prevention of Pollution from Ships, designed to reduce marine pollution. Due to these increasingly restrictive regulations, several pollution reduction methods have been developed in recent years. In recent years, many primary and secondary reduction techniques have been proposed and employed in marine engines. Some methods directly improve combustion such as exhaust gas recirculation, water addition, modification of the injection process, etc. Other methods are based on exhaust gas after treatments, such as selective catalytic reduction systems. Nevertheless, the increasingly restrictive legislation makes it very difficult to continue developing efficient reduction procedures at competitive prices, and alternative fuels become another possible solution.

The performance of marine engines is vital for the efficiency, environment, and safety. Besides, it is very important to reduce emissions as well as dependency on fossil fuels. Current engines require specific knowledge to reduce, as soon as possible, consumption and emissions. Innovative solutions are being increasingly developed in the recent years. The improvement of both computational and experimental techniques makes it possible to develop new solutions

This Special Issue contains 12 peer-review scientific papers about developments in the research of marine engines performance and emissions. These papers were developed by 39 authors from The Netherlands, Finland, Sweden, Korea, Slovakia, China, Chile, Ukraine, Poland, and Spain. Most papers discuss emissions from marine engines [1–9], and others discuss performance, such as control technology [10], and turbocharger compression [11,12]. Puškár et al. [1] and Sui et al. [2] analyzed alternative fuels. Perez and Reusser [3] optimized the emissions profile using a shaft generator with optimum tracking-based control scheme. Winnes et al. [4] analyzed a scrubber. Kim et al. [5] analyzed the intake and exhaust system. Lamas et al. [6] analyzed NOx reduction using ammonia injection and compared with water injection. Witkowski [7] analyzed several methods to reduce emissions. Lehtoranta et al. [8] analyzed a methane oxidation catalyst for LNG (Liquefied Natural Gas) ships. Lamas et al. [10] analyzed the injection system. Homišin et al. [10] analyzed an electronic constant twist angle control system suitable for torsional vibration tuning of propulsion systems. Shen et al. [11] developed a marine two-stroke diesel engine MVEM (Mean Value Engine Model) with in-cylinder pressure trace predictive capability and a novel compressor model. Finally, Píštěk et al. [12] developed a mistuning identification method of integrated bladed discs of marine engine turbochargers.

Funding: This research received no external funding.

Institutional Review Board Statement: Not applicable.

Informed Consent Statement: Not applicable.

Conflicts of Interest: The author declares no conflict of interest.

References

1. Puškár, M.; Kopas, M.; Sabadka, D.; Kliment, M.; Šoltésová, M. Reduction of the Gaseous Emissions in the Marine Diesel Engine Using Biodiesel Mixtures. *J. Mar. Sci. Eng.* **2020**, *8*, 330. [CrossRef]
2. Sui, C.; de Vos, P.; Stapersma, D.; Visser, K.; Ding, Y. Fuel Consumption and Emissions of Ocean-Going Cargo Ship with Hybrid Propulsion and Different Fuels over Voyage. *J. Mar. Sci. Eng.* **2020**, *8*, 588. [CrossRef]
3. Perez, J.R.; Reusser, C.A. Optimization of the Emissions Profile of a Marine Propulsion System Using a Shaft Generator with Optimum Tracking-Based Control Scheme. *J. Mar. Sci. Eng.* **2020**, *8*, 221. [CrossRef]
4. Winnes, H.; Fridell, E.; Moldanová, J. Effects of Marine Exhaust Gas Scrubbers on Gas and Particle Emissions. *J. Mar. Sci. Eng.* **2020**, *8*, 299. [CrossRef]
5. Kim, K.-H.; Kong, K.-J. One-Dimensional Gas Flow Analysis of the Intake and Exhaust System of a Single Cylinder Diesel Engine. *J. Mar. Sci. Eng.* **2020**, *8*, 1036. [CrossRef]
6. Lamas Galdo, M.I.; Castro-Santos, L.; Rodriguez Vidal, C.G. Numerical Analysis of NOx Reduction Using Ammonia Injection and Comparison with Water Injection. *J. Mar. Sci. Eng.* **2020**, *8*, 109. [CrossRef]
7. Witkowski, K. Research of the Effectiveness of Selected Methods of Reducing Toxic Exhaust Emissions of Marine Diesel Engines. *J. Mar. Sci. Eng.* **2020**, *8*, 452. [CrossRef]
8. Lehtoranta, K.; Koponen, P.; Vesala, H.; Kallinen, K.; Maunula, T. Performance and Regeneration of Methane Oxidation Catalyst for LNG Ships. *J. Mar. Sci. Eng.* **2021**, *9*, 111. [CrossRef]
9. Lamas, M.I.; Castro-Santos, L.; Rodriguez, C.G. Optimization of a Multiple Injection System in a Marine Diesel Engine through a Multiple-Criteria Decision-Making Approach. *J. Mar. Sci. Eng.* **2020**, *8*, 946. [CrossRef]
10. Homišin, J.; Kaššay, P.; Urbanský, M.; Puškár, M.; Grega, R.; Krajňák, J. Electronic Constant Twist Angle Control System Suitable for Torsional Vibration Tuning of Propulsion Systems. *J. Mar. Sci. Eng.* **2020**, *8*, 721. [CrossRef]
11. Shen, H.; Zhang, J.; Yang, B.; Jia, B. Development of a Marine Two-Stroke Diesel Engine MVEM with In-Cylinder Pressure Trace Predictive Capability and a Novel Compressor Model. *J. Mar. Sci. Eng.* **2020**, *8*, 3020. [CrossRef]
12. Píštěk, V.; Kučera, P.; Fomin, O.; Lovska, A. Effective Mistuning Identification Method of Integrated Bladed Discs of Marine Engine Turbochargers. *J. Mar. Sci. Eng.* **2020**, *8*, 379. [CrossRef]

Journal of
Marine Science and Engineering

Article

Fuel Consumption and Emissions of Ocean-Going Cargo Ship with Hybrid Propulsion and Different Fuels over Voyage

Congbiao Sui [1,2], Peter de Vos [2], Douwe Stapersma [2], Klaas Visser [2] and Yu Ding [1,*]

[1]   College of Power and Energy Engineering, Harbin Engineering University, Harbin 150001, China;
      suicongbiao@hrbeu.edu.cn
[2]   Faculty of Mechanical, Maritime and Materials Engineering, Delft University of Technology,
      2628 CD Delft, The Netherlands; P.deVos@tudelft.nl (P.d.V.); D.Stapersma@tudelft.nl (D.S.);
      K.Visser@tudelft.nl (K.V.)
*   Correspondence: dingyu@hrbeu.edu.cn; Tel.: +86-451-8258-9370

Received: 13 July 2020; Accepted: 3 August 2020; Published: 6 August 2020

Abstract: Hybrid propulsion and using liquefied natural gas (LNG) as the alternative fuel have been applied on automobiles and some small ships, but research investigating the fuel consumption and emissions over the total voyage of ocean-going cargo ships with a hybrid propulsion and different fuels is limited. This paper tries to fill the knowledge gap by investigating the influence of the ship mission profile, propulsion modes and effects of different fuels on the fuel consumption and emissions of the ship over the whole voyage, including transit in open sea and manoeuvring in close-to-port areas. Results show that propulsion control and electric power generation modes have a notable influence on the ship's fuel consumption and emissions during the voyage. During close-to-port manoeuvres, propelling the ship in power-take-in (PTI) mode and generating the electric power by auxiliary engines rather than the main engine will reduce the local NOx and HC (hydrocarbons) emissions significantly. Sailing the ship on LNG will reduce the fuel consumption, CO_2 and NOx emissions notably while producing higher HC emissions than traditional fuels. The hybridisation of the ship propulsion and using LNG together with ship voyage optimisation, considering the ship mission, ship operations and sea conditions, will improve the ship's fuel consumption and emissions over the whole voyage significantly.

Keywords: ship propulsion system; electric power generating system; hybrid propulsion; propulsion control; LNG; mission profile; power take off/in

---

1. Introduction

Shipping, which is a relatively energy-efficient, environment-friendly and sustainable mode of mass transport of cargo [1], is the dominant and will remain the most important transport mode for world trade [2]. However, the shipping industry consumes more fuel in comparison with other transport modes [3] and shipping-related emissions contribute significantly to the global air pollution and long-term global warming [4,5]. Correlated with fuel consumption, shipping is responsible for approximately 3.1% of annual global CO_2 and approximately 2.8% of annual GHGs (greenhouse gases) on a CO_2e (CO_2 equivalent) basis [6]. Approximately 15% and 13% of global human-made NOx and SOx emissions come from the shipping industry. It is projected that maritime CO_2 emissions will increase significantly by 50% to 250% in the period up to 2050 [6]. Moreover, as fuel cost accounts for approximately 50% to 60% of the total operational cost of a ship [7], a significant fuel consumption reduction will contribute to a considerable save of a ship's operational cost. Consequently, the shipping industry is striving to reduce its fuel consumption and emissions due to the increasingly high fuel price, social concerns on the environmental impact and the resulting mandatory and strict emission control regulations worldwide [8].

Ship mission profile during the voyage has a significant influence on the fuel consumption and exhaust emissions of ships [9]. Therefore, the ship mission profiles should be taken into consideration when evaluating ship transport performance [10]. However, one of the major drawbacks of the present IMO (International Maritime Organization)'s EEDI (Energy Efficiency Design Index) is that it only considers one operating point without taking the ship's representative mission profiles into account. On the contrary, the EEOI (energy efficiency operational indicator), which is also developed by IMO, is calculated for a voyage or a number of voyage legs based on real operating conditions [11]. In [12], based on a case study of a handy size Chemical/Product Tanker of 38,000 DWT (Deadweight tonnage), Acomi, et al. investigate the voyage energy efficiency by calculating the EEOI of the ship using both commercial software and onboard measures. In [13], in the case study of a RoPax vessel, Coraddu et al. estimate the ship operational performance of the ship voyage using the EEOI as the measure by real data statistics and numerical simulations. In [14], Hou et al. optimise the vessel speed of an ice zone ship to find a minimum EEOI in an ice zone. In [15], in the case study of a VLCC (Very Large Crude Carrier) tanker, Safaei et al. address the reduction in fuel consumption of the ship voyage using route optimisation considering ship profile and sea conditions. In [16], in the case study of a bulk carrier, Zaccone et al. develop a 3D dynamic programming optimisation method to select the optimal path and speed profile for the ship voyage aiming to minimise the voyage fuel consumption and taking also into account ship safety and comfort. However, most of the research focuses on route planning and ship speed profile when studying the ship voyage optimisation. During ship operation, propulsion control [17,18], power management [19–21] and ship operational speeds [22,23] will significantly influence the fuel consumption and emissions performance of ships. However, quantitative and systematic investigations on the influence of various ship operations, including propulsion control, power management and operational speeds, on the ship performance of the whole voyage are still limited.

Hybrid propulsion, which is a combination of mechanical and electrical propulsion, is a promising option to improve the economic, environmental and operational performance of ships [24,25]. In the basic form of the hybrid propulsion system, the propeller can be mechanically driven by an internal combustion engine and/or electrically driven by an electric motor, which may also be able to work as an electric generator. If the electric motor is powered by a hybrid power supply, such as diesel generator(s), natural gas generator(s), fuel cells and/or batteries, it will be a hybrid propulsion with a hybrid power supply system [26]. The operation modes of a hybrid propulsion system include power take off (PTO); slow power take in (PTI); boost power take in [27]. Among others, the benefits of a hybrid propulsion include reduced fuel consumption; reduced CO_2 emissions and other pollutants; possibility to sail and operate with zero emission in coastal and port areas; greater redundancy; noise reduction; lower maintenance [24,28]. However, different ship types can benefit differently from the hybrid propulsion due to their diverse operational profiles [28,29]. In [29], Jafarzadeh and Schjølberg study the operational profiles of eight different ship types, including tankers, bulk carriers, general cargo ships, container ships, Ro-Ro ships, reefers, offshore ships and passenger ships, aiming to identify what ship types are able to benefit from hybrid propulsion. Hybrid ship propulsion is typically applied on naval vessels, towing vessels, offshore vessels and passenger ships including ferries. However, the current applications and research of hybrid propulsion are mainly limited on small ships, while the applications and research on large ocean-going vessels are rare.

To improve the safety and operability of ocean-going cargo ships and to reduce their global greenhouse gas emissions and the local pollutant emissions in coastal and port areas, few studies on the potential applications of hybrid propulsion and power supply system on the big ocean-going cargo ships can be found. In [30], based on a multi-physical domain model, a conceptual hybrid propulsion system for a very large crude oil carrier (VLCC) has been studied for the potential benefit of improving the ship's safety and operability in heavy sea conditions without reducing the system efficiency. In [31] and [32], the potential benefits of hybrid propulsion for large ocean-going cargo vessels to increase fuel efficiency and reduce greenhouse gas emissions and pollutant emissions are investigated as well. In [33], the impact of battery–hybrid propulsion on the fuel consumption and emissions of an ocean-going chemical tanker when sailing in coastal and port areas during port approaches has been investigated. However, it is concluded that the battery–hybrid propulsion for ocean-going cargo ships, even when only sailing at low ship speed in close-to-port areas for a short time, is still not a realistic option nowadays even though it can produce zero local emissions; the main reason is that the required battery capacity is very large and the weight of the battery becomes unacceptable.

Using LNG (liquefied natural gas) as the alternative marine fuel is another promising and attractive solution to reducing the local and regional environmental impact and operational costs of ships [34,35]. Compared to using conventional marine fuels, using LNG produces significantly less pollutant emissions, such as NOx, SOx and PM (particle matter), and CO_2 emissions will also be reduced as well [24,36]. Another driver for using LNG as a marine fuel is the current favourable fuel price compared to the increasing price of conventional fuel oil [37]. However, one of the disadvantages in the use of LNG as marine fuel is that it may have a worse impact on climate change (global warming) than using conventional fuels, when taking the life-cycle emissions of methane (CH_4), which is a worse greenhouse gas than CO_2, into consideration [35,37]. Currently, a relatively small number of ships run on LNG and adopting LNG as a fuel is attractive for ships sailing on fixed routes and large ships sailing in short sea and coastal areas, especially in emission control areas (ECAs) [38–40]. With more stricter emissions regulations coming into force and more infrastructure of LNG fuel growing worldwide, larger ocean-going vessels are expected to select LNG as a fuel in the foreseeable future [39]. There are many publications indicating the potential benefits of using LNG as a marine fuel, however, quantitative investigations on the impact of using LNG as a fuel on the fuel consumption and emissions of ships over the whole voyage, taking the ship's operational profile into consideration, are limited.

This paper will therefore investigate the potential influence of the application of the hybrid ship propulsion and electric power generation system with different fuels as well as various propulsion control and power management strategies on the ocean-going cargo ship in reducing the fuel consumption and emissions over the whole voyage.

The main goals and outline of this paper are:

(1) To introduce the conceptual hybrid propulsion and electric power generation system of a benchmark ocean-going chemical tanker (Section 2).
(2) To explain the "average" indicators of the fuel consumption and emissions performance of the ship taking the ship mission profiles of both the transit voyage in open sea and manoeuvre in close-to-port areas into consideration (Section 3).
(3) To present the fuel consumption and emissions models of both the main engine (two-stroke diesel engine) and the auxiliary engines (four-stroke diesel engine) as well as the model corrections on different fuel types (Section 4).
(4) To introduce the ship mission profiles of the transit voyage sailing in open sea and the close-to-port manoeuvre in coastal and port areas (Section 5).
(5) To quantitatively and systematically investigate the influence of the ship operational speeds, propulsion control modes, electric power generation modes, sailing on different fuel types and PTI propulsion mode on the fuel consumption and emissions performance over the whole voyage, including the transit in open sea and manoeuvre in close-to-port areas (Section 6).

In Section 7, the conclusions, limitations and uncertainties of the present paper and the recommendations for future work will be provided.

2. Hybridisation of the Benchmark Chemical Tanker

The ocean-going 13,000 DWT chemical tanker introduced in [41] has been chosen to investigate the fuel consumption and emissions performance of the ship when transiting in open sea and manoeuvring in coastal and port areas. The propulsion system of the chemical tanker consists of a two-stroke diesel engine working as the main engine, a controllable pitch propeller (CPP) and the shafting system. The electric power generation system includes a shaft generator (power take off, PTO) driven by the main engine and three auxiliary generators. The ship particulars of the chemical tanker are given in Table 1 and the particulars of the main engine (two-stroke diesel engine) and auxiliary engines (four-stroke diesel engine) are given in Table 2.

Table 1. Ship particulars of the benchmark chemical tanker.

Length Between Perpendiculars (m)	113.80
Breadth Molded (m)	22.00
Depth Molded (m)	11.40
Design Draught (m)	8.50
Design Displacement (m^3)	16,988
Dead Weight Tonnage (ton)	13,000
Design Ship Speed (kn)	13.30

Table 2. Particulars of the main engine and auxiliary engines.

Parameters	Main Engine	Auxiliary Engine
Engine Type	MAN 6S35ME-B9.3-TII (2-stroke)	DAIHATSU 6DE-18 (4-stroke)
Number of engines (-)	1	3
Rated Power (kW)	4170	750
Rated Speed (rpm)	167	900
Stroke (m)	1.55	0.28
Bore (m)	0.35	0.185
Mean Effective Pressure (MPa)	1.67	2.21

In the benchmark ship, originally, the shaft generator can only work in PTO mode. In order to investigate the potential fuel consumption and emissions performance of a hybrid ocean-going cargo ship especially when sailing in coastal and harbour areas, in this paper, the original propulsion system and electric power generation system of the benchmark chemical tanker have been conceptually hybridised. In the conceptual hybrid ship propulsion and electric power generation system (Figure 1), the shaft generator can also work as a shaft motor in PTI (power take in) mode. To investigate the influence of sailing on different fuels on the fuel consumption and emissions of the ship, both the main engine and auxiliary engines have been assumed (conceptually updated) so that they can also use LNG (liquefied natural gas) as their fuels, without considering the details of how the engines will be updated and how different fuels will be stored and managed onboard the ship, which are out of the scope of this paper. So, after the updates, the benchmark ocean-going chemical tanker will have a hybrid ship propulsion and electric generation system.

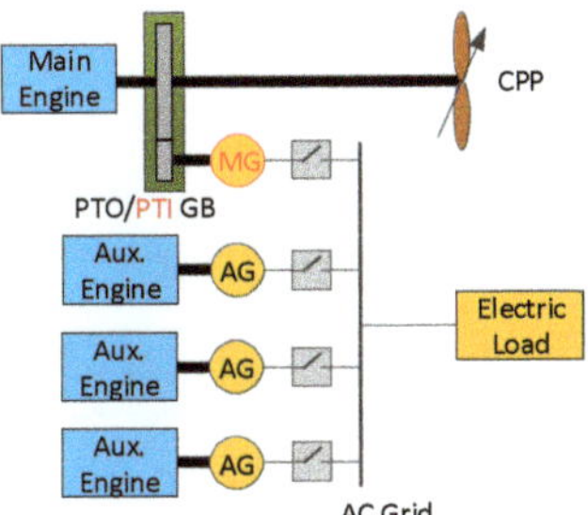

Figure 1. Layout of the updated chemical tanker propulsion system and electric generating system.

3. Mean Value Indicators of Fuel Consumption and Emissions

The mean value indicators of fuel consumption and emissions presented and used in this paper are defined based on the theories introduced in [10,33,41]. When taking the ship mission profile into account and in order to express the ship performance as a single value, an operational average value of energy effectiveness and energy (fuel) index has been introduced in [10].

The mean energy conversion effectiveness $\bar{\varepsilon}_{EC}$ over voyage, which is the weighted average value over the mission profile of the ship that will be defined later in the following chapter, is defined in Equation (1).

$$\bar{\varepsilon}_{EC} = \frac{\sum_i W_{D,i} \cdot V_i \cdot \Delta t_i}{\sum_i \left(\Phi_{FE,main,i} + \Phi_{FE,aux,i} \right) \cdot \Delta t_i} \tag{1}$$

where $W_{D,i}$, V_i, $\Phi_{FE,main,i}$, $\Phi_{FE,aux,i}$ and Δt_i are the ship dead weight (N), ship speed (m/s), energy flow into main engine (J/s), energy flow into auxiliary engines (J/s) and time of duration in each part of the voyage (h).

Same as for the definition of the mean energy conversion effectiveness, the mean fuel index $\overline{FI}$ (g/(ton·mile)) and mean emission index $\overline{EI}$ (g/(ton·mile)) averaged over the whole voyage of the ship are defined by Equations (2) and (3), respectively [33].

$$\overline{FI} = \frac{\sum_i \left(\Phi_{Fuel,main,i} + \Phi_{Fuel,aux,i} \right) \cdot \Delta t_i}{\sum_i M_{D,i} \cdot V_i \cdot \Delta t_i} \tag{2}$$

$$\overline{EI} = \frac{\sum_i \left(\Phi_{Emission,main,i} + \Phi_{Emission,aux,i} \right) \cdot \Delta t_i}{\sum_i M_{D,i} \cdot V_i \cdot \Delta t_i} \tag{3}$$

where $\Phi_{Fuel,main,i}$, $\Phi_{Fuel,aux,i}$, $\Phi_{Emission,main,i}$, $\Phi_{Emission,aux,i}$ and $M_{D,i}$ are the fuel mass flow into the main engine (g/h), fuel mass flow into auxiliary engines (g/h), emission mass flow generated by the main engine (g/h), emission mass flow generated by auxiliary engines (g/h) and dead weight tonnage of the ship (t) in each part of the voyage, respectively.

4. Mathematical Models

The ship propulsion and electric generation system model of the benchmark chemical tanker introduced in [41] has been updated in this paper and the model structure is shown in Figure 2. The models of fuel consumption and emissions of diesel engines will be briefly presented in this paper. The models of ship resistance, propeller open water characteristics, wake factor, thrust deduction factor and relative rotative efficiency can be found in [41] for detail. The onboard electric loads are modelled by a constant value, which is set as 350 kW. The shaft generator/motor and the auxiliary generators are modelled by constant energy conversion efficiencies (which is consistent with the constant auxiliary power assumption and the practice of switching on/off auxiliary generators to ensure proper loading of the engines). The energy conversion efficiencies of both the generator(s) and motor are set as 95% in the model.

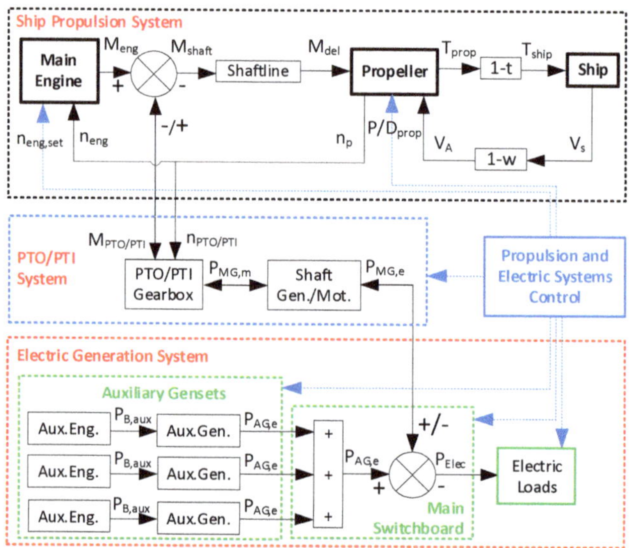

Figure 2. Model structure of the integrated ship propulsion and electric generation systems.

4.1. Models of Fuel Consumption and Emissions of Diesel Engines

The fuel consumption is calculated using the engine torque model introduced in [3], in which the engine torque is modelled as a function of the fuel consumption and engine speed in the form of second order Taylor expansion of two variables, including the cross product, as shown in Equation (4). Similarly, the emissions are modelled as functions of the engine torque and engine speed [33] as shown in Equation (5). The constant coefficients in the models can be fitted using engine test data.

$$M^* = f_1\left(m_f^*, N^*\right)$$
$$= 1 - a \cdot \left(1 - m_f^*\right) + b \cdot \left(1 - m_f^*\right)^2 - c \cdot (1 - N^*) + d \cdot (1 - N^*)^2 + 2 \cdot e \cdot \left(1 - m_f^*\right) \cdot (1 - N^*) \tag{4}$$

$$\Phi_{em}^* = f_2(M^*, N^*)$$
$$= 1 - a_{em} \cdot (1 - N^*) + b_{em} \cdot (1 - N^*)^2 - c_{em} \cdot (1 - M^*) + d_{em} \cdot (1 - M^*)^2 + 2 \cdot e_{em} \cdot (1 - N^*) \cdot (1 - M^*) \tag{5}$$

where M^*, N^*, m_f^* and Φ_{em}^* are the normalised engine torque, engine speed, fuel mass injected per cycle, emission mass flow, which are normalised by dividing the relevant variables using the corresponding nominal value of the variables.

Only the emissions of the carbon dioxide CO_2, NOx and HC (hydrocarbons) are investigated in this paper. The carbon dioxide is the direct product of the complete combustion of the fuel. So, the CO_2 emission is directly determined by the fuel consumption. Note that, as the test data of the auxiliary engines installed in the benchmark ship is not available, the fuel consumption and emissions models of the auxiliary engines are calibrated using the average data of (a small number of) similar engines that are available in the internal dataset. The calibration of the model of the main engine using the engine test data has been introduced in detail in [41]. More details on the calibration of the NOx and HC emissions models of the main engine and auxiliary engines can be found in Appendix B.

4.2. Corrections for Different Fuel Types

The test results of fuel consumption and emissions for developing and calibrating the models of both the main engine and the auxiliary engines have been corrected at ISO (International Organization for Standardization) standard reference conditions using the standard LHV (Lower Heating Value) of the fuel oil (42,700 kJ/kg), referring to ISO 15550:2016 and ISO 3046-1:2002. However, in this paper, the influence of sailing on different fuel types on the ship fuel consumption and emissions will be investigated. For instance, when the ship is in operation, the fuel type for the main engine can be HFO (heavy fuel oil), MDF (marine diesel fuel) or LNG (liquefied natural gas); while the fuel type for the auxiliary engines can be MDF (marine diesel fuel) or LNG (liquefied natural gas). Therefore, the fuel consumption, NOx emission and HC emission during ship operations need to be corrected accordingly by Equation (6) using the multiplying correcting factors shown in Table 3 (and uncertainty in these factors which is in between brackets).

Table 3. Correcting factors of fuel consumption and NOx and HC emissions.

Fuel Type	ISO *	HFO *	MDF *	LNG *
LHV (kJ/kg)	42,700	41,500	42,000	48,000 (+/− 2000)
C_{fuel} (-)	1	1.0289	1.0167	0.8896 (+/− 0.047)
C_{NOx} (-)	1	1.2	1.0	0.2 (+/− 0.1)
C_{HC} (-)	1	1.5	1.0	10.0 (+/− 6)

* ISO, ISO standard reference conditions, referring to ISO 15550:2016 and ISO 3046-1:2002. HFO, heavy fuel oil; MDF, marine diesel fuel; LNG, liquefied natural gas; LHV, lower heat values.

The underlying assumption for correcting the fuel consumption is that the efficiencies of the engines remain the same when the fuel types are changed. The correcting factors for NOx and HC emissions when using HFO and MDF are set based on the internal dataset. When using LNG as the fuel, the NOx emission will be significantly reduced by approximately 80% compared to diesel fuel [36]; while the HC emission will be much higher than the diesel fuel because of the methane slip during engine operation [34,42,43]. So, for a simple assumption, the correcting factors for NOx and HC emissions when using LNG are set as 0.2 and 10, respectively, based on the information in the available literature and engine specifications. Note that the formation mechanisms and the environmental and human health impacts of HC (hydrocarbons) emissions from diesel fuel and LNG are different although they are all called hydrocarbons in this paper as well as in other literature. The HC emissions from diesel fuel, in general, are the consequence of incomplete combustion and they are hazardous to human health (e.g., carcinogenic). However, the HC emissions from LNG are mainly methane emissions caused by "methane slip" or unburnt methane and it is mainly a greenhouse gas; it has no direct health effects on humans (in modest concentrations) [44], but it may cause suffocation if the concentration of methane in the air is too high [45]. It is assumed that the non-dimensional conversion factor between fuel consumption and CO_2 emission is 3.206 kg/kg for diesel fuels and 2.750 kg/kg for LNG [46].

$$\Phi_x = C_x \cdot \Phi_{x,ISO} \tag{6}$$

where Φ_x is the fuel consumption or the emissions of HFO, MDF and LNG (kg/s); $\Phi_{x,ISO}$ is the fuel consumption or emissions of fuel at ISO (kg/s); C_x is the correcting factors of fuel consumption and emissions for different fuel types represented in Table 3.

5. Ship Mission Profile

The sea condition in which the ship sails will be first defined. The added ship resistance in service due to sea condition, which is relative to that in sea trial condition (calm sea condition), is quantified by the sea margin (SM), which is defined by Equation (7).

$$SM = \frac{P_{E,service} - P_{E,trial}}{P_{E,trial}} \tag{7}$$

where $P_{E,service}$ is the ship effective power in service conditions and $P_{E,trial}$ is the ship effective power in sea trial condition.

The sea margin is modelled as the function of the fouling of hull and propeller, displacement, sea state and water depth. It is assumed that the ship displacement is the design displacement and does not change during the whole voyage including both transit in open sea and manoeuvre in close-to-port areas. In normal sea condition, when sailing in open sea, the ship sails in deep water and the ship resistance addition (10%) is mainly due to the sea state; while, when sailing in coastal and port areas, the ship resistance addition (10%) is mainly due to the shallow water and the effect of sea state is neglected. According to the above assumptions, in normal sea condition, the sea margins when sailing in both open sea and close-to-port area are 15% due to the combined effects of the ship fouling, displacement, sea state and water depth.

5.1. Transit in Open Sea

A combination of ship mission and sea condition profiles for three different voyages sailing at open sea of the chemical tanker has been defined as shown in Table 4. Each voyage is divided into three parts, namely Transit A, Transit B and Transit C. In different parts of the voyage the ship speed, transport distance and sea condition, which is represented by the sea margin (SM), are different. Transit A is in a calm sea condition (low sea margin, 5%), Transit B is in heavy weather (high sea margin, 30%) and Transit C is in a normal sea condition (15% sea margin). The three voyages (I, II and III) have the same total distance (650 n miles) and the same average sea margin (15%), which is defined by Equation (8), but have different average ship speeds from fast (13.5 kn) to slow steaming (10 kn).

$$\overline{SM} = \frac{\sum_i SM_i \cdot P_{E,i} \cdot \Delta t_i}{\sum_i P_{E,i} \cdot \Delta t_i} \tag{8}$$

where $\overline{SM}$ is the average sea margin of each voyage obtained by averaging the power over the whole voyage; SM_i is the sea margin of each part of the voyage; $P_{E,i}$ is the ship effective power at design draft with clean hull and calm weather; Δt_i is the time of duration in each part of the voyage.

Table 4. Ship mission profiles when sailing in open sea.

Ship Missions and Sea Conditions		Voyages		
		I	**II**	**III**
Transit A (Calm Sea State: SM = 5%)	Ship Speed V_A (kn)	13.90	12	10
	Time T_A (h)	4.27	7.89	7.89
	Sea Margin SM_A (-)	1.05	1.05	1.05
Transit B (Heavy Sea State: SM = 30%)	Ship Speed V_B (kn)	12	12	10
	Time T_B (h)	5.26	5.26	5.26
	Sea Margin SM_B (-)	1.3	1.3	1.3
Transit C (Normal Sea State: SM = 15%)	Ship Speed V_C (kn)	13.66	12	10
	Time T_C (h)	38.62	41.02	51.85
	Sea Margin SM_C (-)	1.15	1.15	1.15
The whole transit voyage	Average Ship Speed (kn)	13.50	12.00	10.00
	Total Transit Time (h)	48.15	54.17	65.00
	Total Transit Distance (n mile)	650	650	650
	Average Sea Margin (-)	1.15	1.15	1.15

The average ship speeds for the three voyages are systematically going down from voyage I (13.5 kn) to voyage III (10 kn) and transit time is going up accordingly. However, to make it more realistic, for voyage I with average ship speeds of 13.5 kn, the ship speeds of Transit B (where the ship is in heavy weather and the sea margin is 30%) are reduced to 12 kn, because of the high sea state, while the speed loss is assumed to be recovered in part C of the transit. For the voyage II and voyage III in which the average speeds are 12kn and 10kn, respectively, the ship speed during the whole voyage remains the same. The detailed determination of the mission profiles when the ship transits in open sea can be found in Appendix A.

For each ship voyage, the influence of different ship propulsion control modes and electric power generation modes on the fuel consumption and emissions of the ship over the whole voyage will be investigated. The ship propulsion control modes (Constant Revolution Mode and Constant Pitch Mode) and electric power generation modes (PTO Mode and Aux Mode) have been introduced in [41] in details and they are briefly summarised in Table 5. The combinator curves for the two different propulsion control modes are shown in Figure A5 in Appendix C.

Table 5. Ship propulsion control modes and power generating modes.

Propulsion Control Modes and Power Generating Modes		Descriptions
Propulsion control modes	Constant Revolution Mode	CONSTANT propeller revolution and CHANGING propeller pitch (Generator Law)
	Constant Pitch Mode	CONSTANT propeller pitch and CHANGING propeller revolution until minimum revolution is reached (Propeller Law)
Electric power generation modes	PTO Mode *	Shaft generator ON and auxiliary generator OFF
	Aux Mode	Shaft generator OFF and auxiliary generator ON

* PTO, power take off.

Note that, if the commanded ship speed cannot be reached within the power limit of the main engine because of providing PTO power, the shaft generator will be shut down and the electric power needed by the ship will be supplied by auxiliary generators, such as Transit A and Transit C of voyage I during which the PTO switch is turned off and the main engine only provides power for propulsion due to the demanded high ship speeds under the corresponding sea margin shown in Table 4.

5.2. Manoeuvring in Coastal and Port Areas

The ship mission profile during the close-to-port manoeuvre is shown in Table 6. The ship speed is 7 knots in coastal areas and 5 knots in port areas. The sailing time and sailing distance of the ship in the same area are combined together, respectively, in the ship mission profile in Table 6. In coastal areas, the total sailing distance when approaching and leaving the harbour is 40 nautical miles and the total sailing time is 5.71 h. In port areas, the total sailing distance is 10 nautical miles and the total sailing time is 2 h. The sea margin when the ship is sailing in coastal and port areas is assumed to be normal, i.e., 15% sea margin, and the added ship resistance is because of the smaller depth in coastal and port areas compared with the open sea, where the sea state is the main reason for the added ship resistance.

Five different operation cases of the ship are studied to investigate the influence of the ship propulsion and the electric generation modes of a hybrid propulsion ship on the fuel consumption and emissions performance during the close-to-port manoeuvre.

Table 6. Ship mission profile sailing in coastal and port areas.

Sailing Area	Ship Mission Profile	
Coastal Area	Ship Speed (kn)	7
	Sailing Distance (n mile)	40
	Sailing Time (h)	5.71
Port Area	Ship Speed (kn)	5
	Sailing Distance (n mile)	10
	Sailing Time (h)	2
The whole harbour approaching and leaving manoeuvre	Average speed (kn)	6.49
	Total Time (h)	7.71
	Total Distance (n mile)	50

- In case I, the main engine burning heavy fuel oil (HFO) provides power for the ship propulsion and onboard electric loads through shaft generator in PTO mode, while the auxiliary engines are shut down.
- In case II, it is the same as case I except that the fuel burnt by the main engine is changed from heavy fuel oil (HFO) to marine diesel fuel (MDF).
- In case III, it is also the same as case I except that the fuel for the main engine is changed from HFO to LNG (liquefied natural gas).
- In case IV, the auxiliary engines burning marine diesel fuel (MDF) provide power for the ship propulsion through shaft motor in PTI mode and onboard electric loads, while the main engine is shut down.
- In case V, it is the same as case IV except that the fuel for the auxiliary engines is changed from MDF to LNG.

So, in case I, case II and case III, the ship propulsion system works in PTO mode but on different fuels; while in case IV and case V, the ship propulsion system works in PTI mode but on different fuels. Only the constant revolution mode, in which the ship speed is controlled by changing the propeller pitch and the propeller revolution is kept constant, will be studied during the close-to-port manoeuvre.

6. Results and Discussions

6.1. Average Ship Performance over Transit Voyage in Open Sea

6.1.1. Influence of Different Operation Modes

The influence of different propulsion control modes and electric power generation modes on the ship performance when sailing in the open sea has been investigated from the voyage perspective. In this section, the fuels for the main engine and the auxiliary engines are set as HFO and MDF, respectively. The average energy conversion effectiveness of the ship over the voyage is shown in Figure 3. In voyage I, where the average ship speed is 13.5 kn, different propulsion control and electric power generation modes do not make much difference on the average energy conversion effectiveness (around 100). The reason is that the propeller pitch of the two control modes is almost the same to reach that high ship speed. However, in voyage II and III, where the average ship speeds are 12 kn and 10 kn, respectively, the differences are much more obvious. The average energy conversion effectiveness will increase with the decrease in the average ship speed. Taking the constant revolution control and PTO electric generation modes, for example, the energy conversion effectiveness will increase from 100 to 130 and 150 when the average ship speed decreases from 13.5 kn (voyage I) to 12 kn (voyage II) and 10 kn (voyage III). The major reason is the increase in the ship weight/resistance ratio when reducing the ship speed and it is also the main reason why ship "slow steaming" can save fuel consumption over the voyage.

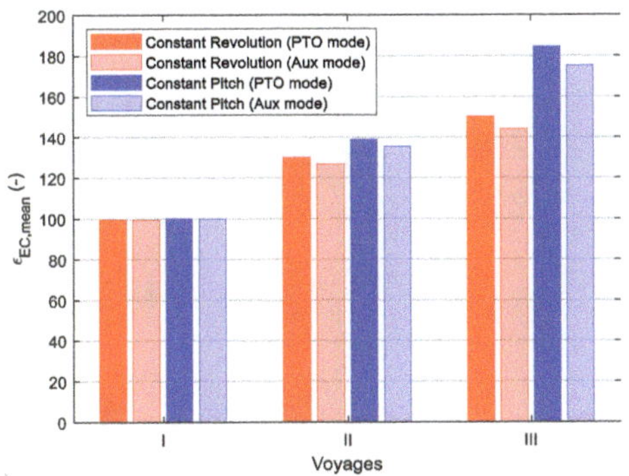

Figure 3. Mean value of energy conversion effectiveness.

The "average" ship performance in terms of the fuel and emissions indices during the whole voyage of the ship are presented in Figure 4. The average fuel index (Figure 4a) and the average CO_2 emission index (Figure 4b) of the ship over the voyage are in fact the inverse of the average energy conversion effectiveness, so they have the inverse trends. For example, in constant pitch control mode and PTO electric generation mode, the fuel index will decrease from 4.37 to 3.15 and 2.37 (g/(ton·mile)), and the CO_2 emission index will decrease from 14.03 to 10.11 and 7.61 (g/(ton·mile)) when the average ship speed reduces from 13.5 to 12 and 10 (kn). Moreover, during the ship voyage, controlling the ship speed in constant pitch mode rather than the constant revolution mode, and providing the electric power by the shaft generator instead of the auxiliary generator will also reduce the fuel consumption and CO_2 emission over the voyage.

When the average ship speed decreases, the average NOx and HC emission index over the whole voyage will also reduce as shown in Figure 4c,d. For instance, under constant pitch and PTO electric generation mode, the NOx emission index decreases from 0.33 to 0.24 and 0.17 (g/(ton·mile)) and the HC index reduces from 0.0118 to 0.0098 and 0.0081 (g/(ton·mile)). The constant pitch mode has a higher average NOx emission index (Figure 4c) than the constant revolution mode. Generating the electric power in PTO mode during the ship voyage will also reduce the NOx emission especially in constant pitch operating mode. The constant revolution mode has lower HC emission index (Figure 4d) than the constant pitch mode especially at low average ship speeds. Unlike the average fuel consumption, CO_2 emission and NOx emission, the average HC emission over the voyage will increase when the electric power is provided by the shaft generator (PTO mode). According to the simulation results of the defined three voyages, an efficient way to reduce the fuel consumption and emissions indexes is to reduce the average ship speed of the voyage, i.e., slow steaming. Different propulsion control modes and power generation modes also make some differences on the average fuel consumption and emissions indexes of the voyage especially at low ship speeds.

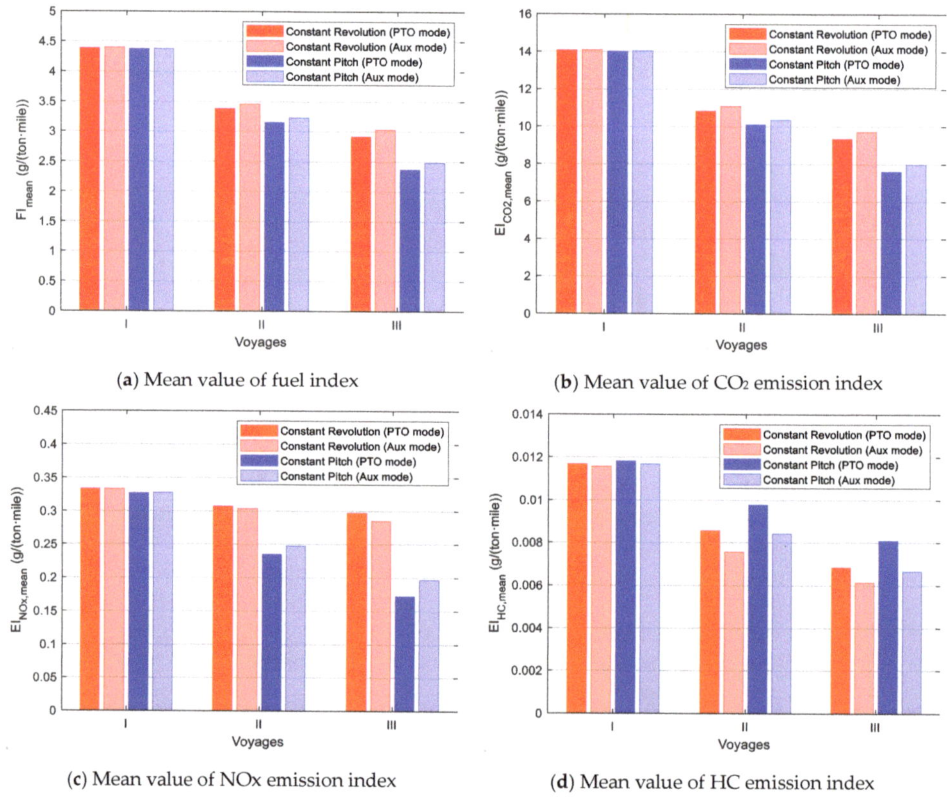

(**a**) Mean value of fuel index

(**b**) Mean value of CO₂ emission index

(**c**) Mean value of NOx emission index

(**d**) Mean value of HC emission index

Figure 4. Mean value of fuel and emissions indices.

6.1.2. Influence of Sailing on Different Fuels

In this section, when investigating the influence of different fuel types on the fuel consumption and emissions over the whole transit voyage, only voyage II, in which the ship sails at 12 knots while at sea, will be looked into. The propulsion control mode is set as constant pitch mode and the electric generation mode is set as PTO mode. The average fuel and emissions indices of the ship over the whole transit voyage when sailing on different fuels, i.e., HFO, MDF and LNG, are shown in Figure 5. The average fuel index of the ship (Figure 5a) when sailing on LNG (2.727 g/(ton·mile)) is about 13.5% less compared with sailing on HFO (3.154 g/(ton·mile)) and 12.5% less than MDF (3.117 g/(ton·mile)). The average CO_2 emission index of the ship (Figure 5b) when sailing on LNG (7.50 g/(ton·mile)) is about 25.8% less compared with sailing on HFO (10.11 g/(ton·mile)) and 25% less than MDF (10.00 g/(ton·mile)). Note that, the lower heat values (LHV) and the conversion factors between fuel consumption and CO_2 emissions are different for different fuels. The average NOx emission index of the ship (Figure 5c) when sailing on MDF (0.196 g/(ton·mile)) is about 17% less than sailing on HFO (0.236 g/(ton·mile)), while sailing on LNG (0.039 g/(ton·mile)) can further reduce the NOx emission index by 80% compared with sailing on MDF. However, the average HC emission index (Figure 5d) of the ship when sailing on LNG (0.065 g/(ton·mile)) is much higher than sailing on HFO (0.0098 g/(ton·mile)) and MDF (0.0065 g/(ton·mile)), which is one of the major disadvantages of using LNG as the marine fuel. This is primarily caused by methane slip and unburnt methane during engine operations.

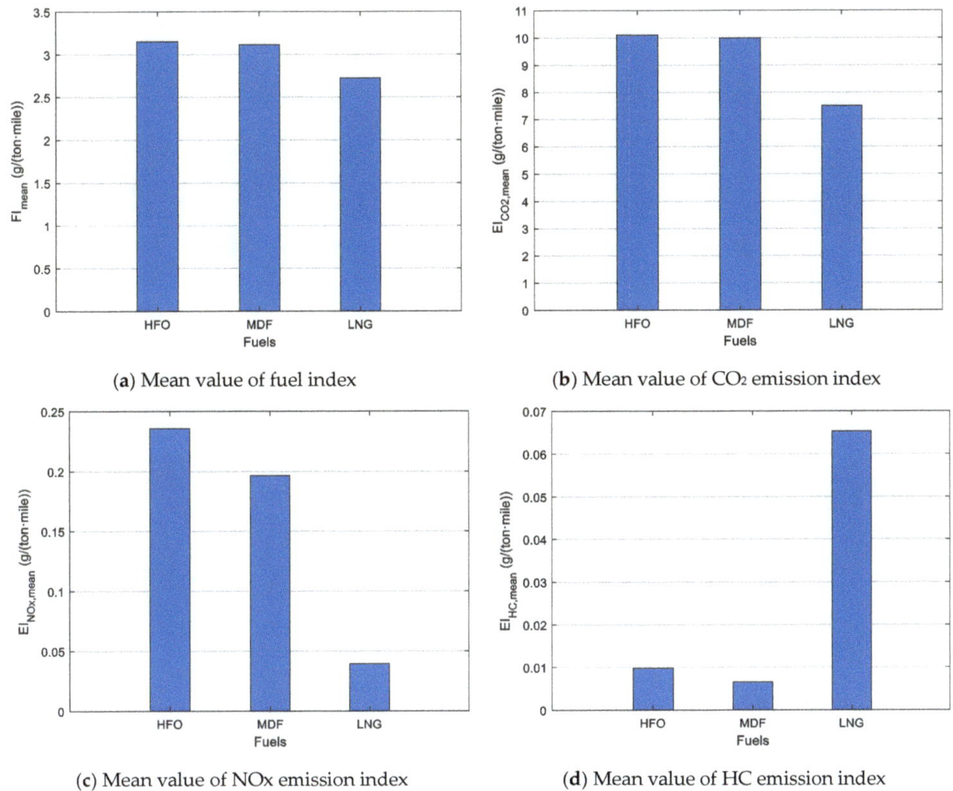

(**a**) Mean value of fuel index

(**b**) Mean value of CO₂ emission index

(**c**) Mean value of NOx emission index

(**d**) Mean value of HC emission index

Figure 5. Mean value of fuel and emissions indices when using different fuels.

6.2. Average Ship Performance over Manoeuvring in Close-to-Port Areas

The fuel consumption and emissions performance of the ship during manoeuvre in close-to-port areas under five different operation cases introduced in Section 5.2 are investigated. The fuel consumption and emissions of the ship during the whole close-to-port manoeuvre are shown in Figure 6. The average fuel indices (Figure 6a) and CO_2 emission indices (Figure 6b) of the ship when sailing on main engine in PTO mode (Case I, II and III) are lower than sailing on auxiliary engines in PTI mode (Case IV and V). However, sailing on auxiliary engines in PTI mode can reduce the local NOx (Figure 6c) and HC emissions indices (Figure 6d) significantly compared with sailing on the main engine in PTO mode.

When sailing on conventional fuels (Case I, II and IV), the fuel index and CO_2 emission index in case I, which are 3.10 and 9.93 (g/(ton·mile)), respectively, are slightly higher than those in Case II, which are 3.06 and 9.81 (g/(ton·mile)); but they are notably lower than those in Case IV, which are 4.25 and 13.62 (g/(ton·mile)), respectively. However, the NOx and HC emission indices in Case I, which are 0.35 and 0.0074 (g/(ton·mile)), respectively, are much higher than those in Case II, which are 0.28 and 0.0049 (g/(ton·mile)), respectively; the NOx and HC emission indices in Case IV, which are 0.17 and 0.0029 (g/(ton·mile)), respectively, are further lower than Case II. The reason is that the fuel consumption performance of the main engine (two-stroke) is better than that of the auxiliary engine (four-stroke) burning MDF while the NOx and HC emission performance of the main engine is worse especially when burning HFO compared with the auxiliary engine.

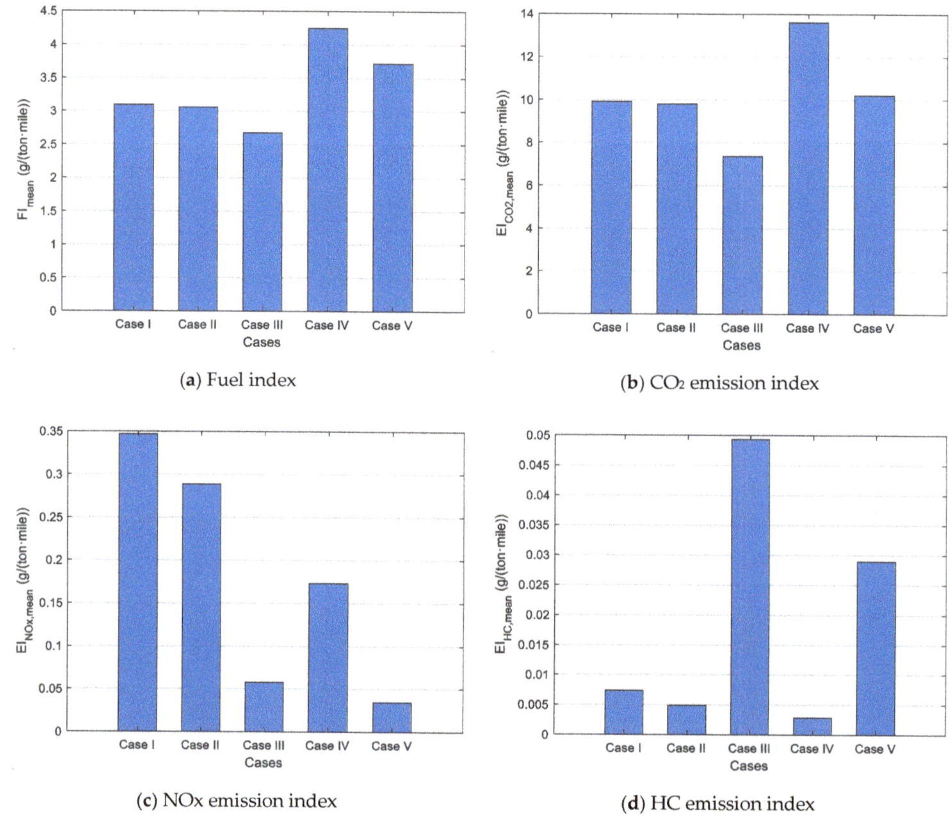

Figure 6. Fuel and emissions indices in different cases.

Sailing on LNG (Case III and V) instead of conventional fuels in coastal and port areas can both reduce fuel consumption and CO_2 emission indices of the ship, in particular the local NOx emission index (0.058 g/(ton·mile) in Case III and 0.035 g/(ton·mile) in Case V) will decrease significantly. However, the local HC emission index (0.049 g/(ton·mile) in Case III and 0.029 g/(ton·mile) in Case V) is much higher when sailing on LNG than sailing on HFO and MDF.

Therefore, comparing the five cases, in order to reduce the ship emissions significantly when manoeuvring in close-to-port areas, the ship should be driven by the auxiliary engines through PTI mode. However, a balance between CO_2 and NOx emissions on the one hand and HC emissions on the other needs to be made when selecting to sail the ship on LNG rather than the conventional fuels. Note that, as mentioned earlier, the methane emissions of LNG have no direct health effects on humans. At the same time, it is actually a more potent greenhouse gas than CO_2. So, reducing the local pollution point of view when driving the ship in PTI mode and LNG fuel when manoeuvring in close-to-port areas are better choices compared to the other cases.

6.3. Fuel Consumption and Emissions of the Whole Voyage

In summary, the average fuel and emissions indexes of the ship over the whole voyage, including transit at open sea, approaching and leaving harbour manoeuvre, are shown in Table 7. According to the previous discussions in Sections 6.1 and 6.2, there are many different combinations of ship operation cases during the whole voyage. For simplicity, only two cases sailing the ship on two different fuels, i.e., HFO and LNG, have been selected and are shown in Table 7. The propulsion control modes for transit in open sea and manoeuvring in close-to-port areas are set as constant pitch mode and constant revolution mode, respectively; the electric power generation modes for the whole voyage are set as PTO mode. Compared to transit in open sea, harbour approaching and leaving manoeuvres only take a small part of the total voyage, so, the results of the total voyage shown in Table 7 are mainly determined by the fuel and emissions indexes of the voyage when sailing at open sea.

Table 7. Average fuel and emissions indexes of the whole voyage.

Fuel Type	HFO	LNG
$\overline{FI}$ (g/(ton·mile))	3.15	2.72
$\overline{EI}_{CO2}$ (g/(ton·mile))	10.10	7.49
$\overline{EI}_{NOx}$ (g/(ton·mile))	0.24	0.041
$\overline{EI}_{HC}$ (g/(ton·mile))	0.0096	0.0642

7. Conclusions and Recommendations

In this paper, the influences of the ship propulsion control modes, electric power generation modes, ship operational speeds, propulsion modes as well as sailing on different fuels on the fuel consumption and emissions of an ocean-going benchmark chemical tanker have been investigated taking the ship's operational profiles into account. The current IMO's EEDI considers only one operating point when estimating ship energy efficiency, however, a lower EEDI does not necessarily mean less ship fuel consumption and emissions when sailing with a certain mission profile over the whole voyage. So, the mean value indicators weighted over the ship mission profile should be used when estimating the fuel consumption and emissions performance of the ship over the voyage.

When transiting in open sea, reducing the ship average operational speed will effectively reduce both the fuel consumption and emissions of the ship over the voyage. To reduce the ship operational speed, reducing the propeller revolution rather than the propeller pitch is more preferable, as pitch reduction will reduce the propeller efficiency and consequently increase the fuel consumption especially for voyage where the ship speed is reduced. Generating the electric power by the shaft generator (PTO mode) rather than the auxiliary generator (Aux mode) will further reduce the fuel consumption while the NOx and HC emissions could increase. However, compared to the propulsion control modes, the electric generation modes have relatively minor influence on fuel consumption and emissions of the ship.

When the ship is sailing and manoeuvring in the coastal and port areas, changing the fuel for the main engine from heavy fuel oil (HFO) to marine diesel fuel (MDF) will reduce the NOx and HC emissions significantly while slightly reducing the fuel consumption and CO_2 emissions. Providing the power for ship propulsion (PTI mode) and onboard electric loads by the auxiliary engines and shutting down the main engine will further reduce the local NOx and HC emissions significantly while the fuel consumption and CO_2 emission will increase notably mainly due to the lower engine efficiency of the auxiliary engines.

Using LNG (liquefied natural gas) as the fuel for both the main and auxiliary engines will reduce the NOx emission significantly compared to using HFO (heavy fuel oil) or MDF (marine diesel fuel). So, sailing the ship on LNG in close-to-port areas will produce much less local environmental impact due to the much less local pollutant emissions. In particular, sailing the ship in PTI mode on LNG will further reduce the local pollutant emissions in coastal and port areas. The fuel consumption and CO_2 emission of the ship will also decrease notably over the whole voyage when sailing on LNG instead of HFO and MDF. However, the hydrocarbon (HC) emission is much higher when using LNG as a marine fuel than traditional diesel fuel due to the methane (CH_4) slip and unburnt methane during engine operations and although it has no direct effects on human health, it may have a worse impact on climate change (global warming) when taking the life-cycle emissions of natural gas into consideration (although the lifetime of the emitted substance should then also be taken into account, which is outside the scope of this paper). It is clear either way that methane emissions from LNG engines should be minimised as much as possible.

Last but not least, there are still some limitations and uncertainties existing in this paper and these will be further studied in future work. One of the limitations is that only CO_2, NOx and HC emissions are investigated and the other emissions, such as sulphur oxides (SOx), particulate matter (PM) and soot (C), etc., generated by the ship are not considered. Another limitation is that only fuel consumption and emissions are investigated in this paper while the ship capital expenditure (CAPEX), operating expenditure (OPEX) and the operational safety of both the engine and ship have been left outside the scope. The uncertainty is in the simplistic assumption on the fuel consumption and emission performance of the engines when using LNG as the fuel. In future work, in addition to improving the fuel consumption and emissions models, the potential influence of the application of the hybrid propulsion and alternative fuels on the CAPEX and OPEX of the ship, and the operational safety of both the engine and ship especially in adverse sea conditions will be investigated. The trade-off relationships between the ship CAPEX and OPEX, and between the energy effectiveness of the ship and the operational safety, need to be investigated.

Author Contributions: Conceptualisation, C.S. and D.S.; Data curation, C.S.; Formal analysis, C.S. and P.d.V.; Investigation, C.S. and P.d.V.; Methodology, C.S. and D.S.; Project administration, K.V.; Resources, Y.D.; Software, C.S.; Supervision, P.d.V. and K.V.; Validation, P.d.V., D.S. and Y.D.; Visualisation, C.S.; Writing–original draft, C.S.; Writing–review and editing, P.d.V., D.S., K.V. and Y.D. All authors have read and agreed to the published version of the manuscript.

Funding: This project partly is financially supported by the International Science and Technology Cooperation Program of China, 2014DFA71700; Marine Low-Speed Engine Project-Phase I; China High-tech Ship Research Program.

Acknowledgments: The authors would like to thank CSSC Marine Power Co., Ltd. (CMP) for providing the ample data of the benchmark ship and the engines, and the support for the measurements of the two-stroke marine diesel engines.

Conflicts of Interest: The authors declare no conflict of interest.

Abbreviations

Abbreviations

AC	alternating current
AG	auxiliary generator
Aux.Eng.	auxiliary engine
Aux.Gen.	auxiliary generator
CPP	controllable pitch propeller
GB	gearbox
HFO	heavy fuel oil
LNG	liquefied natural gas
MCR	maximum continuous rating
MDF	marine diesel fuel
MG	(shaft) motor/generator
PTO	power take off
PTI	power take in

Symbols

C_{fuel}	correcting factor of fuel consumption (-)
C_{NOx}	correcting factor of NOx emission (-)
C_{HC}	correcting factor of HC emission (-)
$\overline{EI}$	mean value emission index (g/(ton·mile))
$\overline{FI}$	mean value fuel index (g/(ton·mile))
LHV	lower heating value (kJ/kg)
M^*	normalised engine torque (-)
M_D	dead weight tonnage of the ship (t)
M_{del}	delivered torque to propeller (Nm)
M_{eng}	engine torque (Nm)
m_f	injected fuel mass per cycle (kg)
m_f^*	normalised injected fuel mass per cycle (-)
M_{shaft}	shaft torque (Nm)
N^*	normalised engine speed (-)
n_{eng}	engine rotational speed (r/s)
n_p	propeller rotational speed (r/s)
$P_{AG,e}$	electric power of auxiliary generator (W)
$P_{B,aux}$	power of auxiliary engines (W)
P_E	ship effective power (W)
$P_{E,service}$	ship effective power in service conditions (W)
$P_{E,trial}$	ship effective power in trial conditions (W)
P_{Elec}	electrical power of onboard grid (W)
$P_{MG,e}$	electrical power of shaft motor/generator (W)
$P_{MG,m}$	mechanical power of shaft motor/generator (W)
sfc	specific fuel consumption (g/kWh)
SM	sea margin (-)
$\overline{SM}$	average sea margin over voyage (-)
t	thrust deduction fraction (-)
T_{prop}	propeller thrust (N)
T_{ship}	ship thrust (N)
V	ship speed (m/s)
v_A	propeller advance sped (m/s)
v_s	ship speed (m/s)
w	wake fraction (-)
W_D	deadweight of ship (N)
$\overline{\varepsilon}_{EC}$	mean value energy conversion effectiveness (-)
Φ_{em}^*	normalised emission mass flow (-)
$\Phi_{Emission,main}$	emission mass flow of main engine (g/h)
$\Phi_{Emission,aux}$	emission mass flow of auxiliary engines (g/h)
$\Phi_{FE,aux}$	energy flow of fuel to auxiliary engines (J/s)
$\Phi_{FE,main}$	energy flow of fuel to main engine (J/s)
$\Phi_{Fuel,main}$	fuel mass flow into main engine (g/h)
$\Phi_{Fuel,aux}$	fuel mass flow into auxiliary engines (g/h)

Appendix A

Mission Profile in Open Sea

According to the transit voyage in open sea defined in Section 5.2, the average sea margin over the whole voyage is defined by Equation (A1).

$$\overline{SM} = \frac{SM_A \cdot P_{E,A} \cdot t_A + SM_B \cdot P_{E,B} \cdot t_B + SM_C \cdot P_{E,C} \cdot t_C}{P_{E,A} \cdot t_A + P_{E,B} \cdot t_B + P_{E,C} \cdot t_C} \tag{A1}$$

The average ship speed over the whole voyage is defined by (A2).

$$\overline{V} = \frac{V_A \cdot t_A + V_B \cdot t_B + V_C \cdot t_C}{t_A + t_B + t_C} \tag{A2}$$

The distance of the whole voyage is:

$$V_A \cdot t_A + V_B \cdot t_B + V_C \cdot t_C = 650 \;\; (\text{nmile}) \tag{A3}$$

Voyage I:

The maximum ship speeds in a calm sea state (SM = 5%) and a normal sea state (SM = 15%) in Case I are 13.90 kn and 13.66 kn, respectively. The maximum ship speed in a heavy sea state (SM = 30%) is limited at 12 kn. In order to reach the average ship speed of 13.5 kn over the whole voyage, it is assumed that the ship sails at the maximum speeds at different parts of the voyage defined above. In Voyage I, the average sea margin over the whole voyage is 15%. According to Equations (A1)–(A3), the time the ship sails in different parts of the voyage in Voyage I are: $t_{I,A} = 4.27$ (h), $t_{I,B} = 5.26$ (h), $t_{I,C} = 38.62$ (h).

Voyage II:

The average sea margin of the whole voyage in Voyage II is also 15%. The ship speed over the whole voyage is 12 kn and the corresponding ship effective power over the whole voyage is 1229.59 kW. It is assumed that the time the ship sails in a heavy sea state in Voyage II and Voyage III is the same as that in Voyage I, which is 5.26 h. Then, according to Equations (A1), (A2) and (A3), the time the ship sails in different parts of the voyage are: $t_{II,A} = 7.89$ (h), $t_{II,B} = 5.26$ (h), $t_{II,C} = 41.02$ (h).

Voyage III:

Similar to Voyage II, the time the ship sails in different parts of the voyage in Voyage III, where the average ship speed is 10 kn, are: $t_{III,A} = 7.89$ (h), $t_{III,B} = 5.26$ (h), $t_{III,C} = 51.85$ (h).

Appendix B

Calibration of NOx and HC Emissions Models

The coefficients $a_{em} \sim e_{em}$ ($a_{NOx} \sim e_{NOx}$ and $a_{HC} \sim e_{HC}$) in the NOx and HC emissions models in Equation (5) are constants which can be determined by engine test data using the least square fitting method. The test data provided in the technical file of the engine EIAPP (Engine International Air Pollution Prevention) certificate has been used for modelling the NOx and HC emissions as well as the fuel consumption. When calibrating the engine fuel consumption model and emissions model, the engine test data come from the same operating points (i.e., 25%, 50%, 75%, 100% of rated engine load), which are selected along both the generator curve (E2 cycle) and the propeller curve (E3 cycle). However, the engine installed on the benchmark chemical tanker has been tested only for E2 cycle rather than E3 cycle due to the installed controllable pitch propeller. So, the test data for E3 cycle is obtained based on the mean value of the E3 cycle test data of the other four engines from the same engine family. "The mean value of data of different engines is taken in the following way, firstly, the mean value of data of different engines at nominal points are taken as the new nominal value of the engine; secondly, the mean values of part load percentages, i.e., the ratios of part load value to nominal value, along generator curve (E2 cycle) and propeller curve (E3 cycle) of different engines are taken as the part load percentages of the engine along generator curve and propeller curve, respectively [41]." Note that the specific HC emission data at the point of 75% nominal power of the E2 cycle is believed too high (0.42g/kWh) to be reasonable compared to the data of the other points and there is no physical explanation for the measurement that lies so far outside the line that connects the other data points. Therefore, it is corrected (as a rule of thumb) to a lower value (0.30g/kWh) as shown in Figure A2b to make the trend smoother and the fitting results more acceptable in spite of the fact that we only have E2 cycle data of one two-stroke diesel engine. The modelling result of the NOx and HC emissions of the main engine are shown in Tables A1 and A2, Figures A1 and A2.

Table A1. Coefficients of NOx emission model of main engine.

Nominal Parameters			Coefficients				
$\Phi_{NOx,nom}$ (g/s)	$M_{eng,nom}$ (kNm)	$n_{eng,nom}$ (rpm)	a_{NOx}	b_{NOx}	c_{NOx}	d_{NOx}	e_{NOx}
13.3208	238.4465	167	2.1463	−0.8538	0.4046	−0.5199	1.4678

Table A2. Coefficients of HC emission model of main engine.

Nominal Parameters			Coefficients				
$\Phi_{HC,nom}$ (g/s)	$M_{eng,nom}$ (kNm)	$n_{eng,nom}$ (rpm)	a_{HC}	b_{HC}	c_{HC}	d_{HC}	e_{HC}
0.4170	238.4465	167	−0.0595	−0.0088	1.6009	0.7635	−0.0424

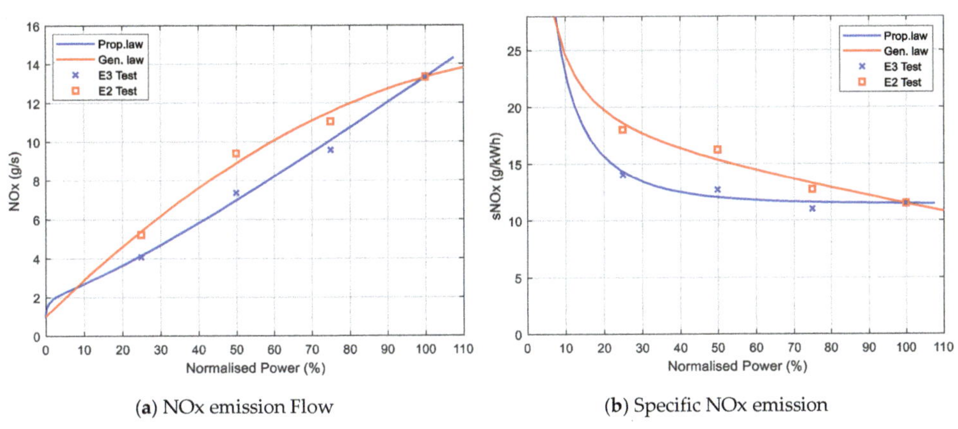

(a) NOx emission Flow

(b) Specific NOx emission

Figure A1. NOx emission.

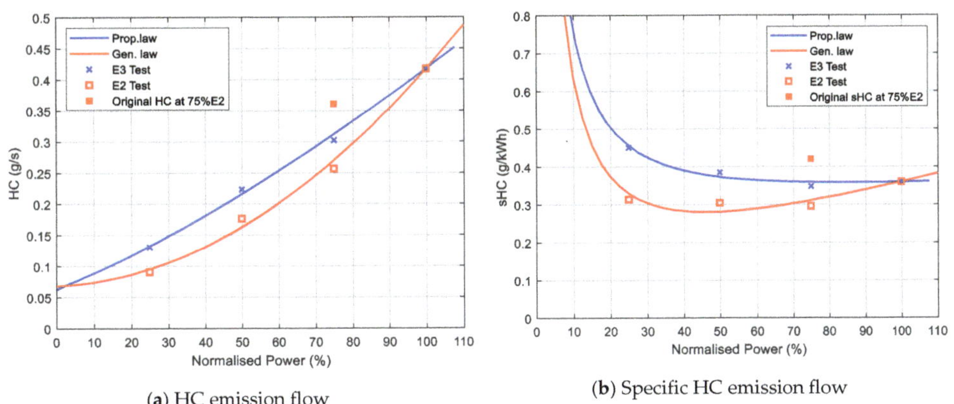

(a) HC emission flow

(b) Specific HC emission flow

Figure A2. HC emission.

The calibrated coefficients of the engine torque model, NOx and HC emissions model of the auxiliary engine, are shown in Tables A3–A5. The modelling results of the fuel consumption and emissions of both the main engine and auxiliary engines are shown in Figures A3 and A4.

Table A3. Coefficients of engine torque model of auxiliary engine.

Nominal Parameters			Coefficients				
$M_{eng,nom}$ (kNm)	$m_{f,nom}$ (g/cyl/cycle)	$n_{eng,nom}$ (rpm)	a	b	c	d	e
7.9577	1.0417	900	-0.0558	-0.6022	0.9446	-0.1548	0.1567

Table A4. Coefficients of NOx emission model of auxiliary engine.

Nominal Parameters			Coefficients				
$\Phi_{NOx,nom}$ (g/s)	$M_{eng,nom}$ (kNm)	$n_{eng,nom}$ (rpm)	a_{NOx}	b_{NOx}	c_{NOx}	d_{NOx}	e_{NOx}
1.8750	7.9577	900	0.6642	0.2174	0.8867	-0.0267	0.3099

Table A5. Coefficients of HC emission model of auxiliary engine.

Nominal Parameters			Coefficients				
$\Phi_{HC,nom}$ (g/s)	$M_{eng,nom}$ (kNm)	$n_{eng,nom}$ (rpm)	a_{HC}	b_{HC}	c_{HC}	d_{HC}	e_{HC}
0.0271	7.9577	900	1.0213	0.1307	0.2600	-0.4000	0.2703

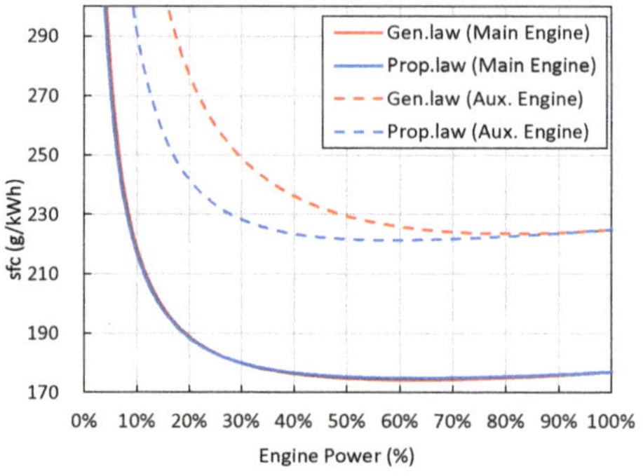

Figure A3. Specific fuel consumption of main engine and auxiliary engines.

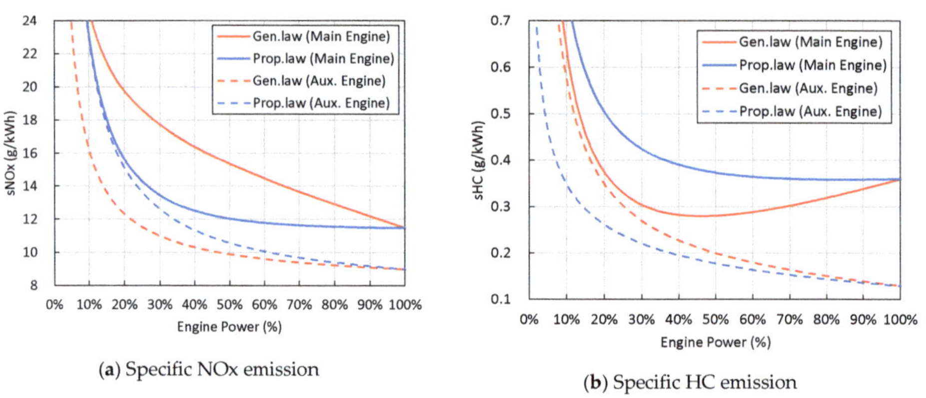

(**a**) Specific NOx emission

(**b**) Specific HC emission

Figure A4. Specific NOx and HC emissions of main engine and auxiliary engines.

Appendix C

Combinator Curves for Ship Propulsion Control

The combinator curves for the ship propulsion control modes (constant revolution mode and constant pitch mode) are shown in Figure A5. Setting the sea margin as 15%, the corresponding main engine power in different propulsion control and electric generation modes is shown in Figure A6.

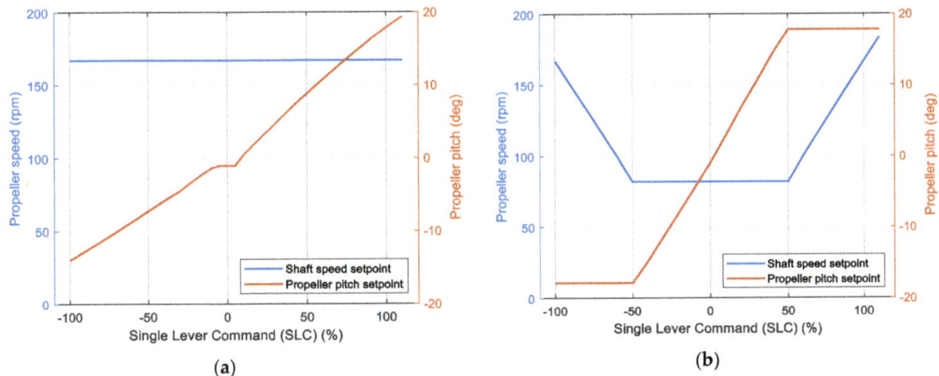

(a)　　　　**(b)**

Figure A5. Combinator curve of different propulsion control modes. (**a**) Constant revolution mode, and (**b**) constant pitch mode.

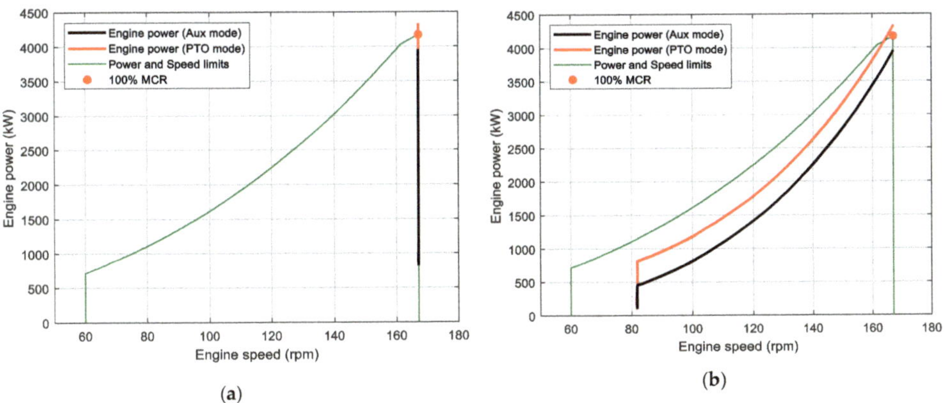

(a)　　　　**(b)**

Figure A6. Main engine power in different propulsion control and electric generation modes. (**a**) Constant revolution mode, and (**b**) constant pitch mode.

References

1. Chapman, L. Transport and climate change: A review. *J. Transp. Geogr.* **2007**, *15*, 354–367. [CrossRef]
2. UNCTAD. *Review of Maritime Transport 2019*; UNITED Nations Publication: Geneva, Switzerland, 2019.
3. Shi, W. Dynamics of Energy System Behaviour and Emissions of Trailing Suction Hopper Dredgers. Ph.D. Thesis, Delft University of Technology, Delft, The Netherlands, 2013.
4. Moreno-Gutiérrez, J.; Pájaro-Velázquez, E.; Amado-Sánchez, Y.; Rodríguez-Moreno, R.; Calderay-Cayetano, F.; Durán-Grados, V. Comparative analysis between different methods for calculating on-board ship's emissions and energy consumption based on operational data. *Sci. Total Environ.* **2019**, *650*, 575–584. [CrossRef]
5. Taljegard, M.; Brynolf, S.; Grahn, M.; Andersson, K.; Johnson, H. Cost-effective choices of marine fuels in a carbon-constrained world: Results from a global energy model. *Environ. Sci. Technol.* **2014**, *48*, 12986–12993. [CrossRef]

6.   IMO. *Third IMO Greenhouse Gas Study 2014, Executive Summary and Final Report*; International Maritime Organisation (IMO): London, UK, 2015.

7.   Perera, L.P.; Mo, B.; Kristjánsson, L.A.; Jønvik, P.C.; Svardal, J.Ø. Evaluations on ship performance under varying operational conditions. In Proceedings of the 34th International Conference on Ocean, Offshore and Arctic Engineering, St. John's, NL, Canada, 31 May–5 June 2015.

8.   Andersson, K.; Baldi, F.; Brynolf, S.; Lindgren, J.F.; Granhag, L.; Svensson, E. *Shipping and the Environment*; Springer: Berlin/Heidelberg, Germany, 2016; pp. 3–27.

9.   Klein Woud, H.; Stapersma, D. *Design of Propulsion and Electric Power Generation Systems*; IMarEST, Institute of Marine Engineering, Science and Technology: London, UK, 2002.

10.   Stapersma, D. *Main Propulsion Arrangement and Power Generation Concepts. Encyclopedia of Maritime and Offshore Engineering*; John Wiley & Sons, Ltd.: Hoboken, NJ, USA, 2017; pp. 1–40.

11.   MEPC. Guidelines for voluntary use of the ship energy efficiency operational indicator (EEOI). In *International Maritime Organization Report*; MEPC.1/Circ.684, 17 August 2009; IMO: London, UK, 2009.

12.   Acomi, N.; Acomi, O.C. Improving the Voyage Energy Efficiency by Using EEOI. *Procedia-Soc. Behav. Sci.* **2014**, *138*, 531–536. [CrossRef]

13.   Coraddu, A.; Figari, M.; Savio, S. Numerical investigation on ship energy efficiency by Monte Carlo simulation. *Proc. Inst. Mech. Eng. Part M J. Eng. Marit. Environ.* **2014**, *228*, 220–234. [CrossRef]

14.   Hou, Y.H.; Kang, K.; Liang, X. Vessel speed optimization for minimum EEOI in ice zone considering uncertainty. *Ocean Eng.* **2019**, *188*, 106240. [CrossRef]

15.   Safaei, A.; Ghassemi, H.; Ghiasi, M. Voyage Optimization for a Very Large Crude Carrier Oil Tanker: A Regional Voyage Case Study. *Sci. J. Marit. Univ. Szczec.* **2015**, *44*, 83–89.

16.   Zaccone, R.; Ottaviani, E.; Figari, M.; Altosole, M. Ship voyage optimization for safe and energy-efficient navigation: A dynamic programming approach. *Ocean Eng.* **2018**, *153*, 215–224. [CrossRef]

17.   Geertsma, R.D.; Negenborn, R.R.; Visser, K.; Loonstijn, M.A.; Hopman, J.J. Pitch control for ships with diesel mechanical and hybrid propulsion: Modelling, validation and performance quantification. *Appl. Energy* **2017**, *206*, 1609–1631. [CrossRef]

18.   Geertsma, R.D.; Visser, K.; Negenborn, R.R. Adaptive pitch control for ships with diesel mechanical and hybrid propulsion. *Appl. Energy* **2018**, *228*, 2490–2509. [CrossRef]

19.   Kalikatzarakis, M.; Geertsma, R.D.; Boonen, E.J.; Visser, K.; Negenborn, R.R. Ship energy management for hybrid propulsion and power supply with shore charging. *Control Eng. Pract.* **2018**, *76*, 133–154. [CrossRef]

20.   Vu, T.L.; Ayu, A.A.; Dhupia, J.S.; Kennedy, L.; Adnanes, A.K. Power Management for Electric Tugboats Through Operating Load Estimation. *IEEE Trans. Control Syst. Technol.* **2015**, *23*, 2375–2382. [CrossRef]

21.   Zhao, F.; Yang, W.; Tan, W.W.; Yu, W.; Yang, J.; Chou, S.K. Power management of vessel propulsion system for thrust efficiency and emissions mitigation. *Appl. Energy* **2016**, *161*, 124–132. [CrossRef]

22.   Psaraftis, H.N.; Kontovas, C.A. Speed models for energy-efficient maritime transportation: A taxonomy and survey. *Transp. Res. Part C Emerg. Technol.* **2013**, *26*, 331–351. [CrossRef]

23.   Lee, C.; Lee, H.L.; Zhang, J. The impact of slow ocean steaming on delivery reliability and fuel consumption. *Transp. Res. Part E Logist. Transp. Rev.* **2015**, *76*, 176–190. [CrossRef]

24.   Carlton, J.; Aldwinkle, J.; Anderson, J. Future ship powering options: Exploring alternative methods of ship propulsion. *Lond. R. Acad. Eng.* **2013**, *2013*, 1–95.

25.   Bouman, E.A.; Lindstad, E.; Rialland, A.I.; Strømman, A.H. State-of-the-art technologies, measures, and potential for reducing GHG emissions from shipping—A review. *Transp. Res. Part D Transp. Environ.* **2017**, *52*, 408–421. [CrossRef]

26.   Geertsma, R.D.; Negenborn, R.R.; Visser, K.; Hopman, J.J. Design and control of hybrid power and propulsion systems for smart ships: A review of developments. *Appl. Energy* **2017**, *194*, 30–54. [CrossRef]

27.   Kwasieckyj, B. Efficiency Analysis and Design Methodology of Hybrid Propulsion Systems. Master's Thesis, Delft University of Technology, Delft, The Netherlands, 2013.

28.   Bennabi, N.; Charpentier, J.F.; Menana, H.; Billard, J.Y.; Genet, P. Hybrid propulsion systems for small ships: Context and challenges. In Proceedings of the 2016 XXII International Conference on Electrical Machines (ICEM), Lausanne, Switzerland, 4–7 September 2016; pp. 2948–2954.

29.   Jafarzadeh, S.; Schjølberg, I. Operational profiles of ships in Norwegian waters: An activity-based approach to assess the benefits of hybrid and electric propulsion. *Transp. Res. Part D Transp. Environ.* **2018**, *65*, 500–523. [CrossRef]

30. Yum, K.K.; Skjong, S.; Tasker, B.; Pedersen, E.; Steen, S. Simulation of a Hybrid Marine Propulsion System in Waves. In Proceedings of the 28th CIMAC World Congress, Helsinki, Finland, 6–10 June 2016.

31. Dedes, E.K.; Hudson, D.A.; Turnock, S.R. Investigation of Diesel Hybrid systems for fuel oil reduction in slow speed ocean going ships. *Energy* **2016**, *114*, 444–456. [CrossRef]

32. Kern, C.; Tvete, H.A.; Alnes, Ø.Å.; Sletten, T.; Hansen, K.R.; Knaf, A.; Sames, P.; Puchalski, S. Battery Hybrid Oceangoing Cargo Ships. In Proceedings of the CIMAC Congress 2019, Vancouver, BC, Canada, 10–14 June 2019.

33. Sui, C.; Stapersma, D.; Visser, K.; Ding, Y.; de Vos, P. Impact of Battery-Hybrid Cargo Ship Propulsion on Fuel Consumption and Emissions during Port Approaches. In Proceedings of the CIMAC Congress 2019, Vancouver, BC, Canada, 10–14 June 2019.

34. Wei, L.; Geng, P. A review on natural gas/diesel dual fuel combustion, emissions and performance. *Fuel Process Technol.* **2016**, *142*, 264–278. [CrossRef]

35. Brynolf, S.; Fridell, E.; Andersson, K. Environmental assessment of marine fuels: Liquefied natural gas, liquefied biogas, methanol and bio-methanol. *J. Clean Prod.* **2014**, *74*, 86–95. [CrossRef]

36. Burel, F.; Taccani, R.; Zuliani, N. Improving sustainability of maritime transport through utilization of Liquefied Natural Gas (LNG) for propulsion. *Energy* **2013**, *57*, 412–420. [CrossRef]

37. Thomson, H.; Corbett, J.J.; Winebrake, J.J. Natural gas as a marine fuel. *Energy Policy* **2015**, *87*, 153–167. [CrossRef]

38. Acciaro, M. Real option analysis for environmental compliance: LNG and emission control areas. *Transp. Res. Part D Transp. Environ.* **2014**, *28*, 41–50. [CrossRef]

39. Schinas, O.; Butler, M. Feasibility and commercial considerations of LNG-fueled ships. *Ocean Eng.* **2016**, *122*, 84–96. [CrossRef]

40. AEsoy, V.; Einang, P.M.; Stenersen, D.; Hennie, E.; Valberg, I. LNG-fuelled engines and fuel systems for medium-speed engines in maritime applications. *SAE Tech. Pap.* **2011**. [CrossRef]

41. Sui, C.; Stapersma, D.; Visser, K.; de Vos, P.; Ding, Y. Energy effectiveness of ocean-going cargo ship under various operating conditions. *Ocean Eng.* **2019**, *190*, 106473. [CrossRef]

42. Lehtoranta, K.; Aakko-Saksa, P.; Murtonen, T.; Vesala, H.; Kuittinen, N.; Rönkkö, T.; Ntziachristos, L.; Karjalainen, P.; Timonen, H.; Teinilä, K. Particle and Gaseous Emissions from Marine Engines Utilizing Various Fuels and Aftertreatment Systems. In Proceedings of the CIMAC Congress 2019, Vancouver, BC, Canada, 10–14 June 2019.

43. Anderson, M.; Salo, K.; Fridell, E. Particle- and Gaseous Emissions from an LNG Powered Ship. *Environ. Sci. Technol.* **2015**, *49*, 12568–12575. [CrossRef]

44. Stapersma, D. Diesel Engines—A Fundamental Approach to Performance Analysis, Turbocharging, Combustion, Emissions and Heat Transfer. In *Emissions and Heat Transfer, Part II: Diesel Engines B—Combustion, Emissions and Heat Transfer*; Lecture Notes NLDA/TUDelft April 2010; NLDA & Delft UT: Delft, The Netherlands, 2010.

45. Jo, J.Y.; Kwon, Y.S.; Lee, J.W.; Park, J.S.; Rho, B.H.; Choi, W. Acute respiratory distress due to methane inhalation. *Tuberc. Respir. Dis.* **2013**, *74*, 120–123. [CrossRef]

46. IMO. *Annex 5 Resolution Mepc.308(73) (Adopted on 26 October 2018) 2018 Guidelines on the Method of Calculation of the Attained Energy Efficiency Design Index (eedi) for New Ships*; IMO: London, UK, 2018.

Journal of
*Marine Science
and Engineering*

Article

Electronic Constant Twist Angle Control System Suitable for Torsional Vibration Tuning of Propulsion Systems

Jaroslav Homišin, Peter Kaššay, Matej Urbanský, Michal Puškár *, Robert Grega and Jozef Krajňák

Faculty of Mechanical Engineering, Technical University of Košice, Letná 9, 042 00 Košice, Slovakia;
Jaroslav.Homisin@tuke.sk (J.H.); Peter.Kassay@tuke.sk (P.K.); Matej.Urbansky@tuke.sk (M.U.);
Robert.Grega@tuke.sk (R.G.); Jozef.Krajnak@tuke.sk (J.K.)
* Correspondence: Michal.Puskar@tuke.sk; Tel.: +421-55-602-2360

Received: 28 August 2020; Accepted: 17 September 2020; Published: 18 September 2020

Abstract: Currently, great emphasis on reducing energy consumption and harmful emissions of internal combustion engines is placed. Current control technology allows us to customize the operating mode according to the currently required output parameters, while the tuning of mechanical systems in terms of torsional vibration is often ignored. This article deals with a semi-active torsional vibroisolation system using pneumatic flexible shaft coupling with constant twist angle control. This system is suitable, as it is specially designed, for the tuning of mechanical systems where the load torque has fan characteristics (fans, ship propellers, pumps). The main goal of this research is to verify the ability of an electronic control system developed by us to maintain the pre-set constant twist angle of the used pneumatic flexible shaft coupling during operation. The constant twist angle control function was tested on a laboratory torsional oscillating mechanical system. Presented results show that the proposed electronic control system meets the requirements for its function, namely that it can achieve, sufficiently accurately and quickly, the desired constant twist angle of the pneumatic flexible shaft coupling. It is possible to assume that the presented system will increase the technical level of the equipment where it will be applied.

Keywords: marine propulsion system; pneumatic flexible shaft coupling; pneumatic tuner of torsional oscillations; torsional vibration; semi-active vibroisolation; constant twist angle control; fan characteristics; model-based control; pneumatic bellows

1. Introduction

Nowadays, reducing energy consumption and harmful emissions of internal combustion engines is a very important issue addressed by both research institutions and manufacturers of internal combustion engines. Development of control technology allows us to customize the operating mode of the device according to the currently required output parameters, while often the tuning of mechanical systems in terms of torsional vibration is ignored [1–6]. Tuning current mechanical drives in terms of torsional vibration by conventional (passive) vibroisolation methods is increasingly problematic. This is mainly due to operation in a wide range of speeds, uneven operation cylinders (deactivated cylinders, uneven fuel supply to the cylinders) and also the increased value of excitation amplitudes.

Vibroisolation methods can be divided according to what extent there is a controlled change of system parameters and according to energy supply requirements during operation of [7,8]: *passive vibroisolation*, where the dynamic parameters of the mechanical system cannot be actively changed and these solutions do not require additional energy for their function; *semi-active vibroisolation*, also called as *adaptive vibroisolation*, where it is possible to change basic parameters such as torsional stiffness, damping coefficient and mass moment of inertia, and these vibroisolation systems need energy during the change of their parameters; and *active vibroisolation*, where an additional dynamic torque component is introduced into the system, and these vibroisolation systems need constant power supply during the operation of the mechanical system.

In the case of passive vibroisolation, the system is pre-tuned in terms of torsional vibration in such a way that the system parameters such as torsional stiffness of shafts (depending on their diameter) or mass moment of inertia (e.g., by adding a flywheel) are suitably adjusted. Another way is to use special devices such as flexible shaft couplings, vibration absorbers and dampers. In this method of torsional oscillating mechanical systems tuning, the properties of the system are predetermined. However, it should be noted that the properties of these elements (apart from flywheels and pendulum vibration absorbers) may change over time, causing the mechanical system to be not tuned properly. Viscous dampers contain fluids such as silicon oils whose properties change over time. Due to high temperatures, the used liquid may solidify, causing the damper to lose its function completely [9]. Flexible couplings with rubber elastic elements can significantly change their properties depending on the static torque, operating temperature, the number of operating cycles during operation and also due to the aging of the used rubber material [10–12].

Semi-active vibroisolation systems use devices with the possibility to change their mechanical parameters affecting the size of torsional vibration (torsional stiffness, damping coefficient and moment of inertia). Devices using a change of torsional stiffness include some types of already manufactured shaft couplings such as pneumatic flexible shaft couplings [13–15] and magnetic shaft couplings [16,17]. Moreover, in the field of robotics, attention is currently paid to elements with variable torsional stiffness (variable stiffness actuators, variable stiffness joints) and variable damping devices too. Although the use of these devices is expected mainly where there may be an unexpected collision with surrounding objects or humans (e.g., household robots), there is a possibility to utilize their principles in the field of mechanical drives too. A fairly comprehensive overview of these elements can be found in [18]. Devices with variable mass moment of inertia include flywheels containing weights, whose center of gravity in the radial direction relative to the axis of rotation can be shifted [19]. Alternatively, they contain a fluid mechanism [20].

Active vibroisolation systems are developed mostly in the form of design concepts (patents), though some were verified theoretically by dynamic model simulations or by simplified physical models in laboratory conditions, but gradually they are already beginning to be verified in real ship propulsion systems as well [8]. To eliminate torsional vibrations in real time, it is proposed to use devices such as electrodynamic brakes [21–23], piezoelectric dampers [24] and inertial mass actuators [8], which introduce additional torque to the system. The time course of the additional torque is proposed mostly as harmonic [7,8] or as periodic pulses [21,23].

This article deals with a semi-active vibroisolation system using pneumatic flexible shaft coupling with constant twist angle control. This system is suitable, as it is specially designed, for tuning mechanical systems where the load torque has a fan characteristic when the load torque is approximately proportional to the square of the rotational speed (fans, ship propellers, centrifugal pumps) [25].

The main goal of this research is to verify the ability of an electronic constant twist angle control system (ECTACS), developed by us, to maintain a pre-set constant twist angle of the used pneumatic flexible shaft coupling during operation of an experimental torsional oscillating mechanical system (TOMS) in laboratory conditions.

2. Materials and Methods

2.1. Pneumatic Flexible Shaft Couplings

At our department, new types of pneumatic flexible shaft couplings are being developed [15]. The dynamic torsional stiffness of these couplings can be continuously changed by changing the air pressure in their pneumatic flexible elements. They also show less changes in properties due to the aging of the material of the elastic elements, because the elastic elements have low rigidity and most of the load is transmitted by the compressive force of the air [13]. Pneumatic flexible shaft couplings can be used for tuning torsional oscillating mechanical systems in two main ways [13–15]:

(1) *Tuning*, where the value of pressure is pre-set to a suitable value out of operation. In this case, the pneumatic flexible shaft coupling works as a classical shaft coupling with option to adapt its torsional stiffness out of operation. However, it still represents a passive vibroisolation system.

(2) *Continuous tuning*, where the pressure is adjusted (via the control system) to current operating conditions directly during operation. Thus, the pneumatic flexible shaft coupling in this case acts not as a classical shaft coupling, but it can be considered as a *pneumatic tuner of torsional oscillations (PTTO)* used as an element for the realization of semi-active vibroisolation.

Several methods of continuous tuning using pneumatic tuners of torsional oscillations were proposed and patented [26]. The two main continuous tuning methods, which were also practically implemented in laboratory conditions, are *extremal control* [27] and *constant twist angle control* [13].

2.1.1. Used Pneumatic Tuner of Torsional Oscillations

In the presented research, a tangential pneumatic tuner of torsional oscillations with fully interconnected flexible elements with a 4–2/70–T–C-type designation was used (Figure 1). This type of designation means that the PTTO has 4 double-bellows flexible elements with an outer diameter of 70 mm, the elements are placed tangentially and their compression spaces are fully interconnected. This PTTO consists of two identical hubs (1) connected by pneumatic flexible elements (2). The individual pneumatic flexible elements are mounted between triangular consoles (3). Under loading torque, one pair of opposing pneumatic flexible elements is stretched and at the same time the other pair is equally compressed, allowing the PTTO to transmit torque in both directions of rotation. The compression space of the PTTO is fully interconnected by polyamide hoses with a diameter of 6 mm (4). Its filling with gaseous medium is realized through a rotary air supply, which is a part of one of the connecting flanges of the PTTO. Each flexible element includes two pneumatic screw connections (5).

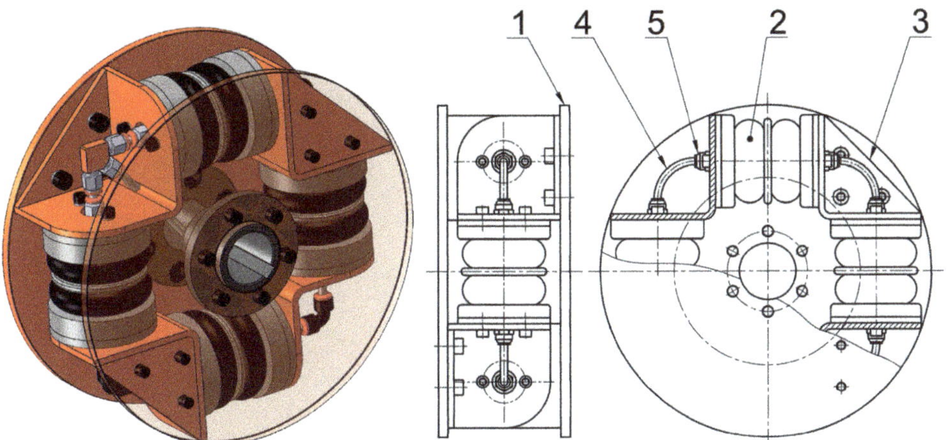

Figure 1. Pneumatic tuner of torsional oscillations type 4–2/70–T–C.

In order to react flexibly with the variable load torque generated directly during the operation of the mechanical system, it is necessary to ensure that the pressure in its entire compression space changes as quickly and evenly as possible. This is ensured by the full interconnection of pneumatic elements compression spaces. The speed and uniformity of the filling of the compression space can be influenced by increasing the inner diameter of the connecting hoses too, but at the expense of reducing the resistance to the air flow between the pneumatic flexible elements. In practice, however, this means a reduction in the damping properties of the PTTO itself [28,29].

Generally, the static load torque of pneumatic couplings M_{stat} [N·m] at a twist angle φ [rad] can be expressed as [30]

$$M_{stat} = M_{G(\varphi)} + p_T \cdot (S_e \cdot r)_{(\varphi)}, \tag{1}$$

where $M_{G(\varphi)}$ [N·m] is the pneumatic flexible element rubber shell torque, p_T [Pa] is the overpressure in the pneumatic flexible elements of the coupling, S_e [m^2] is the effective area of the coupling's compression space and r [m] is the distance of the center of the effective area S_e from the coupling's axis. Expression $Se \cdot r$ [m^3] is then the static moment of the effective area to the coupling's axis. Rubber shell torque and static moment of effective area are expressed as a function of the twist angle φ. The parameters M_G and $S_e \cdot r$ can be determined from measured static load torque—static load characteristics (Figure 2a) and overpressure (Figure 2b) depending on the twist angle φ at different initial overpressures. The full procedure of obtaining these parameters is described in [30].

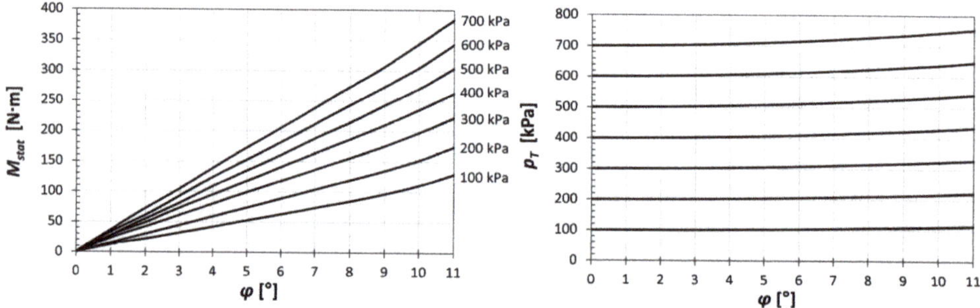

Figure 2. Results of *4–2/70–T–C*-type pneumatic tuner of torsional oscillations static measurements: (**a**) static load characteristics; (**b**) overpressure.

In Figure 3, the resulting rubber shell torque M_G (Figure 3a) and effective area of the coupling compression space $Se \cdot r$ (Figure 3b) in graphical and equation form are shown [28].

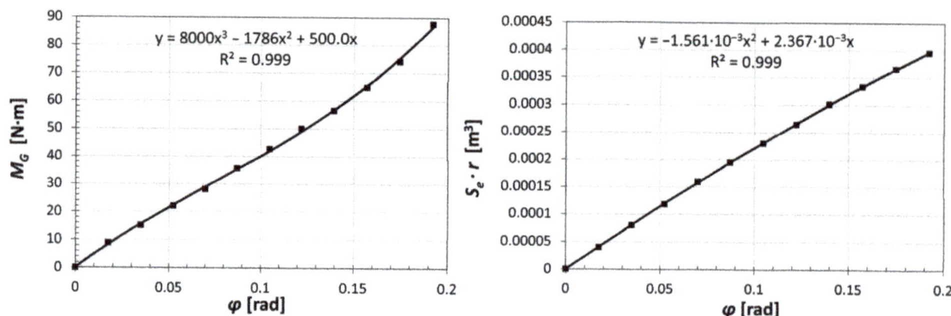

Figure 3. Static parameters of *4–2/70–T–C*-type pneumatic tuner of torsional oscillations: (**a**) rubber shell torque; (**b**) effective area of the coupling compression space.

The model shows good agreement with the actual PTTO, as the difference between the calculated and measured values of the static load torque does not exceed 5% [28]

2.2. Constant Twist Angle Control

The principle of constant twist angle control is that it maintains a pre-set constant static twist angle of the PTTO [31]. This is ensured by a control system which adapts the overpressure of the gaseous medium in the PTTO to the static load torque in order to maintain the pre-set constant twist angle. Constant twist angle control is suitable for tuning mechanical systems with fan characteristics. The principle of operation is explained in more detail below [32].

The transmitted static load torque M_{stat} and dynamic stiffness k_{dyn} [N·m·rad^{-1}] of the PTTO by a given constant twist angle φ_{const} (for simplicity neglecting the rubber shell torque M_G which is relatively small compared to the total transmitted torque [13]) can be generally considered as proportional to the overpressure p_{TC} in the compression space of the PTTO corresponding to the given constant twist angle (see Equation (1)).

Thus, the static load torque can be expressed as

$$M_{stat} = a_{(\varphi_{const})} \cdot p_{TC} \tag{2}$$

and dynamic torsional stiffness as

$$k_{dyn} = b_{(\varphi_{const})} \cdot p_{TC}, \tag{3}$$

where $a_{(\varphi_{const})}$ [N·m·Pa^{-1}] and $b_{(\varphi_{const})}$ [N·m·rad^{-1}·Pa^{-1}] are factors whose values depend on the constant twist angle φ_{const} only.

The transmitted static load torque corresponding to the fan characteristics (of the propeller) can be expressed as

$$M_{stat} = c_f \cdot \omega^2, \tag{4}$$

where c_f [N·m·rad^{-2}·s^2] is a constant and ω [rad·s^{-1}] is the angular speed of the propeller shaft.

From the equality of Equations (2) and (4) for the static load torque M_{stat}, then expressing p_{TC} and putting it into Equation (3), the dynamic torsional stiffness will be

$$k_{dyn} = b_{(\varphi_{const})} \cdot \frac{c_f}{a_{(\varphi_{const})}} \cdot \omega^2. \tag{5}$$

Considering a two-mass torsional oscillating mechanical system with dynamic torsional stiffness k_{dyn} and equivalent mass moment of inertia I_{RED} [kg·m^2], the natural angular frequency of this system will be

$$\Omega_0 = \sqrt{\frac{k_{dyn}}{I_{RED}}} = \sqrt{\frac{b_{(\varphi_{const})} \cdot c_f}{a_{(\varphi_{const})} \cdot I_{RED}}} \cdot \omega = C_{(\varphi_{const})} \cdot \omega, \tag{6}$$

where $C_{(\varphi_{const})}$ [-] is a factor depending on the given constant twist angle φ_{const} only. This means that the natural frequency of the mechanical system will be proportional to the angular speed of the propeller shaft, and the value of factor $C_{(\varphi_{const})}$ can be properly set by selecting a suitable constant twist angle φ_{const}. This is illustrated with an interference diagram of a two-mass torsional oscillating mechanical system using the PTTO with linear characteristics in Figure 4, where ω is the angular speed of the shaft, ω_{e1} and ω_{e2} are the excitation frequencies resulting from the periodically alternating load torque and $\Omega_0(\varphi_{const1 \ldots 8})$ are the natural frequencies corresponding to the pre-set constant twist angles $\varphi_{const1} < \varphi_{const2} < \ldots < \varphi_{const8}$. The intersections of the excitation frequencies with the natural frequency represent resonances at which the transmitted load torque reaches its maximum value. The case of resonance should be avoided in the operating speed range, as it can be dangerous in terms of a high alternating load torque.

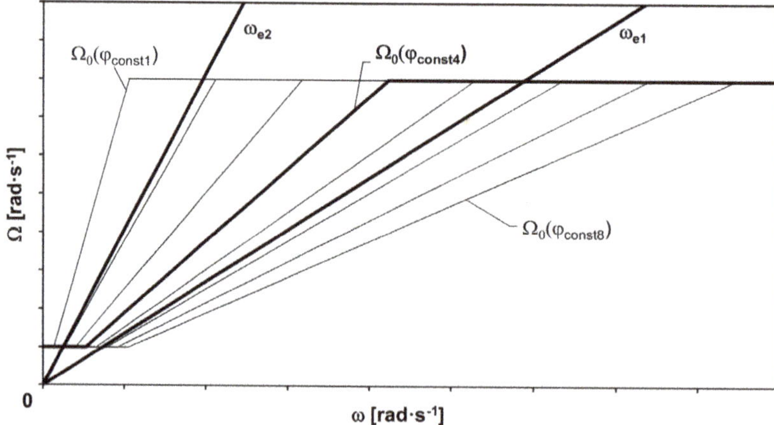

Figure 4. Interference diagram of a mechanical system with constant twist angle control.

According to Figure 4, the course of natural frequencies consists of three parts. The first part, the *sub-regulatory area*, is a horizontal line beginning at zero speed when the load torque is zero (resulting from the fan characteristics) and the PTTO is inflated to the minimum operating pressure. At a certain speed, the torque twists the PTTO to the selected constant twist angle, and this is the start of the second part representing the *regulatory area* of the PTTO. In the regulatory area, the pressure in the PTTO is regulated to a value where the PTTO has the desired constant twist angle. The regulatory area ends at a speed when the pressure reaches the maximum operating value. This point is the start of the *over-regulation area*. After this point, the pressure has its maximum operating value. In the presented case, the optimum constant twist angle is φ_{const4}, where the torsional natural frequency is farthest from the excitation frequencies, i.e., farthest from the resonance in the regulatory area of the PTTO.

Results of theoretical analyses in marine propulsion mechanical systems using constant twist angle control presented, for example, in [31,33] show that this type of control ensures proper tuning in terms of torsional vibration and has considerable potential for future application.

Constant twist angle control can be ensured by a constant twist angle regulator which is a part of the PTTO [13,26], or ECTACS can be used.

2.3. Experimental Torsional Oscillating Mechanical System

In order to verify the ability of our ECTACS to keep the twist angle of a PTTO at a pre-set constant value, the ECTACS was applied into an experimental torsional oscillating mechanical system.

The TOMS of the piston compressor drive (Figure 5) is made up of a 3-phase asynchronous electromotor, *Siemens 1LE10011DB234AF4-Z* (11 kW, 1500 RPM) *(1)*. Rotational speed of this electromotor is continuously vector-controlled by a frequency converter, *Sinamics G120C*. The electromotor drives a 3-cylinder piston compressor, *ORLIK 3JSK-75 (3)*, through a PTTO type *4-2/70-T-C (2)* (Figure 1). The compressor in this system acts as a load and torsional vibration exciter too. For proper tuning of the system, it is necessary to know the dynamic behavior under different operating conditions, which was experimentally investigated in [34].

Under standard conditions, the load torque of the system has no fan characteristics, and this means that this TOMS is not very suitable for tuning torsional vibrations with constant twist angle control, so our goal is not the tuning itself but only the verification of the ability to maintain the desired constant twist angle of the PTTO.

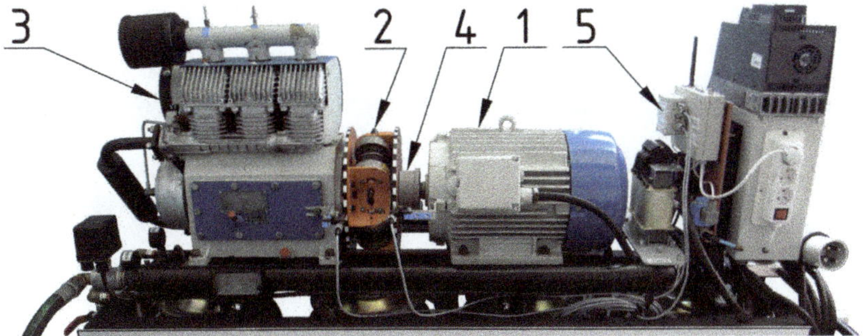

Figure 5. Experimental torsional oscillating mechanical system.

The whole drive is mounted on a rigid sprung frame. The compressed air from the compressor flows into an air pressure tank with a volume of 20 l. The air pressure in the tank is controlled by a throttling valve. So, the load of the TOMS can be adjusted. Maximum compressor output air overpressure is 800 kPa and its value is measured by a pressure sensor, *Danfoss MBS 3000*, with an overpressure measuring range of 0–1 MPa. The same type of pressure sensor is used for measuring the air pressure value in the compression space of the PTTO. The accuracy of the *MBS 3000* sensor with a metal membrane is 0.5% of its measuring range, i.e., 5 kPa (combined fault–nonlinearity, hysteresis and reproducibility). The supply of compressed air into the PTTO is ensured by a rotation supply *(4)*. The mechanical part of the experimental TOMS is described in more detail in [35].

The multifunctional electronic module (MFEM), marked with the number *(5)*, which is a part of an electronic constant twist angle control system, developed by us, has the following functions: Page: 7 Is the italics necessary?

(1)   Power supply for the optoelectronic sensors, pressure sensors and electromagnetic valves;
(2)   Measurement of the black-to-white stripe edge-crossing times for both hubs of the PTTO;
(3)   Measurement of the air pressure value in the compression space of the PTTO and the compressor output air pressure value;
(4)   Communication with the software part (running in a PC) of our ECTACS. The measured data are sent to the PC in order to be further processed in real-time;
(5)   Setting of the needed value of the air pressure in the compression space of the PTTO, which is computed by the software part of our ECTACS. The quick and very accurate pressure setting is carried out by electromagnetic valves, one for the inflation and one for the deflation of the compression space of the PTTO.

2.4. Data Measuring and Processing

The dynamic load torque transmitted by the PTTO causes mutual dynamic angular twisting of the driving and driven hubs of the PTTO. The measurement is based on determining the PTTO's twist angle time course. The PTTO is equipped with black–white tape, which is stuck to the circumference of the driving and driven hubs of the PTTO and they are scanned by a pair of optoelectronic sensors (Figure 6). According to our specific requirements, a pair of *Dewetron* optoelectronic sensors of type *SE-TACHO-PROBE-01* [36] (Figure 6) was used.

These sensors detect the reflection from the reflective black–white moving tape (Figure 6b). The sensors react to the edges between the black and white stripes. There is a distinct change of electrical output voltage at the moment the edge is crossed. These sensors can work with a maximum frequency of 10 kHz but the cleanness of the reflective tape, cleanness of the optical parts of the sensors and the sharpness of the edges between the black and white stripes must be excellent [36].

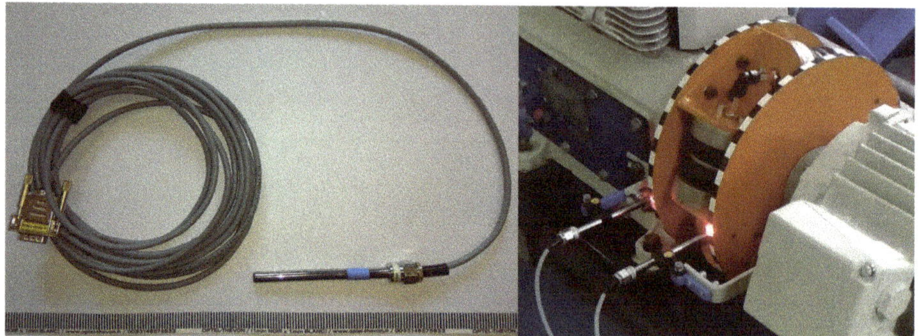

Figure 6. Optoelectronic sensors *Dewetron SE-TACHO-PROBE-01*: (**a**) sensor attached to a data cable; (**b**) pair of sensors mounted in the experimental torsional oscillating mechanical system.

The number of black and white stripe pairs for the driving and driven hubs should be equal and chosen with respect to the maximum twist angle of the PTTO and to the character of the transmitted load torque (especially its dynamic component). It is also important that the length of all the black and white stripes should be equal. The angle corresponding to length of a black and white stripe pair must be larger than the maximum twist angle of the PTTO. For the PTTO type *4-2/70-T-C*, the maximum twist angle is 11° which corresponds to a maximum of 32 pairs of stripes. Further, as the major harmonic component for a three-piston compressor is the 3rd harmonic component, it is advantageous to select the number of black and white stripe pairs as integer multiples of 3 to obtain the best results for the twist angle time course by the equal operation of all cylinders (theoretically "3 identical recorded curve portions in one revolution"). Therefore, 30 black–white stripe pairs for each hub of the PTTO have been selected. The edge-crossing times t_i need to be measured, where the index i stands for the sample order number. In our case, only the black-to-white stripe edges are considered. The times t_i are computed from the counted number of impulses from the counter of the MFEM microprocessor. One impulse represents a time of 1/14745600 s.

Considering the times t_{1i} for the driving hub and the times t_{2i} for the driven hub of the PTTO, the time delays Δt_i [s] can be computed using the following equation:

$$\Delta t_i = t_{2i} - t_{1i}. \tag{7}$$

In the next step, the total twist angle φ_T [rad] of the PTTO can be computed according to the following equation:

$$\varphi_{Ti} = \frac{\pi \cdot n \cdot \Delta t_i}{30}, \tag{8}$$

where n [min^{-1}] is the immediate rotational speed of the mechanical system.

From the total twist angle of the PTTO φ_T, its static component φ_{stat} can be computed as the mean value:

$$\varphi_{stat} = \frac{\sum_{i=1}^{k} \varphi_{Ti}}{k}, \tag{9}$$

where k [-] is the number of samples, which has to be an integer multiple of the black–white stripe pairs number on the hub. In practice, the floating average method is used for this computation.

The computations according to Equations (7)–(9) are performed by the software part of our ECTACS in PC in real time. This way, the controlled variable (φ_{stat}) for the ECTACS can be computed. It is very advantageous because the torsional vibration does not directly affect the control device like in the case of regulators directly built into the PTTO.

2.5. Description of the Constant Twist Angle Control System Function

The goal of constant twist angle control is to maintain a pre-set static (mean) value of the twist angle by any given static load torque M_{stat} resulting from the current operating mode. Although it is possible to use a closed-loop PID control system with a static twist angle as a controlled variable and overpressure as a manipulated variable, based on the fact that the mathematical and physical model of the PTTO (presented in Section 2.1.1) is well known, it was decided to use a model-based adaptive control system [37]. This approach allows us to reach the desired value of the static twist angle more quickly, which is very important in terms of passing through the resonance as quickly as possible.

The static load torque of the PTTO at a periodically alternating load torque can be computed as

$$M_{stat} = M_{G(\varphi_{stat})} + p_T \cdot (S_e \cdot r)_{(\varphi_{stat})}, \tag{10}$$

where p_T is the mean overpressure in the PTTO, and the values of the rubber shell torque M_G and static moment of effective area $S_e \cdot r$ are computed from the static twist angle φ_{stat}. The value of the twist angle is continuously measured and its mean value is computed by the control system.

The rubber shell torque $M_{G(\varphi stat)}$ and the static moment of effective area $S_e \cdot r_{(\varphi stat)}$ are computed as fifth degree polynomials:

$$M_{G(\varphi_{stat})} = \sum_{i=0}^{5} a_i \cdot \varphi_{stat}^i, \tag{11}$$

$$(S_e \cdot r)_{(\varphi_{stat})} = \sum_{i=0}^{5} b_i \cdot \varphi_{stat}^i. \tag{12}$$

After computing the static load torque according to Equation (10), the value of overpressure p_{TC} needed for obtaining the desired constant twist angle φ_{const}, using Equations (11)–(12), where the desired constant twist angle φ_{const} is set instead of the static twist angle φ_{stat}, can be expressed as follows:

$$p_{TC} = \frac{M_{stat} - M_{G(\varphi_{const})}}{(S_e \cdot r)_{(\varphi_{const})}}. \tag{13}$$

After setting the new value of overpressure according to Equation (13), the value of the actual static twist angle is measured. The difference between the actual static twist angle φ_{stat} and the desired static twist angle φ_{const} can be expressed as

$$\Delta\varphi = \varphi_{const} - \varphi_{stat}. \tag{14}$$

In the case where the achieved value of the mean twist angle lies outside the insensitivity range φ_{ins}, $|\Delta\varphi| > \varphi_{ins}$, but inside the fine tuning range φ_{FT}, $|\Delta\varphi| \leq \varphi_{FT}$, fine tuning is used. The value of the needed overpressure is then computed as

$$p_{TC} = p_T + X \cdot \Delta\varphi \cdot c, \tag{15}$$

where X [Pa·rad^{-1}] is a derivation of p_{TC} according to Equation (13) by angle φ_{const} and c [-] is a constant factor. The value of the constant factor c should be selected in the range $(0; 1)$.

Derivation X is then computed as

$$X = -\frac{\sum_{i=1}^{5} i \cdot a_i \cdot \varphi_{stat}^{i-1}}{(S_e \cdot r)_{(\varphi_{stat})}} - \frac{\left(M_{stat} - M_{G(\varphi_{stat})}\right) \cdot \sum_{i=1}^{5} i \cdot b_i \cdot \varphi_{stat}^{i-1}}{\left((S_e \cdot r)_{(\varphi_{stat})}\right)^2}. \tag{16}$$

The flowchart of the constant twist angle control algorithm is shown in Figure 7.

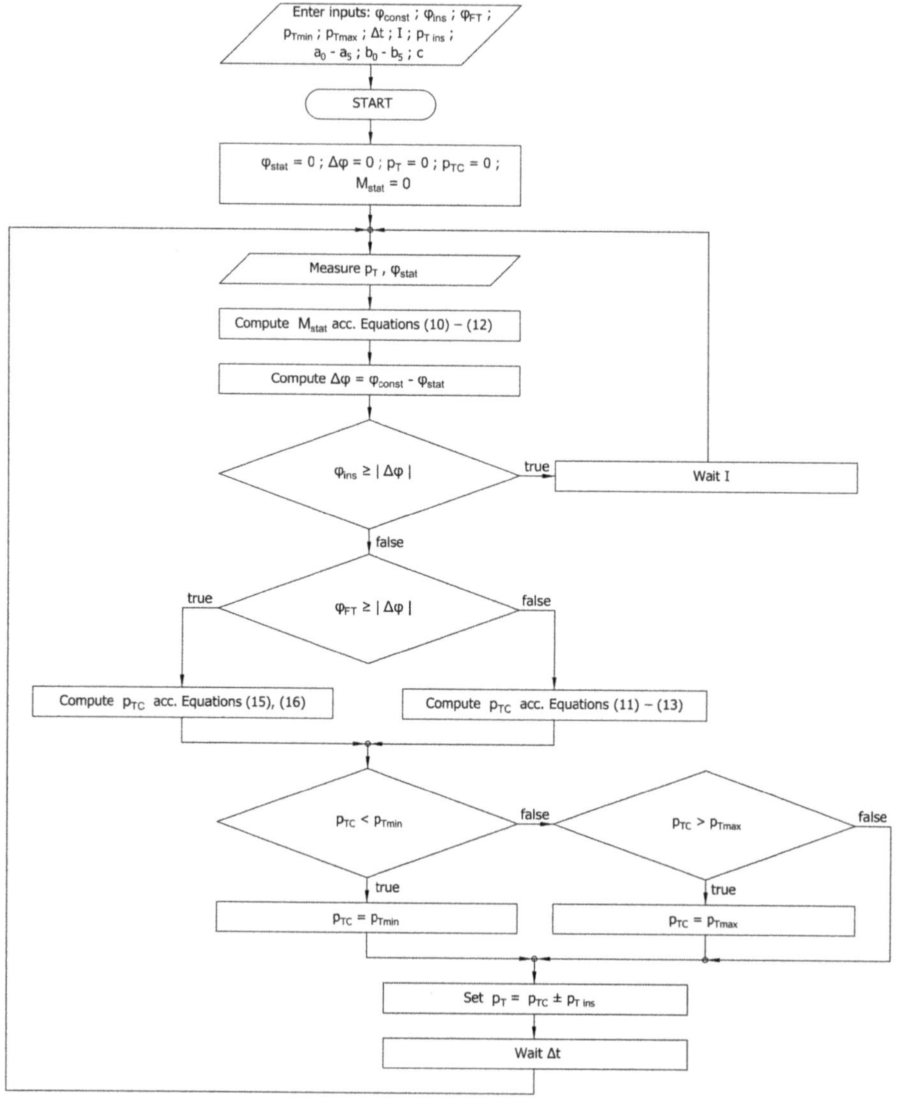

Figure 7. Flowchart of constant twist angle control algorithm.

The parameters of the algorithm can be set via the software graphic user interface of the ECTACS on a PC.

3. Results and Discussion

In this research, as the focus is placed on verifying the ability of our ECTACS to maintain a defined constant value φ_{const} of the PTTO's twist angle static component, and torsional vibration in the TOMS does not directly affect the process of twist angle control (is not present in the algorithm), only the static component M_{stat} of the load torque transmitted by the PTTO is shown in the presented results.

In Figure 8, the time courses of the static component of the load torque transmitted by the PTTO M_{stat}, compressor output air overpressure p_C and rotational speed n of the TOMS during the experimental measurement are displayed. Three different operating modes (marked with numbers 1–3 in Figure 8) were chosen for testing:

(1) Minimum rotational speed, negligible compressor output air overpressure (caused only by flow resistance in the compressor output piping);

(2) Maximum rotational speed, negligible compressor output air overpressure (caused by flow resistance in the compressor output piping);

(3) Maximum rotational speed, maximum compressor output air overpressure set by the throttling valve.

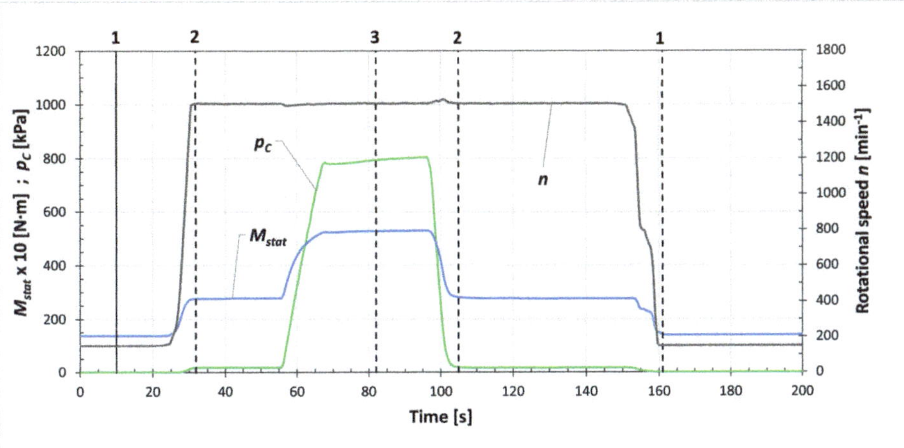

Figure 8. Operating modes of the mechanical system used during testing.

The vertical dashed lines in Figure 8 mark the times where the operating modes begin to be steady state (the rotational speed n and compressor output air overpressure p_C do not change).

The sequence of mechanical system operating modes was chosen so that the mechanical system is initially minimally loaded (operating mode 1), then it is partially loaded (operating mode 2) and then it is fully loaded (operating mode 3). The sequence subsequently continues with the unloading of the mechanical system (operating modes 2 and 1).

The aim of our ECTACS is to keep the static component $\varphi_{stat} = \varphi_{const}$ regardless of the operating mode. Therefore, the φ_{stat} is the controlled variable. The manipulated variable is the overpressure p_T in the compression space of the PTTO whose operating range was set to 0–800 kPa. Since the φ_{stat} is used directly as an input variable, it is a feedback control system. Since our system uses a very accurate mathematical and physical model of the PTTO for the computation of the needed value of p_T, it also allows us to set the φ_{stat} very accurately (±0.1 degree without difficulty) during the operation of the mechanical system (Figures 9 and 10).

In Figure 9, the control process by mechanical system loading is shown. The course of the φ_{stat} is very close to the defined constant value φ_{const} (in our case $\varphi_{const} = 2°$) of the PPTO's twist angle static component after the change of operating mode 1 to operating mode 2 and subsequently operating mode 2 to operating mode 3. In the first case, the set point $\varphi_{const} \pm 0.05°$ tolerance was reached in two steps of p_T change, and in the second case in three steps of p_T change (although it was very close after two steps of p_T change).

There are certain idle intervals after the changes of the manipulated variable p_T. The intervals are necessary in order to stabilize and measure the controlled variable correctly because transitional effects arise after the change of p_T in the mechanical system. The needed stabilization time depends on the type, character and dynamical behavior of the mechanical system. The selection of a stabilization time for a specific mechanical system requires an individual approach based on theoretical or experimental investigation of the transitional effects caused especially by the rapid coupling pressure changes [38–45].

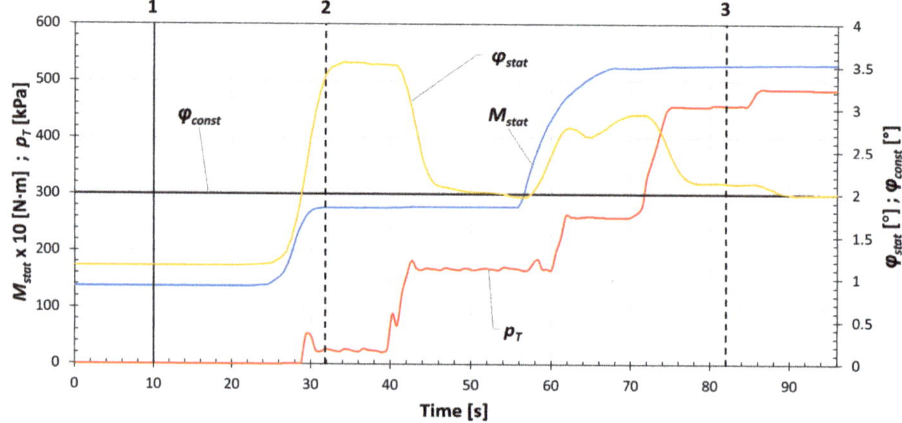

Figure 9. Control process by loading the mechanical system.

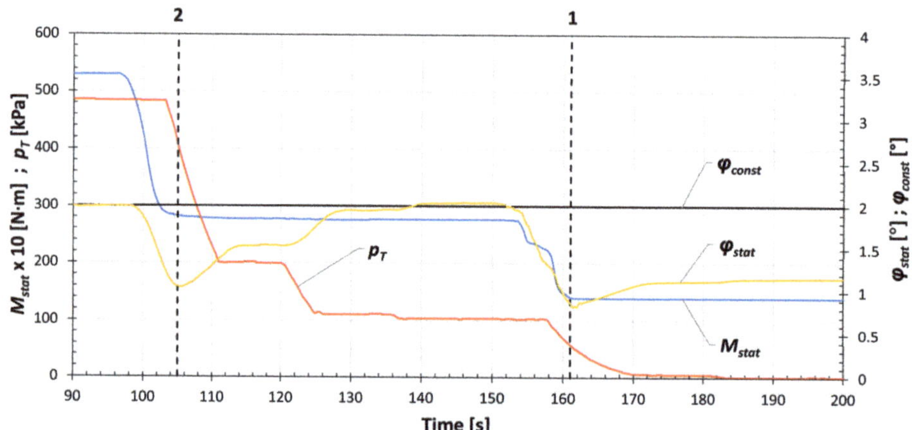

Figure 10. Control process by unloading the mechanical system.

It is also important to notice that the control system reacts immediately to the change of the φ_{stat}, and it does not wait for the steady state. Since operation mode 1 is characterized by the minimum rotational speed and negligible compressor output air overpressure, the transmitted load torque is also the minimum and therefore unable to twist the PTTO to the desired φ_{const} even by zero overpressure p_T in the compression space of the PTTO.

In Figure 10, the control process by mechanical system unloading is shown. The course of the φ_{stat} is very close to the defined constant value $\varphi_{const} = 2°$ after the change of operational state 3 to operational state 2. The set point $\varphi_{const} \pm 0.05°$ tolerance was reached in three steps of p_T change (although it was very close again after two steps of p_T change). Therefore, it is very important to select the set point tolerance reasonably (the wider the set point tolerance, the shorter the control process). From our existing research, e.g., [31,33,46–54], it can be said that the set point tolerance of 0.1° meets our requirements for practical applications of the PTTO with constant twist angle control. Again, regarding the change to operating mode 1, the transmitted load torque is the minimum and therefore it is unable to twist the PTTO to the desired constant twist angle φ_{const}.

The customizable graphical user interface of the software of our ECTACS is shown in Figure 11.

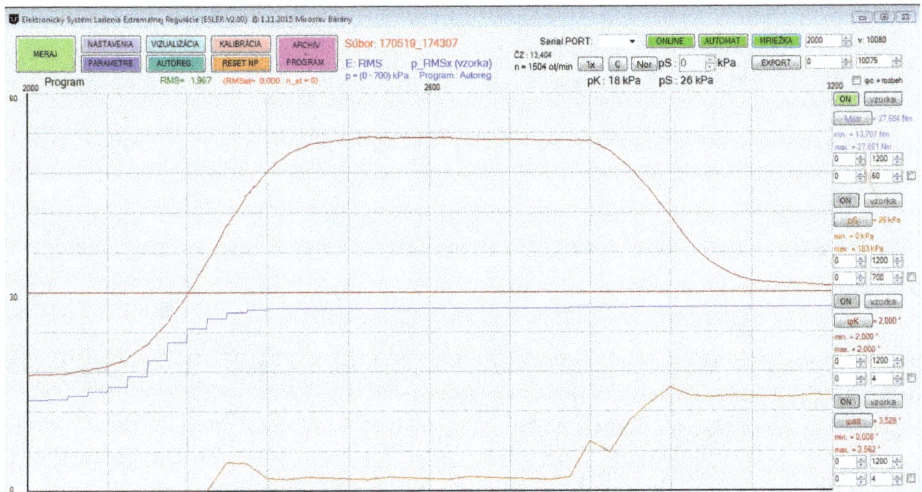

Figure 11. The graphical user interface of the electronic constant twist angle control system.

The software is developed by us and it is programmed in C++. It allows us to set, via the graphical user interface, all needed parameters, for example, the φ_{const}, tolerances, stabilization times, parameters of the PTTO's mathematical and physical model, etc., even during the operation of the mechanical system. By monitoring the time courses of the selected parameters, the whole control process and the response of the mechanical system to the changes of the parameters or operating modes in real time can be observed. Furthermore, the data can be stored and viewed or exported for further analysis in post-processing mode.

4. Discussion

From the presented results, it is obvious that the proposed electronic constant twist angle control system meets the requirements for its function, namely that it can achieve, sufficiently accurately and quickly, the desired constant twist angle of the pneumatic tuner of torsional oscillations. It is possible to assume that the proposed system will increase the technical level of the equipment where it will be applied.

However, it is well known that every technical solution has advantages and disadvantages. The use of the presented electronic constant twist angle control system provides the following general advantages:

- The system provides a quick and very accurate setting of a constant twist angle of the pneumatic tuner, thanks to the mathematical and physical model of the pneumatic tuner used for the twist angle computations;
- Torsional vibration in the mechanical system does not directly affect the control device like in the case of regulators directly built into the pneumatic tuner;
- Dynamic mass properties of the mechanical system are not influenced by additional masses because there are no regulators built into the pneumatic tuner;
- It allows the use of any torsional oscillating mechanical system (regardless of size and transmitted load torque);

- There is no friction between the sensors and the hubs of the pneumatic tuner;
- It is possible to quickly replace the broken, damaged or malfunctioning sensor.
- It is also necessary to mention the general disadvantages of the presented electronic system:
- The function of the system is sensitive to dirt on the optical part of the optoelectronic sensors or the reflective black–white tape. This issue could be fixed using sensors with a similar working principle, for example, proximity probes and a toothed wheel instead of black and white stripes;
- The system needs a power supply and in its present state and also a PC for the software part of the system;
- The pneumatic tuner's mathematical and physical model parameters, which need to be set in the software, have to be known.

In our further research, we are planning to verify the proper tuning of a torsional oscillating mechanical system with fan load characteristics using our electronic constant twist angle control system.

5. Patents

Homišin, J. *Control system for continuous tuning of pneumatic coupling's angular twist.* Utility model SK7497Y1, Industrial Property Office of Slovak Republic, Banská Bystrica 2016. 1 August 2016. Available online: https://wbr.indprop.gov.sk/WebRegistre/Tlac/Download?fileName=COO.2161.100.7.3398451 (accessed on 18 September 2020). (In Slovak)

Homišin J. *Continuously tuned mechanical system.* Patent PL 216901 B1, The Patent Office of the Republic of Poland, Warszawa 24 October 2013. Available online: https://api-ewyszukiwarka. pue.uprp.gov.pl/api/collection/68227816fb32c5ac9046f508f5d65afa (accessed on 18 September 2020). (In Polish)

Author Contributions: Conceptualization, P.K. and M.U.; methodology, J.H.; software, R.G.; validation, M.U. and P.K.; formal analysis, J.H.; investigation, R.G., M.P. and J.K.; resources, M.P. and J.K.; data curation, M.U.; writing—original draft preparation, P.K. and M.U.; writing—review and editing, P.K., M.U. and M.P.; visualization, M.U.; supervision, J.H.; project administration, J.H.; funding acquisition, J.H. All authors have read and agreed to the published version of the manuscript.

Funding: Research was funded by grant project *VEGA 1/0528/20 "Solution of new elements for mechanical system tuning."* "This work was supported by the Slovak Research and Development Agency under the Contract no. APVV-19-0328."

Acknowledgments: The authors would like to thank Ing. Miroslav Bárány for the technical support.

Conflicts of Interest: The authors declare no conflict of interest.

References

1. Zoul, V. Dynamic of propulsion, present situation and trends. *Trans. Univ. Košice* **2014**, *2*, 101–106.
2. Czech, P. Application of probabilistic neural network and vibration signals for gasket under diesel engine head damage. *Sci. J. Sil. Univ. Technol. Ser. Transp.* **2013**, *78*, 39–45.
3. Czech, P. Determination of the course of pressure in an internal combustion engine cylinder with the use of vibration effects and radial basis function—Preliminary research. In *Communications in Computer and Information Science*; Mikulski, J., Ed.; Springer: Berlin, Germany, 2012; Volume 329, pp. 175–182. [CrossRef]
4. Puškár, M.; Bigoš, P. Output performance increase of two-stroke combustion engine with detonation combustion optimization. *Strojarstvo* **2010**, *52*, 577–587.
5. Sinay, J.; Puškár, M.; Kopas, M. Reduction of the NOx emissions in vehicle diesel engine in order to fulfill future rules concerning emissions released into air. *Sci. Total Environ.* **2018**, *624*, 1421–1428. [CrossRef] [PubMed]
6. Czech, P.; Wojnar, G.; Burdzik, R.; Konieczny, L.; Warczek, J. Application of the discrete wavelet transform and probabilistic neural networks in IC engine fault diagnostics. *J. Vibroeng.* **2014**, *16*, 1619–1639.
7. Liang, X.; Shu, G.; Dong, L.; Wang, B.; Yang, K. Progress and Recent Trends in Torsional Vibration of Internal Combustion Engines. In *Advances in Vibration Analysis Research*; Ebrahimi, F., Ed.; InTech: Rijeka, Croatia, 2011; pp. 245–272. [CrossRef]

8. Bartel, T.; Herold, S.; Infante, F.; Käsgen, J.; Matthias, M.; Millitzer, J.; Perfetto, S. Active Vibration Reduction of Ship Propulsion Systems. In Proceedings of the 2018 Joint Conference—Acoustics, Ustka, Poland, 11–14 September 2018; Marszal, J., Kochańska, I., Eds.; Polish Acoustical Society: Gdańsk, Poland, 2018; pp. 15–20. [CrossRef]

9. Feese, T.; Hill, C. Prevention of Torsional Vibration Problems in Reciprocating Machinery. In Proceedings of the 38th Turbomachinery Symposium, Houston, Texas, 14–17 September 2009; Turbomachinery Laboratories, Texas A&M University: College Station, TX, USA, 2018; pp. 213–238. [CrossRef]

10. Gurský, P. Identification of the Influence of Work Cycles on the Basic Properties of Different Types of Flexible Couplings and Their Mutual Comparison. Ph.D. Thesis, Technical University of Košice, Košice, Slovakia, 2011. (In Slovak).

11. Han, H.S.; Lee, K.H.; Park, S.H. Parametric study to identify the cause of high torsional vibration of the propulsion shaft in the ship. *Eng. Fail. Anal.* **2016**, *59*, 334–346. [CrossRef]

12. Han, H.; Lee, K.; Park, S. Evaluation of the increased stiffness for the elastic coupling under the dynamic loading conditions in a ship. *Eng. Fail. Anal.* **2016**, *68*, 254–262. [CrossRef]

13. Homišin, J. *New Types of Flexible Shaft Couplings: Development, Research, Application*; Vienala: Košice, Slovakia, 2002. (In Slovak)

14. Neupauerová, S. Investigation of stress allocation in the rubber-cord flexible element by FEF. *Acta Mech. Slov.* **2005**, *9*, 27–30. (In Slovak)

15. Homišin, J. Contribution and perspectives of new flexible shaft coupling types—Pneumatic couplings. *Sci. J. Sil. Univ. Technol. Ser. Transp.* **2018**, *99*, 65–77. [CrossRef]

16. Lubin, T.; Mezani, S.; Rezzoug, A. Experimental and theoretical analyses of axial magnetic coupling under steady-state and transient operations. *IEEE Ind. Electron. Mag.* **2014**, *61*, 4356–4365. [CrossRef]

17. Sudano, A.; Accoto, D.; Zollo, D.; Guglielmelli, E. Design, Development and Scaling Analysis of a Variable Stiffnes Magnetic Torsion Spring. *Int. J. Adv. Robot. Syst.* **2013**, *10*, 1–11. [CrossRef]

18. VanderBorght, B.; Albu-Schaeffer, A.; Bicchi, A.; Burdet, E.; Caldwell, D.; Carloni, R.; Catalano, M.G.; Eiberger, O.; Friedl, W.; Ganesh, G.; et al. Variable impedance actuators: A review. *Robot. Auton. Syst.* **2013**, *61*, 1601–1614. [CrossRef]

19. Yang, T. The Principles and Structure of Variable-Inertia Flywheels. European Patent EP 0508790 (A1), 19 February 1997. Available online: https://patents.google.com/patent/EP0508790A1/zh (accessed on 18 September 2020).

20. Lewis, O.G. Variable Inertia Liquid Flywheel. U.S. Patent 3248967, 3 May 1966. Available online: http://www.freepatentsonline.com/3248967.pdf (accessed on 18 September 2020).

21. Walkowc, J. Active Torsional Vibration Damper. U.S. Patent 5678460A, 21 October 1997.

22. Kadomukai, Y.; Yamakado, M.; Nakamura, Y. Torque Controlling Apparatus for Internal Combustion Engine. U.S. Patent 4922869, 8 May 1990. Available online: https://patents.google.com/patent/US4922869 (accessed on 18 September 2020).

23. Vadamalu, R.S.; Beidl, C.; Hohenberg, G.; Muehlbauer, K. Active torsional vibration reduction: Potential analysis and controller development for a belt-driven 48 V system. *Automot. Engine Technol.* **2019**, *4*, 139–151. [CrossRef]

24. Przybyłowicz, P.M. Torsional vibration control by active piezoelectric system. *J. Theor. Appl. Mech.* **1995**, *33*, 809–823.

25. Hughes, A. *Electric Motors and Drives: Fundamentals, Types and Applications*, 3rd ed.; Elsevier: Oxford, UK, 2006.

26. Homišin, J. Characteristics of pneumatic tuners of torsional oscillation as a result of patent activity. *Acta Mech. Autom.* **2016**, *10*, 316–323. [CrossRef]

27. Homišin, J. Static optimisation of mechanical systems based on the method of extremal regulation. *Sci. J. Sil. Univ. Technol. Ser. Transp.* **2019**, *103*, 15–29. [CrossRef]

28. Čopan, P. Application of new Tuning Method of Torsional Oscillating Mechanical Systems. Ph.D. Thesis, Technical University of Košice, Košice, Slovakia, 2014. (In Slovak).

29. Pešík, L.; Skarolek, A.; Kohl, O. Vibration Isolation Pneumatic System with a Throttle Valve. In *The Latest Methods of Construction Design*; Dynybyl, V., Berka, O., Petr, K., Lopot, F., Dub, M., Eds.; Springer: Cham, Switzerland, 2016; pp. 75–79. [CrossRef]

30. Kaššay, P.; Homišin, J.; Urbanský, M. Formulation of Mathematical and Physical Model of Pneumatic Flexible Shaft Couplings. *Sci. J. Sil. Univ. Technol. Ser. Transp.* **2012**, *76*, 25–30. [CrossRef]

31. Homišin, J.; Kaššay, P.; Puškár, M.; Grega, R.; Krajňák, J.; Urbanský, M.; Moravič, M. Continuous tuning of ship propulsion system by means of pneumatic tuner of torsional oscillation. *Int. J. Marit. Eng.* **2016**, *156*, A231–A238. [CrossRef]

32. Kaššay, P. Modeling, Analysis and Optimization of Torsional Oscillating Mechanical Systems. Ph.D Thesis, Technical University of Košice, Košice, Slovakia, 2014. (In Slovak).

33. Homišin, J.; Kaššay, P. Optimal tuning method of ships system by means of pneumatic tuner of torsional oscillations. *Acta Mech. Slov.* **2009**, *13*, 38–47. [CrossRef]

34. Homišin, J.; Urbanský, M. Partial results of extremal control of mobile mechanical system. *Diagnostyka* **2015**, *16*, 35–39.

35. Urbanský, M.; Kaššay, P. The new realized mobile device for extremal control research and presentation. *Sci. J. Sil. Univ. Technol. Ser. Transp.* **2015**, *89*, 173–178. [CrossRef]

36. DEWETRON Homepage. Available online: https://www.dewetron.com/products/components-and-sensors/rpm-angle-sensors/ (accessed on 17 July 2020).

37. Zhang, P. *Advanced Industrial Control Technology*, 1st ed.; Elsevier: Oxford, UK, 2010.

38. Urbanský, M.; Homišin, J.; Čopan, P. Examination of mechanical system response to gaseous media pressure changes in the pneumatic coupling. *Sci. J. Sil. Univ. Technol. Ser. Transp.* **2013**, *81*, 143–149.

39. Uradnicek, J.; Kraus, P.; Musil, M.; Bachraty, M. Modeling of frictional stick slip effect leading to disc brake noise vibration and harshness. In Proceedings of the 23rd International Conference, Svratka, Czech Republic, 15–18 May 2017; pp. 1002–1005.

40. Kraus, P.; Uradnicek, J.; Musil, M.; Bachraty, M.; Hulan, T. Thermo-Structural brake squeal fem analysis considering temperature dependent thermal expansion. *Eng. Mech.* **2018**, *24*, 429–432.

41. Musil, M.; Suchal, A.; Uradnicek, J.; Kraus, P. The Complex eigenvalue analysis of brake squeal using finite element method. In Proceedings of the 22nd International Conference, Svratka, Czech Republic, 9–12 May 2016; pp. 406–409.

42. Vinas, J.; Brezinova, J.; Guzanova, A. Tribological properties of selected ceramic coatings. *J. Adhes. Sci. Technol.* **2013**. [CrossRef]

43. Hudák, R.; Šarik, M.; Dadej, R.; Živčák, J.; Harachová, D. Material and Thermal Analysis of Laser Sinterted Products. *Acta Mech. Autom.* **2013**, *7*, 15–19. [CrossRef]

44. Živčák, J.; Šarik, M.; Hudák, R. FEA Simulation of Thermal Processes during the Direct Metal Laser Sintering of Ti64 Titanium Powder. *Measurement* **2016**, *94*, 893–901. [CrossRef]

45. Tlach, V.; Cisár, M.; Kuric, I.; Zajačko, I. Determination of the Industrial Robot Positioning Performance. Modern Technologies in Manufacturing. Available online: https://www.matec-conferences.org/articles/matecconf/abs/2017/51/matecconf_mtem2017_01004/matecconf_mtem2017_01004.html (accessed on 22 June 2020).

46. Brezinova, J.; Guzanova, A. Friction Conditions during the Wear of Injection Mold Functional Parts in Contact with Polymer Composites. *J. Reinf. Plast. Compos.* **2010**, *29*, 1712–1726. [CrossRef]

47. Toth, T.; Zivcak, J. A Comparison of the Outputs of 3D Scanners. *Procedia Eng.* **2014**. [CrossRef]

48. Kučera, P.; Píštěk, V. Testing of the Mechatronic Robotic System of the Differential Lock Control on a Truck. *Int. J. Adv. Robot. Syst.* **2017**, *14*, 1–7. [CrossRef]

49. Kniewald, D.; Guzanova, A.; Brezinova, J. Utilization of Fractal analysis in strength prediction of adhesively-bonded joints. *J. Adhes. Sci. Tech.* **2008**, *22*, 1–13. [CrossRef]

50. Kučera, P.; Píštěk, V.; Prokop, A.; Řehák, K. Measurement of the powertrain torque. In Proceedings of the Engineering Mechanics, Svratka, Czech Republic, 14–17 May 2018; pp. 449–452.

51. Kuric, I. New Methods and Trends in Product Development and Planning. In Proceedings of the 1st International Conference on Quality and Innovation in Engineering and Management (QIEM), Cluj Napoca, Romania, 17–19 March 2011; pp. 453–456.

52. Kučera, P.; Píštěk, V. Prototyping a System for Truck Differential Lock Control. *Sensors* **2019**, *19*, 3619. [CrossRef] [PubMed]

53. Puškár, M.; Kopas, M.; Puškár, D.; Lumnitzer, J.; Faltinová, E. Method for reduction of the NOx emissions in marine auxiliary diesel engine using the fuel mixtures containing biodiesel using HCCI combustion. *Mar. Pollut. Bull.* **2018**. [CrossRef] [PubMed]
54. Jasminská, N.; Brestovič, T.; Puškár, M.; Grega, R.; Rajzinger, J.; Korba, J. Evaluation of hydrogen storage capacities on individual adsorbents. *Measurement* **2014**. [CrossRef]

Journal of
Marine Science and Engineering

Article

Optimization of the Emissions Profile of a Marine Propulsion System Using a Shaft Generator with Optimum Tracking-Based Control Scheme

Joel R. Perez [1,*,†] and Carlos A. Reusser [2,†]

1   Naval and Maritime Sciences Institute, Faculty of Engineering Sciences, Universidad Austral de Chile, 5111187 Valdivia, Chile
2   Department of Electronics, Universidad Tecnica Federico Santa Maria, 2390123 Valparaiso, Chile; carlos.reusser@usm.cl
*   Correspondence: joelperez@uach.cl; Tel.: +56-9-79791242
†   These authors contributed equally to this work.

Received: 15 January 2020; Accepted: 6 February 2020; Published: 20 March 2020

Abstract: Nowadays, marine propulsion systems based on thermal machines that operate under the diesel cycle have positioned themselves as one of the main options for this type of applications. The main comparative advantages of diesel engines, compared to other propulsion systems based on thermal cycle engines, are the low specific fuel consumption of residual fuels, and their higher thermal efficiency. However, its main disadvantage lies in the emissions produced by the combustion of the residual fuels, such as carbon dioxide (CO_2), sulfur oxide (SO_x), and nitrogen oxide (NO_x). These emissions are directly related to the operating conditions of the propulsion system. Over the last decade, the International Maritime Organization (IMO) has adopted a series of regulations to reduce CO_2 emissions based on the introduction of an Energy Efficiency Design Index (EEDI) and an Energy Efficiency Operational Indicator (EEOI). In this context, adding a Shaft Generator (SG) to the propulsion system favoring lower EEDI and EEOI values. The present work proposes a selective control system and optimization scheme that allows operating the shaft generator in Power Take Off (PTO) or Power Take In (PTI) mode, ensuring that the main engine operates, always, at the optimum fuel efficiency point, thus ensuring minimum CO_2 emissions.

Keywords: marine propulsion system; shaft generator; power take-in; power take-off; energy efficiency design index; energy efficiency operational indicator; gradient vector optimization; power converter; torque oriented control

---

1. Introduction

Even though marine propulsion systems have been in constant development since the 18th century, nowadays the most common system used on board large carriers, i.e., container ships and tankers, is a system considering a diesel engine as the prime mover. Most of these engines are of the crosshead type, operating on the two-stroke cycle at low speed having long strokes, turbocharged, and directly coupled to a single fixed-pitch propeller. The power installed for these configurations vary from 10 MW and up to 80 MW [1]. These type of propulsion systems include the use of a Waste Heat Recovery System (WHRS), which relates taking the remaining heat of the exhaust gases generated from the combustion process of the diesel engine. This system was previously known as an economizer and was used to generate steam for heating processes only. Recently, in some cases, the WHRS uses heat to generate steam to be used in turbo-generators [2–4].

Diesel engines for marine applications have many advantages when compared to other prime movers such as turbines. They have a higher thermal efficiency and a low fuel oil consumption of low-cost residual fuels. The disadvantage of consuming residual fuels is the high amount of CO_2, SO_x, and NO_x emissions, which are related to the ship's operational condition [5,6]. Alternative fuels with low carbon and sulfur content have been considered to be used to replace these residuals fuels; however, its use in large carriers is not cost efficient and still presents a poor environmental performance. In this regard, alternative fuels such as Liquefied Natural Gas (LNG), biofuels and hydrogen are some of the most promising alternatives in study as replacement, but still some concerns about their storage, technological maturity and safety mitigating measures [7,8].

The ship's power design requirement represents the main constrain when the operational condition of the ship is assessed and compared to the power demands at normal operating conditions. The prime mover is forced, most of the time, to operate under underrated conditions increasing its fuel oil consumption therefore the amount of emissions. Diesel engines, as the one described, found their optimum fuel oil consumption point at ~75–80% of the MCR [9].

High fuel oil consumptions lead to an increase of emissions, e.g., CO_2 emissions, which have been monitored and since 2011 have been measured using an index called EEDI. This index is part of a mandatory regulation coming from the IMO and applicable to every new ship since 2013 [10,11]. An operational indicator called EEOI has also been considered but this is not mandatory yet, although is integral part of the Ship Energy Efficiency Monitoring Plant (SEEMP), which is mandatory to be implemented on board ships but that is not auditable by any means yet. The EEOI is used as a measuring tool to voluntary assess the efficiency of an existing ship.

The EEDI encourage the use of technologies such as shaft generator to reduce the use of the power installed on board through auxiliary generator sets [12]. This reduction of the use reduces the fuel oil consumption, therefore, as was mentioned before, reduces the amount of CO_2 emissions. The shaft generator force to use the diesel engine in a loading range, quite close to the optimum fuel oil consumption point [13,14]. The use of shaft generators above this point has been considering unjustified because of the possibility to overload the diesel engine leading to increase the fuel oil consumption.

The present work presents a marine diesel engine propulsion system with a direct driving shaft generator and a back to back converter based on the use of a selective control scheme. This scheme enables for the diesel engine to operate at its optimum fuel oil consumption point, which has been renamed as its Minimum Emissions Operating Point (MEOP). The scheme considers the use of the shaft generator as a Power Take Off (PTO) drive when the diesel engine operates below the MEOP and as a Power Take In (PTI) when the diesel engine operates above the MEOP. The shaft generator, at PTO, generates enough power to turn-off the generator set of the ship. These operational conditions have a repercussion on the EEOI, which is to be estimated and analyzed to prove the positive influence to lower the amount of emissions based on the reduction of the specific fuel oil consumption of the diesel engine. After EEOI results, the selective control scheme is going to be used to evaluate its influence over the EEDI of a new design looking for the development of an efficient propulsion system that ensures the compliance with the IMO regulations.

2. Background

Efficiency can be defined as the ratio of the useful work performed by a vessel to the total energy expended, but also can be expressed as actions designed to achieve efficiency [15]. Under these definitions, first, we can consider the vessel efficiency as the amount of fuel consumed, as the energy source to be expended, over the transport work performed by the vessel, as the process of carrying cargo, and second, we can consider vessel efficiency as the implementation of technological and operational means to a vessel to achieve higher levels of efficiency. Both definitions can be applied but for this research, the first is going to be applied over the design of new vessels, and the second over existing vessels. Both definitions can be worked together when analyzing the efficiency of a vessel, as to improve its efficiency in the design stage, it is required to know its current operational efficiency, from the EEOI. The current efficiency can be estimated when evaluating the amount of fuel that is consumed by the diesel engines of the main propulsion system and the auxiliary systems when navigating. This estimation of vessel efficiency provides the baseline from where it is possible to improve it when considering the implementation of technological means, i.e., shaft generators and operational means, i.e., selective control schemes as the described in this research. The amount of fuel saved by the implementation of technological and operational means translated as the improvement into the efficiency of a vessel. Because the consumption of fuel generates emissions, any reduction of fuel consumption leads to lower emission levels. The use of the fuel consumption a as state variable provides the baseline of considering the use of EEDI and EEOI as means of evaluation of vessel design and operational efficiency.

2.1. Emissions from the Combustion Process

Emissions are generated during the process of converting the chemical energy of the fuel into mechanical work, Equation (1) represents the stoichiometric reaction of the fuel and the consequence emissions generation, CO_2 emissions are the higher amount of all of them.

$$C_mH_n + \left(m + \frac{n}{4}\right) O_2 + pN_2 \rightarrow m\,CO_2 + \frac{n}{2}H_2O + pN_2 \tag{1}$$

2.2. Fuel Oil Consumption and Diesel Engine and Shaft Generator Operation

The stoichiometric air to fuel ratio (AFR_{st}) is the minimum amount of air required to burn a kilogram of fuel and, when compared to the actual Air to Fuel Ratio (AFR), the stoichiometric ratio λ, presented in Equation (2), can be found [15]. The AFR can be considered at any engine load for the purposes of analysis.

$$\lambda = \frac{AFR}{AFR_{st}} \tag{2}$$

The engine's output power suffers when operating at lower loads condition, because at this condition, less fuel is available and the engine while trying to achieve a higher load demands more fuel to be provided to overcome the demand. The engine trying to maintain the power output at the desired operational condition increases the specific fuel oil consumption until it reaches the desired engine load, which is related to its speed as can be seen in Figure 1. At low engine load $\lambda \approx 4.0$, which decreases as the engine load is continuously increased. When the load is within the range of 75 to 80% of the engine maximum continuous rating (MCR), the value of λ reaches its minimum. When going above 75–80% load, because higher amounts of air and fuel are required, λ increases reaching a value of ≈ 2.0 at the 100% of the MCR.

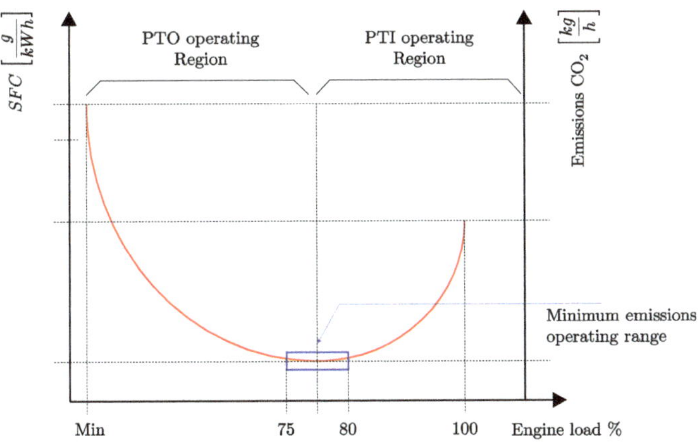

Figure 1. Power Take Off (PTO)/Power Take In (PTI) operating regions.

Figure 1 shows these λ conditions and the optimal fuel oil consumption point or MEOP. The MEOP has been considered in this form because is a representation of the Specific Fuel oil Consumption (SFC), which is one of the factors to evaluate the amount of emissions generated by the engine when using the EEDI and EEOI.

Operationally, Figure 1 differentiate the operating regions of the shaft generator when considering the MEOP over the entire engine load range. Before MEOP the shaft generator operates as PTO and after MEOP operates as PTI. The shaft generator as PTO generates electricity to support the operation of the vessel and reduces the use of the diesel engines of the auxiliary system known as generation set. When operating as PTI, the electrical power to operate the shaft generator as an electric motor is provided by the generator set. Nonetheless, from this assumption, future work will investigate options to improve the efficiency of the generator set operation and the use of electric power sources, i.e., use of batteries and Non-Conventional Renewable Energies (NCRE) sources.

The engine efficiency at the MEOP is the highest and has been used as the evaluation point of the EEDI mandated by the IMO for every new ship constructed. The EEDI started with a minimum value established by 2013 and reduced by a percentage over the next 12 years [10,11], a low EEDI value means a more efficient ship, in terms of its design (hydrodynamics, propulsion system, and auxiliaries).

2.3. Energy Efficiency Design Index EEDI

The EEDI can be defined as a technical measure of CO_2 emissions per ship's capacity per nautical mile applied to new ship designs [10]. The EEDI equation is presented in Appendix A. Here, a modified version to be applied for the purposes of this paper is presented in Equation (3). This modified version accounts only for the main engine influence of emissions generation; therefore, it is an approximation that is going to be modified to establish and represent the influence of the shaft generator over the entire main and auxiliary systems of the ship. The auxiliary engines, shaft generator, and WHRS influence over the EEDI equation has been found minimum when compared to the main engine installed power therefore the EEDI value do not get really affected by them as stated in [11]. The influence of these factors is related to the operation of the ship.

$$EEDI = \frac{P_B \, SFC \, C_F}{DWT \, V_S} \tag{3}$$

The P_B corresponds to the 75% rated installed brake power in kW, C_F is the carbon factor in g CO_2 per $g\,fuel$, DWT is the capacity of the ship in tonnes, and Vs is the ship's design speed in knots.

2.4. Energy Efficiency Operational Indicator EEOI

The MEOP has been also used to calculate the EEOI. The EEOI is the monitoring tool supporting the Ship Energy Efficiency Management Plan (SEEMP) applied to new and existing ships to measure the amount, in grams, of CO_2 per tonne cargo transported per nautical mile for a single voyage [11]. The equation to calculate EEOI is presented in Appendix B and here a modified version to be applied for the purposes of this paper is presented in Equation (4). This modified version uses the SFC instead of the total amount of fuel consumed for a single voyage to relate the indicator with the operational performance described to evaluate EEDI and have a comparison point of the ship's design performance and the actual operation at the required MEOP at low and high loads. The m_c factor accounts for the mass of cargo transported in tonnes and D to the distance, in nautical miles, of the cargo transported.

$$EEOI = \frac{SFC \, C_F}{m_C \, D} \tag{4}$$

Having a comparison point between the design and the operational behavior of the ship allows for a better understanding of these tools to evaluate the current efficiency of the ship, also allowing to consider technological and operational options to improve ship's efficiency. The use of a shaft generator is part of these improvements, and its influence into of ship's efficiency is presented when analysing the Shaft Generator/Motors Emissions Factor f_{gef} into the EEDI equation, this factor is presented in Equation (5), where the power generated accounts and is related to the power generated by the auxiliary engines or generator set to support the service of the ship.

$$f_{gef} = \left(f_i P_{PTI} - f_{eff} P_A E \right) C_{FAE} SFC_{AE} \tag{5}$$

Equation (5) can also be contrasted with the Efficiency Technology Factor (ETF) presented in Equation (6), which relates the reduction in power requirements that any technology generates and is used to improve the ship's efficiency.

$$ETF = f_{eff} P_{eff} C_F SFC \tag{6}$$

The efficiency technology factor f_{eff} in Equations (5) and (6) represents the percentage of influence of the power output of the technology and relates to its efficiency. The background presented allows for the introduction of the control scheme selected as the most appropriate to represent the influence of a shaft generator into the propulsion system of a ship, because relates its fuel consumption and its emissions in accordance with known and validated indexes for ship's efficiency evaluation.

3. Hybrid Propulsion System Characterization

The hybrid propulsion system under study consists of low speed 2-stroke diesel engine, driving directly the propulsion shaft. A permanent magnet synchronous machine is coupled to the diesel engine using a single-stage gearbox. This geared mechanical transmission system enables the combination of the mechanical and electrical prime movers in the same kinematic drivetrain, as shown in Figure 2

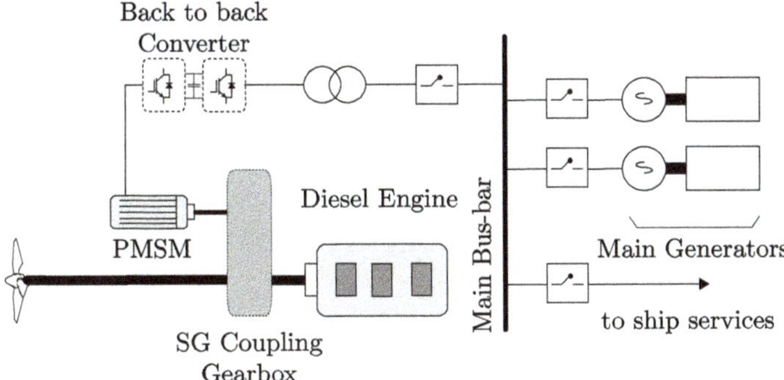

Figure 2. Hybrid propulsion system configuration.

The synchronous machine is connected to the main ship's grid using a back-to-back power converter. This configuration enables bidirectional power flow, between the electric drive and the ship's grid, thus enabling the electric drive to operate in power take-off (PTO) or power take-in (PTI) modes, depending on the direction of the power flow, as shown in Figure 3. Moreover, the use of the back-to-back converter configuration decouples the electric drive and grid control schemes. This enables the grid side control scheme to be synchronized referred to the ship's main busbar frequency, independently of the electric drive operational frequency, and therefore the diesel engine operational speed [16].

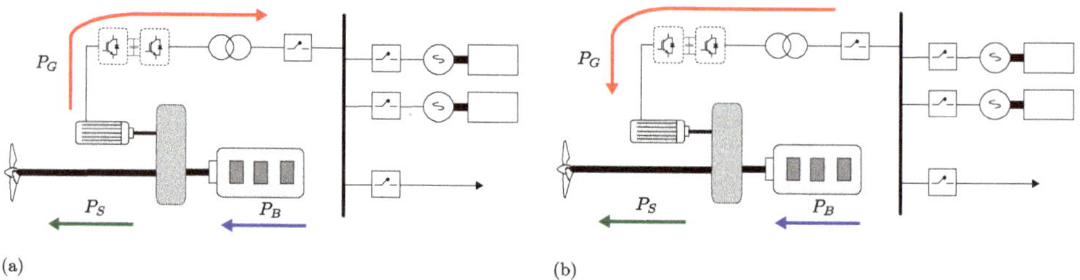

Figure 3. Power flow in (**a**) power take-off operation mode; (**b**) power take-in operation mode.

Power take-off mode: in this mode, the main diesel engine supplies the power needed for the propulsion P_S as well as for the ship's consumers P_G by forcing the electric drive to operate within the generator region. Depending on the ship's load and the required propulsion power, all or some of the generator sets (GS) are turned off, as shown in Figure 3a. The engine developed power $P_B = P_S + P_G$.

Power take in mode: in this mode the electric drive is forced to operate within the motor region, an auxiliary motor, allowing to reduce the main engine's load, as shown in Figure 3b. Depending on the required shaft power, the system operates in booster mode, when both electric and diesel provide deliver power to shaft, or diesel-electric mode, when only the electric drive delivers power. Therefore, the engine developed power $P_B = P_S - P_G$.

3.1. Diesel Engine Model

The diesel engine dynamics is obtained energy conversion principle, by defining the system's Hamiltonian $H(x)$ as given in Equations (7)–(9)

$$H(x) = W_i - \sum_{\ell=1}^{n} Ec_\ell + W_I \tag{7}$$

where W_i is the chemical energy also known as indicated energy, Ec correspond to the system loses, and W_I is the stored energy in the inertia. The indicated energy can be defined as a nonlinear function of the fuel enthalpy h, the fuel flow rate g, and shaft rotational speed $\dot\theta_r$

$$W_i = f_c \left(h, g, \dot\theta_r \right) \tag{8}$$

and the energy stored in the inertia is given as in Equation (9)

$$W_I = \frac{1}{2} L \frac{d}{dt} \theta_r \tag{9}$$

where L stands for the rotational momentum. The total converted energy into mechanical torque is given by Equations (10) and (11).

$$\frac{\partial}{\partial \theta_r} H(x) = 0 \tag{10}$$

$$J \frac{d^2}{dt^2} \theta_r = T_i - T_p - T_f \tag{11}$$

where T_i corresponds to engine's indicated torque, T_p the pumping torque, and T_f the friction torque.

The indicated torque is dependent on the amount of fuel injected into each of cylinders per cycle as given in Equation (12)

$$T_i = \frac{m_{cy}\, n_{cy}\, \rho_h\, \eta_{ig}}{2\,\pi\, n_{cs}} \tag{12}$$

where m_{cy} corresponds to the fuel delivery per cycle per cylinder, n_{cy} to number of cylinders, ρ_h is the heating value of fuel, η_{ig} is the indicated efficiency, and n_{cs} the number of crank revolutions. The indicated efficiency is given as in Equation (13)

$$\eta_{ig} = \eta_{cc} \left(1 - \frac{1}{r_c^{\gamma-1}} \right) \tag{13}$$

where r_c is the compression ratio, γ the gas specific heat capacity ratio in the cylinder, and η_{cc} represents the combustion chamber efficiency.

As presented, the indicated torque is highly dependent on several engine parameters, such as the number of cylinders, the fuel delivery per cycle, the compression ratio, and combustion chamber efficiency. On the other hand, the diesel engine emissions are defined by its residual gas fraction χ_r, which represents a measure of CO_2 concentrations of the working gas in the compression stroke, during the energy conversion process, as defined in Equation (14)

$$\chi_r = \frac{(\tilde{\chi}_{co_2})_C}{(\tilde{\chi}_{co_2})_E} \tag{14}$$

where $(\tilde{\chi}_{co_2})_C$ and $(\tilde{\chi}_{co_2})_E$ stands for the CO_2 fractions during compression and exhaust, respectively.

The torque component corresponding to energy losses T_C is given by Equations (15) and (16)

$$T_C = \frac{\partial}{\partial \theta_r} \sum_{\ell=1}^{n} Ec_\ell \tag{15}$$

$$T_C = T_p + T_f \tag{16}$$

where the pumping torque T_p and friction torque T_f can be expressed as in Equations (17) and (18), respectively.

$$T_p = \frac{V_d}{2\pi n_{cs}} \left(p_{em} - p_{im} \right) \tag{17}$$

$$T_f = \frac{V_d}{2\pi n_{cs}} \left(c_0 + c_1 n_r + c_2 n_r^2 \right) \tag{18}$$

here V_d corresponds to the engine displacement volume, p_{em} is the exhaust pressure to the manifold, and p_{im} the manifold inlet pressure. The friction torque T_f, on the other hand, may be assumed to be a quadratic polynomial depended on the engine revolutions n_r, with c_0, c_1, and c_2 fitting constants.

From Equations (7)–(17) it becomes self-evident that the diesel engine mathematical model is highly nonlinear and dependent on several specific construction parameters. However, a linearized model may be used, considering all important nonlinear characteristics [17–20], which can be modeled as dead-times and time-delays contained in τ_1 and τ_2, respectively, and constant parameters k_1, k_2, as presented in Equations (19)–(20)

$$\frac{d}{dt} y = -\frac{1}{\tau_1} y + \frac{k_1}{\tau_1} u \tag{19}$$

$$J \frac{d}{dt} \omega_r = -B \omega_r + k_2 y (t - \tau_2) - T_L \tag{20}$$

where J stands for the engine's inertia, B is the friction coefficient, ω_r corresponds to the engine rotational speed, u to the speed controller output, y the position of the fuel rail, and T_L the external load torque. Values for k_1, k_2, τ_1, and τ_2 may be found empirically or from the data provided by the manufacturer using a model fitting algorithm, as stated in [21].

Considering the previously made considerations and modeling restrictions, it is possible to build an emissions model, on the basis of the data provided by the manufacturer, using an appropriate polynomial approximation with squares regression.

Given m data points $\{x_i\, y_i\}_{i=1}^{m}$ with x_i given output power and y_i corresponding CO_2 emissions rate; the best fit polynomial for the CO_2 emissions $e(x)$ could be developed using Equation (21)

$$e(x) = \sum_{k=0}^{n} \alpha_k\, x^k \quad n < m - 1 \tag{21}$$

where $\alpha_k\ \forall\, k$ coefficients may be found by minimizing the least square error using Equation (22)

$$A^T A\, a = A^T y \tag{22}$$

with the coefficients vector $a = [\alpha_0\ \dots\ \alpha_n]^T$, the sample value vector $y = [y_0\ \dots\ y_n]^T$, and A the Vandermonde matrix, given as in Equations (23) and (24)

$$A = \begin{bmatrix} 1 & x_1 & x_1^2 & \dots & x_1^n \\ 1 & x_2 & x_2^2 & \dots & x_2^n \\ \vdots & \vdots & \vdots & \vdots & \\ 1 & x_m & x_m^2 & \dots & x_m^n \end{bmatrix} \quad \forall x_i\ i = 1, \dots, m \tag{23}$$

$$a = (A^T A)^{-1} A^T y \tag{24}$$

obtaining finally an n degree polynomial representing the CO_2 emissions profile, as given in Equation (25)

$$y(x) = a_0 + a_1\, x + a_2\, x^2 + \dots + a_n\, x^n \tag{25}$$

3.2. Electric Drive Model and Control

The anisotropic permanent magnet synchronous machine (PMSM) mathematical model in an arbitrary synchronous reference frame $d\,q$ is described as in Equations (26) and (27) [22],

$$v_s^{(dq)} = R_s\, i_s^{(dq)} + \frac{d}{dt}\psi_s^{(dq)} + F\,\psi_s^{(dq)} \tag{26}$$

$$F = \begin{bmatrix} 0 & -\omega_k \\ \omega_k & 0 \end{bmatrix} \tag{27}$$

where v_s corresponds to the stator voltage, R_s to the stator resistance, i_s the stator current, ψ_s the stator flux linkages, and ω_k to the shaft synchronous speed. The stator flux linkages are given as in Equations (28) and (29):

$$\psi_s^{(dq)} = G\, i_s^{(dq)} + P\psi_m \tag{28}$$

$$G = \begin{bmatrix} L_d & 0 \\ 0 & L_q \end{bmatrix} \quad P = \begin{bmatrix} 1 \\ 0 \end{bmatrix} \tag{29}$$

L_d and L_q are the direct and quadrature reference frame inductances, respectively, and ψ_m the permanent magnet flux linkage. The electromechanical torque developed by the PMSM is given in Equations (30) and (31)

$$T_e = \frac{\partial}{\partial \theta_r} W_{fld}\left(\psi_s^{(dq)}, i_s^{(dq)}, \theta_r\right) \tag{30}$$

$$T_e = \frac{3}{2}\, p\left\{\psi_m\, i_s^q + (L_d - L_q)\, i_s^d\, i_s^q\right\} \tag{31}$$

where p are the number of pole pairs.

Despite the classical FOC control scheme, which is used to control the drive shaft speed, in this case, the control objective is to control the torque developed by the electric drive [23], which is achieved by means of the electric torque reference, provided by the optimization algorithm output. The implementation of the electric drive control scheme is shown in Figure 4.

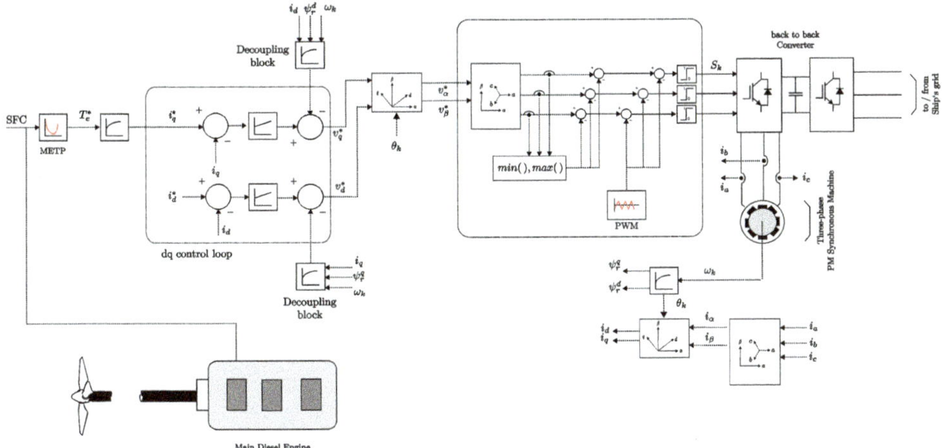

Figure 4. Torque field-oriented control scheme.

3.3. Grid Side Power Flow Control

Grid side control is achieved by implementing active and reactive power control [24,25], using a virtual flux voltage oriented control (VF-VOC) strategy. Active and reactive power, P and Q, respectively, in a synchronous rotating reference frame, grid side-oriented $u\,v$ are given in Equations (32) and (33), as a result of using a voltage orientation in u coordinate.

$$P = \frac{3}{2}\,\mathbb{Re}\left\{v^u\left(i^u + ji^v\right)\right\} \tag{32}$$

$$Q = \frac{3}{2}\,\mathbb{Im}\left\{v^u\left(i^u + ji^v\right)\right\} \tag{33}$$

Thus, by setting the reactive component of the grid current $i^v = 0$ it is possible to maximize the active power flow into the grid. Orientation into the grid side synchronous reference frame $u\,v$ is achieved by extracting the orientation angle θ_p provided by a virtual-flux space vector $\psi^{(xy)}$ referred to the voltage drop in the output inductance $v_o^{(xy)}$ as in Equations (34) and (35). Implementation of the grid side control scheme is provided in Figure 5.

$$\psi^{(xy)} = \int v_o^{(xy)}(t)\,dt \tag{34}$$

$$\theta_p = \text{atan2}\left(\psi^x,\,\psi^y\right) \tag{35}$$

The corresponding grid side dynamic model in the $u\,v$ synchronous reference frame, is given in Equation (36),

$$v^{(uv)} = R\,i^{(uv)} + L\frac{d}{dt}i^{(uv)} + F\,L\,i^{(uv)} + v_g^{(uv)} \tag{36}$$

where $v^{(uv)}$ corresponds to the converter output voltage, $i^{(uv)}$ stands for the grid side current, and $v_g^{(uv)}$ to the main busbar voltage, in the $u\,v$ reference frame. Line parameters of resistance and inductance are given as R and L, respectively, and matrix F has been defined in Equation (27).

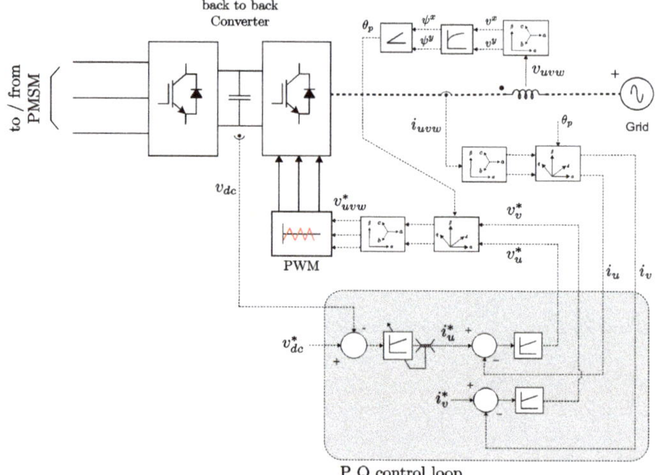

Figure 5. Voltage-oriented control scheme.

4. Optimization Strategy

Let us define an arbitrary optimization problem ϕ as in Equations (37)–(40), given a set of candidate solutions $\mathcal{C}$, a set of solution $\mathcal{S} \subseteq \mathcal{C}$, an objective function $f(x)$, and v the optimization sense.

$$\phi = \langle \mathcal{C}, \mathcal{S}, v, f(x) \rangle \tag{37}$$

$$\mathcal{C} = x \quad ; \quad x = \{ x_0, \ldots, x_n \} \tag{38}$$

$$\mathcal{S} = x \pm \delta \tag{39}$$

$$v = min \{ f(x) |_{x \pm \delta} \} \tag{40}$$

where x corresponds to the system state and δ to the variation of the state introduced by the search direction of the optimization strategy.

The implemented optimization strategy is based on the use of the gradient vector $\nabla f(x)$ as search direction for each iteration. Note that the gradient vector is orthogonal to the plane tangent to the contour surfaces of the function to optimize; $\nabla f(x) = g(x) = [\frac{\partial f}{\partial x_1} \ldots \frac{\partial f}{\partial x_n}]^T$. The gradient vector at a point $g(x_k)$ represents the direction of maximum rate of change, which is given by $| g(x_k) |$

The optimization strategy searches for the point x_k, where $| g(x_k) | \leq \mu_g$, given an initial state x_0; certain convergence parameters μ_g, μ_a, and μ_g; and a normalized search direction p_k; given Equations (41) and (43)

$$p_k = -\frac{g(x_k)}{| g(x_k) |} \tag{41}$$

$$x_{k+1} = x_k + \alpha \, x_k \tag{42}$$

for some α such that satisfies Equation (43)

$$| f(x_{k+1}) - f(x_k) | \leq \mu_a + \mu_r | f(x_k) | \tag{43}$$

The search objective corresponds to the minimum CO_2 emissions operating point of the diesel engine, and the function to optimize $f(x)$, to the diesel engine emissions profile, given a a certain required output power. The result from the optimization problem is used as torque reference for the TFOC electric drive control scheme. Figure 6 shows the algorithm implementation.

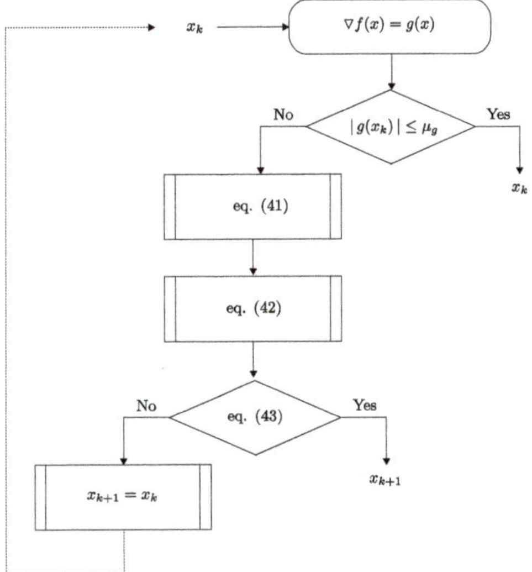

Figure 6. Gradient-based optimization algorithm structure.

5. Simulation Results

In this section, the performance of the diesel engine and the the optimization scheme under different load and operational conditions are presented. The hybrid propulsion system model under study was developed using PLECS. The controllers for the electric drive and the grid side, as well as the optimization algorithm, were developed and implemented in C code.

5.1. Diesel Engine Performance

Results are presented differentiating the performance of the control scheme of the shaft generator and the results of the fuel consumed by the engine when simulating the PTO and PTI conditions. The methodology considers the use of the IMOs EEDI and EEOI tools to show the benefits of the scheme, providing a reference to evaluate the design efficiency and the operational efficiency.

Results are plotted over the two simulation conditions considered but presented over a power range to simulate specific operational conditions such as slow steaming. This operational condition has been considered because represents the ability of the simulation to show the performance of the ship at low ship's speed, which can be used to evaluate a ship design over a higher range of options to get an efficient design. A more accurate evaluation of the EEDI could be necessary but still results are providing a great assertiveness of the methodology selected.

The data used to simulate the performance of the control scheme considers the use of a ship, which has specific information: its capacity, speed, cargo transported, distance navigated, power installed, and the type of fuel consumed. The type of ship considered was selected from a worldwide database of ships [26]. When analyzing the database and the specific information needed to evaluate the design and operational efficiencies of a ship, very large crude oil carriers were the type of ships more reluctant to be used because of the simplicity of their propulsion system and the significance of the amount and type of fuel consumed. The propulsion system consists of a diesel engine directly coupled to a fixed pitch propeller. The auxiliary power installed for this specific type of ship accounts for ~10% of the propulsion power installed [26]. Table 1 presents the open source data used for simulation that were fixed as the initial conditions. The range of data is only a reference of the type of ship found in the database and no consideration to the operational profile of them has been considered for simulation purposes.

Table 1. Operational parameters for simulation.

Type of Ship	Fuel Consumed	Capacity (dwt)	Speed (kn)	Cargo Transported (tonnes)	Installed Power (MW)
Very Large Crude	Heavy Fuel Oil	300,000	15	315,000	25
Oil Carrier (VLCC)	(HFO)	320,000	21	330,000	36

The engine selected to be modeled and evaluated is an engine from MAN [27]. The 7G80ME-C9.2-TII diesel engine was selected having a specified maximum continuous rating (SMCR) power of 33 MW at 72 rpms. The specific fuel oil consumptions (SFC) vary from 187.1 g/kWh at low engine load to 166 g/kWh at high engine load. The normal continuous rating of the engine was considered ~75% of the SMCR. The total power delivered by the auxiliary generators during the navigation has been considered ≈3% of the propulsion engine rated power. When consuming HFO, a carbon factor of 3.114 g CO_2/g Fuel was used to estimate the emissions. With this information, the simulation of the control scheme was carried out and the results are presented next.

5.2. Control Scheme Performance at PTO Operating Region

Results are presented considering the performance of the shaft generator operating as a PTO, including the power and fuel consumption performance and the evaluation of the EEDI and EEOI tools.

The control scheme was evaluated considering a navigation time period long enough to simulate a consistent increase of the brake power of the engine over a step time to reflect its performance and SFC variation to get a steady evaluation of the efficiencies. Blue curves in Figure 7 shows the results of a segment of the PTO operating region while simulating a period of navigation time of 5 h, where, in the left-hand side of the figure, it is possible to appreciate how the brake power lineally increases from 5 MW to 12 MW. The right-hand side of the figure shows the decrease of the SFC from a maximum value at low engine load, the power delivered as PTO is stable enough to allow for the ship to turn-off the generator set. The red curves are the results of not having a shaft generator installed. The increase in the brake power when using a shaft generator is ~5% of the SMCR.

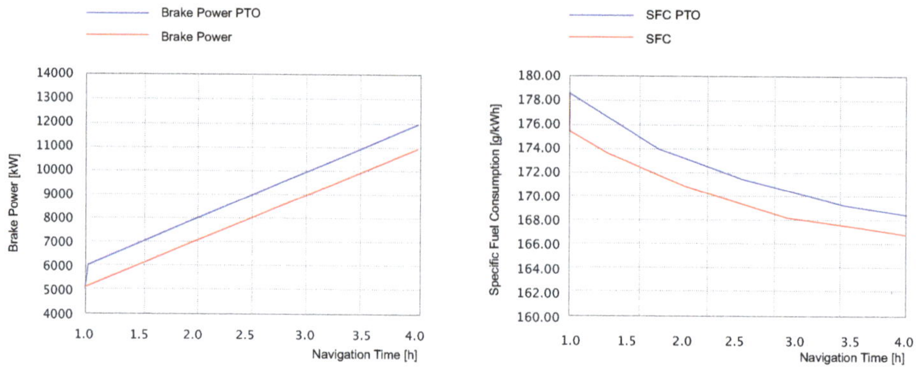

Figure 7. PTO evaluation performance.

When evaluating the EEDI, a specific value of 27,412 g CO_2/g Fuel was calculated. This value represents the amount of CO_2 emissions by design, which is therefore a value that can be modified at design stages only when the main and auxiliary machinery are selected and allocated to the vessel. A better approach could be to install a less powerful engine, but that means a completely different approach of the design spiral of the new ship. Following this evaluation, EEOI has been considered because represents the ability to calculate the emissions of the ship when in service navigating different routes. EEOI allows to check the variations of the same parameters that EEDI uses to be evaluated, therefore provides with a more comprehensive form to understand the operational behavior of a well design ship.

EEDI provides a fixed value at the design conditions, yet EEOI can be used to evaluate the performance at every variation of engine load. EEOI considers the total amount of fuel consumed and mass of cargo transported and distance navigated, the latests being just another representation of cargo capacity and ship's speed, respectively.

One of the main objectives of this work was to evaluate, using EEDI and EEOI, the influence of a shaft generator when applying a control scheme of its operation. The purpose is looking for reduction into the SFC at different engine loads. Also, to prove that the reduction of the SFC compared to the increase into the necessary power to be developed, to overcome the extra necessary brake power to be produced, to propel the ship and to use the shaft generator as PTO and PTI, respectively.

Figure 8 shows the results of the SFC variation at PTO operating region when applying the proposed control scheme, red curve, and the blue curve shows the SFC variation when not having a shaft generator. The SFC differences are between 0.1% to 2% over the whole engine load range plotted, the difference even though can be considered small is quite significant when evaluating the EEOI, having the maximum SFC reduction between 6000 kW and 7000 kW. The difference is barely noticed because of the scale of the plotted results.

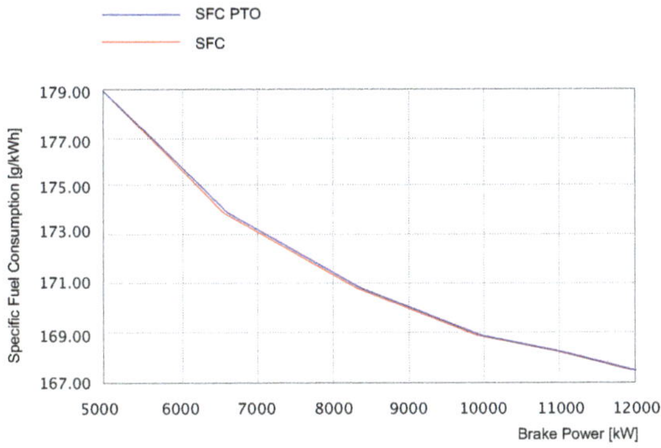

Figure 8. Specific Fuel oil Consumption (SFC) difference at PTO and without shaft generator.

The EEOI values are presented in Figure 9 together with the SFC obtained when using the shaft generator control scheme. The SFC decreases while the engine load increases. The fuel consumed provides EEOI values in accordance with its consumption over the navigation period and the navigated distance, as was described and established before as the initial conditions for simulation. The EEOI increases, but at a low rate over the engine load, which is a consequence of the decrease in the SFC. Although the decrease of the SFC is not quite significant allows for a low increment of the EEOI value, which leads to a overall reduction of the amount of operational CO_2 emissions.

Figure 10 shows the results of comparing the EEOI values of the applied control scheme, red curve, against not having a shaft generator installed, blue curve. Results are showing that the control scheme applied makes the ship to reduce the amount of operational CO_2 emissions even though an extra amount of brake power is necessary to be generated providing great assertiveness of the methodology and the SFC reduction over the period when operating at the PTO operating region.

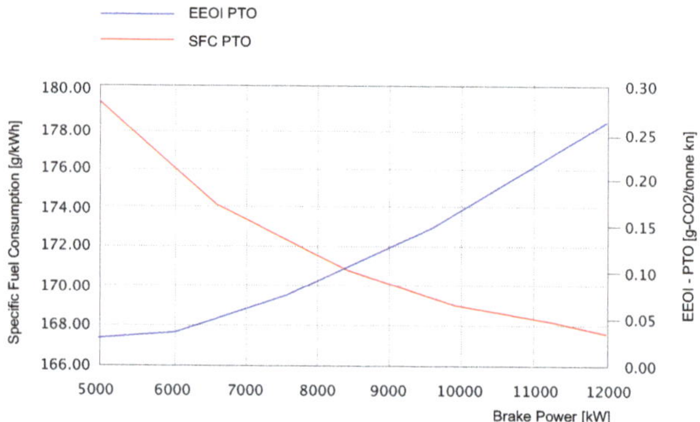

Figure 9. EEOI against SFC–PTO.

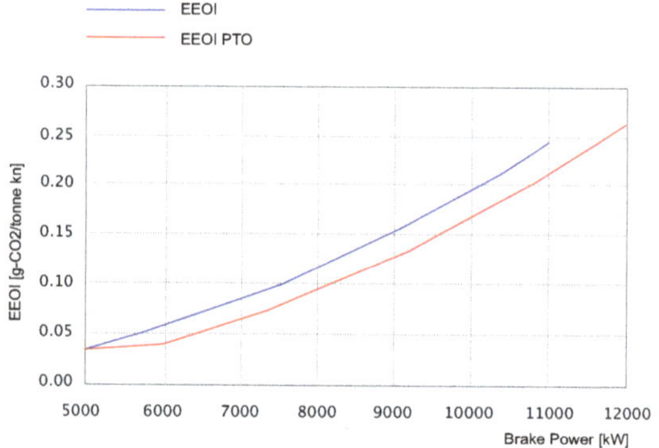

Figure 10. EEOI comparison.

5.3. Control Scheme Performance at PTI Operating Region

Results are presented considering the performance of the shaft generator operating as a PTI including the power and fuel consumption performance and the evaluation of the EEDI and EEOI tools. Following the same procedure to describe the results of the PTO operating region, the PTI operating region results are shown.

The blue curves in Figure 11 show the results of a segment of the PTI operating region while simulating a period of navigation time of 2.5 h, where, in the left hand side of the figure, it is possible to appreciate how the brake power lineally increases from 24 MW to 27 MW when not having a shaft generator, red curve, which also means an increase in the SFC as can be seen in the right-hand side of the figure. When applying the control scheme to operate the shaft generator, the PTI reduces the brake power, blue curve on the left-hand side of the figure therefore, a reduction of the SFC as can be seen in the right-hand side of the figure. The PTI reduces the brake power ~5% of the SMCR.

Figure 12 shows the results of the SFC variation at PTI operating region when applying the proposed control scheme, blue curve, and the red curve shows the SFC variation when not having a shaft generator.

The SFC differences are between −0.4% to 1.5% over the whole engine load range plotted. The difference reflects that even though the engine load increases, for PTI operation, the SFC decreases because the engine goes back to the high efficiency region operation or MEOP described before. On the other hand, the engine without a shaft generator increases the SFC, as expected, because of the operation away of the MEOP.

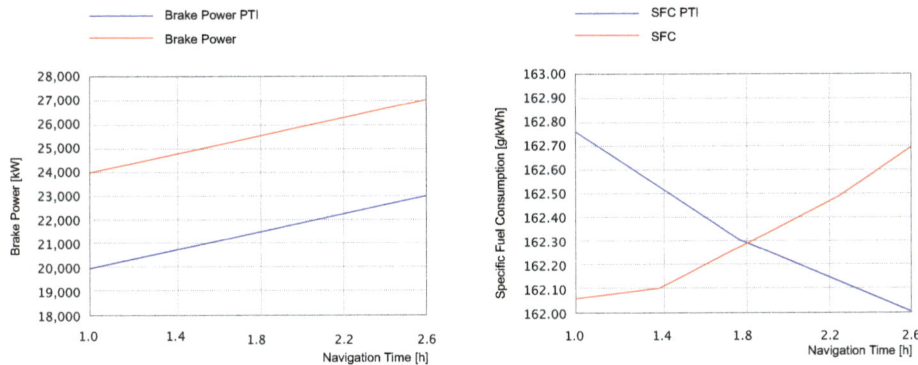

Figure 11. PTI evaluation performance.

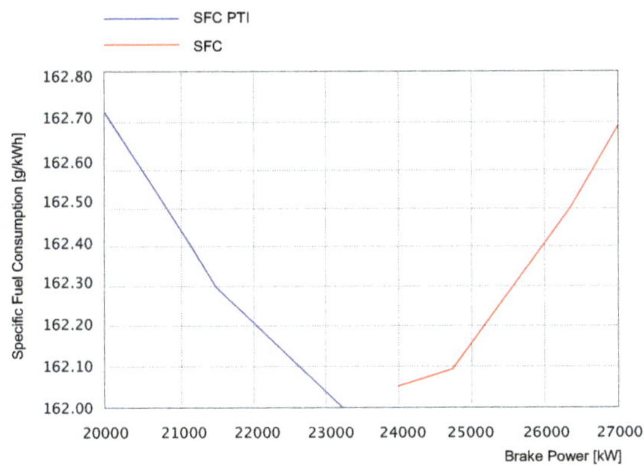

Figure 12. SFC difference at PTI and without shaft generator.

EEOI values are presented in Figure 13 against the SFC obtained when applying the shaft generator control scheme. The SFC decreases while the engine load increases, which gives a set of EEOI values in accordance with the amount of fuel consumed over the navigation period and the navigated distance. The lineal increment of the EEOI is considered low over the engine load and reflects the decrease of the SFC because of the control scheme applied. The operational emissions are reduced in accordance with the reduction of the SFC.

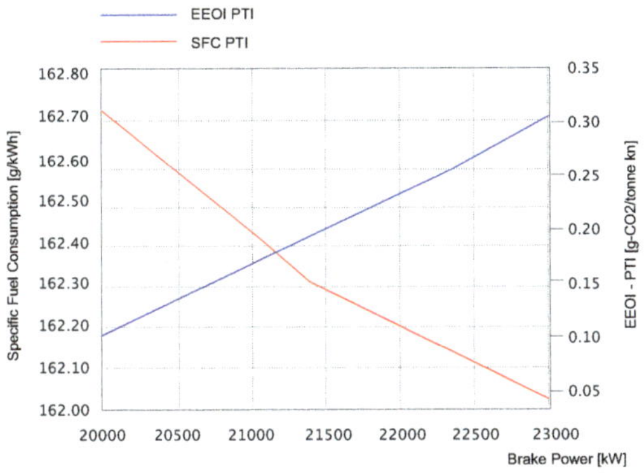

Figure 13. EEOI against SFC at PTI condition.

Figure 14 shows the results of comparing the EEOI values of the applied control scheme, blue curve, against not having a shaft generator installed, red curve.

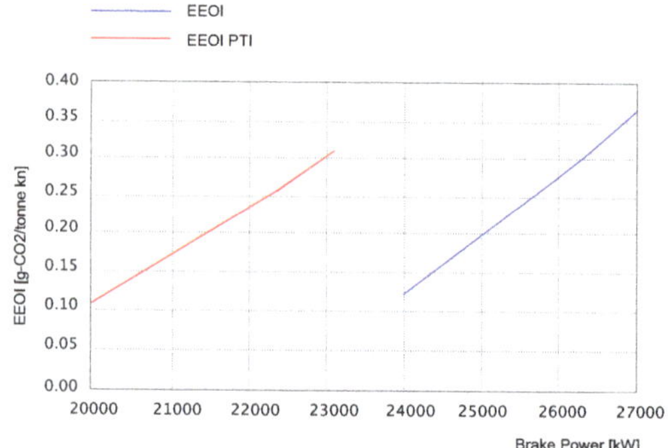

Figure 14. EEOI comparison.

Results are showing that the control scheme applied makes the ship to reduce the amount of operational CO_2 emissions because is making the engine to work closer to MEOP providing great assertiveness of the methodology.

5.4. Electric Drive Performance

In this section, an evaluation of the proposed control scheme, using a Minimum Emissions Point Tracking algorithm, based on a gradient vector optimization technique, is presented. Simulation results include two different torque steps at ① and at ②, both representing different shaft power operational conditions P_s @ 90% rated power as base condition before ①, de-rating to 60% of rated power in ①, and finally going to 80% rated power at ②.

Figure 15 shows the performance of the electric drive (the performance of the PMSM) in terms of its controlled currents. At low operational load below 75% of rated power, the MEPT algorithm sets the electric machine's torque reference in the generator region, therefore the torque producing current $i_q < 0$, whereas the flux producing current $i_d = 0$. On the other hand, when entering into a high load condition above the minimum SFC point, the MEPT, forces the PMSM to operate in the motor region, therefore the torque producing current $i_q > 0$, while the flux producing current $i_d = 0$.

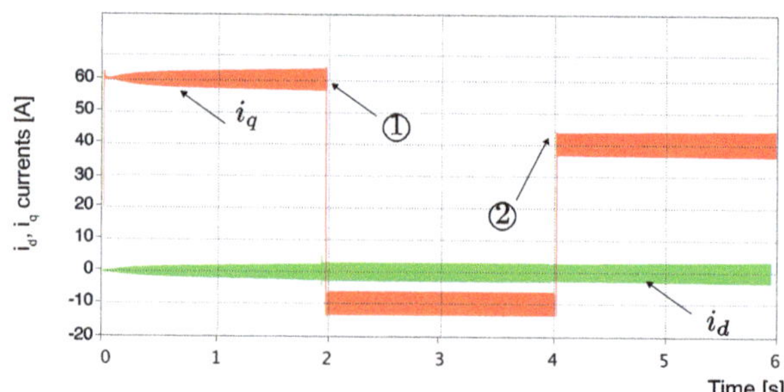

Figure 15. Electric drive performance $i_d\ i_q$ during torque demand step.

In Figure 16 the tracking performance of the optimization algorithm is shown, in the presence of a step change of the diesel engine torque. As shown, the torque electric reference T_e^* presents a fast and accurate response and keeps on tracking the minimum emissions operation point.

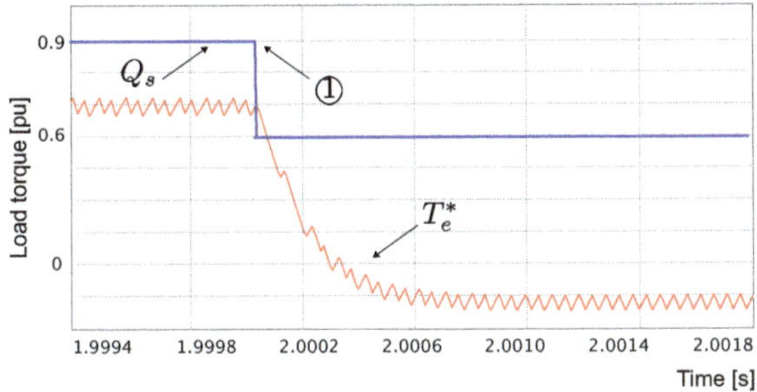

Figure 16. Optimization algorithm step dynamic response.

The grid side currents dynamic performance is shown in Figure 17, showing a sinusoidal behavior and fast dynamic response, during the transition from PTI to PTO mode at ① and from PTO to PTI operation at ②. Moreover, the fast tracking dynamics of the optimization algorithm, ensures minimization on the current wave-form distortion. On the other hand, due the fact that the grid side control scheme is decoupled from the electric drive control scheme, and ensures zero reactive power flow, the amplitude of the grid side currents are dependent on the electric drive's torque reference.

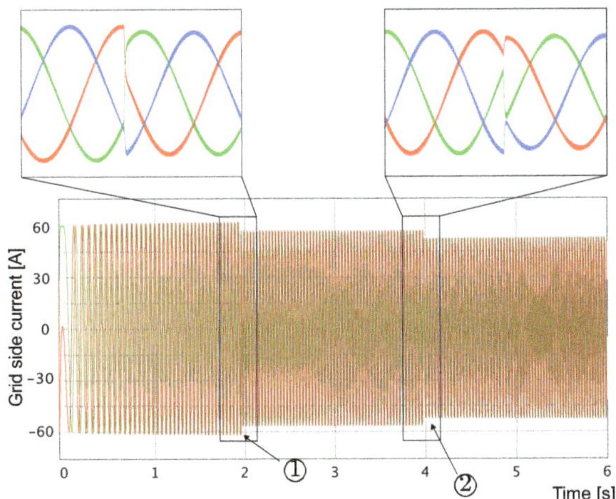

Figure 17. Grid side currents performance.

6. Conclusions and Future Work

Results of the fuel consumed by the engine when simulating the PTO and PTI conditions are not yet significant in helping to reduce EEOI values at both operating regions, which means less operational CO_2 emissions.

The control scheme, when simulating the engine performance at PTI operating region, makes the engine work closer to the MEOP, which leads to low SFC, and therefore low operational emissions.

The control scheme shows the high reliability and accuracy to follow the optimization algorithm with good dynamics, at both operating conditions as PTI and PTO. It also ensures bidirectional power flow, with low distortion of in the grid side currents.

The optimization function presents an accurate performance to obtain a local search for the minimum emissions point, starting at a random state. Future results may include the use of an adaptive perturbation function to ensure full convergence when reaching the minimum emissions point.

Slow steaming was mentioned because of the application of the scheme yet needs to be worked out separately from a design stage to provide more accurate conclusions to this work.

The ship and engine data considered for the simulations is open source and provides great value to continue to be used in this research.

Future work will consider the application of the proposed hybrid propulsion control scheme in a small-scale vessel for experimental validation of the SFC performance, as well as adding a more accurate operational profile of the vessel to work in depth the auxiliary engines and WHRS operational emissions generation influence.

Author Contributions: Conceptualization, J.R.P. and C.A.R.; methodology, J.R.P. and C.A.R.; simulation and programming, C.A.R.; validation, J.R.P. and C.A.R.; formal analysis, J.R.P.; investigation, J.R.P. and C.A.R.; data curation, J.R.P.; writing—original draft preparation, J.R.P. and C.A.R.; writing—review and editing, J.R.P. and C.A.R. All authors have read and agreed to the published version of the manuscript.

Funding: This research received no external funding.

Acknowledgments: The authors wish to thank the financial support from the Chilean Found for Human Resource Development (CONYCIT) through its Ph.D. scholarships (CONICYT/21130448).

Conflicts of Interest: The authors declare no conflicts of interest.

Abbreviations

The following abbreviations are used in this manuscript:

AFR	Air to fuel ratio
EEDI	Energy Efficiency Design Index
EEOI	Energy Efficiency Operational Indicators
FOC	Field oriented control
IMO	International Maritime Organization
LNG	Liquefied Natural Gas
MCR	Maximum continuous rating
MEOP	Minimum Emissions Operating Point
NCRE	Non-conventional renewable energies
PMSM	Permanent magnet synchronous machine
PTI	Power Take-In
PTO	Power Take-Off
SEEMP	Energy Efficiency Monitoring Plant
SFC	Specific fuel oil consumption
SG	Shaft Generator
SMCR	Specified maximum continuous rating
TFOC	Torque field oriented control
VF-VOC	Virtual flux voltage oriented control
WHRS	Waste Heat Recovery System

Appendix A

$$EEDI = \frac{A + B + C - D}{E} \tag{A1}$$

- Main Engines Emissions: Factor A

$$A = \left(\prod_{j=1}^{m} f_j \right) \left(\sum_{i=1}^{n} P_{ME(i)}\, c_{F\,ME(i)}\, SFC_{ME(i)} \right) \tag{A2}$$

f_i Correction factor for ship specific design elements.
P_{ME} Power of main engines.
$C_{F\,ME}$ Main engine conversion factor between fuel consumption and CO_2 emission.
SFC_{ME} Main engine specific fuel consumption.

- Auxiliary Engines Emissions: Factor B

$$B = (P_{AE}\, c_{F\,AE}\, SFC_{AE}) \tag{A3}$$

P_{AE} Power of auxiliary engines.
$c_{F\,AE}$ Auxiliary engine conversion factor between fuel consumption and CO_2 emission.
SFC_{AE} Auxiliary engine specific fuel consumption.

- Generators/Motors Emissions: Factor C

$$C = \left(\left(\prod_{j=1}^{m} f_j \sum_{i=1}^{n} P_{PTI(i)} - \sum_{i=1}^{n} f_{eff(i)}\, P_{AE\,eff(i)} \right) c_{F\,AE}\, SFC_{AE} \right) \tag{A4}$$

f_i Correction factor for ship specific design elements.
P_{PTI} Power of shaft motor divided by the efficiency of shaft generator.
f_{eff} Availability factor of innovative energy efficiency technology.
$P_{AE\,eff}$ Auxiliary power reduction due to individual technologies for electrical energy efficiency.
$c_{F\,AE}$ Auxiliary engine conversion factor between fuel consumption and CO_2 emission.
SFC_{AE} Auxiliary engine specific fuel consumption.

- Efficiency Technologies: Factor D

$$D = \sum_{i=1}^{n} f_{eff(i)}\, P_{eff(i)}\, c_{F\,ME}\, SFC_{ME} \tag{A5}$$

f_{eff} Availability factor of innovative energy efficiency technology.
P_{eff} Output power of innovative mechanical energy efficient technology.
$C_{F\,ME}$ Main engine conversion factor between fuel consumption and CO_2 emission.
SFC_{ME} Main engine specific fuel consumption.

- Transport Work: Factor E

$$E = f_i\, f_c\, C_p\, V_{ref}\, f_w \tag{A6}$$

f_i Capacity factor.
f_c Cubic capacity correction factor.
C_p Capacity or deadweight.
V_{ref} Ship speed.
f_w Weather factor.

Appendix B

$$EEOI = \frac{\sum_j F_j\, c_{F(j)}}{m_c\, D} \tag{A7}$$

j Fuel type.
F_j Mass of consumed fuel j.
$c_{F(j)}$ Fuel mass to CO_2 mass conversion factor for fuel j.
m_c Cargo carried (tonnes).
D Distance in nautical miles corresponding to the cargo carried.

References

1. Shu, G.; Liang, Y.; Wei, H.; Tian, H.; Zhao, J.; Liu, L. A review of waste heat recovery on two-stroke IC engine aboard ships. *Renew. Sustain. Energy Rev.* **2013**, *19*, 385–401. [CrossRef]
2. Singh, D.V.; Pedersen, E. A review of waste heat recovery technologies for maritime applications. *Energy Convers. Manag.* **2016**, *111*, 315–328. [CrossRef]
3. Hossain, S.; Bari, S. Waste Heat Recovery From the Exhaust of a Diesel Generator Using Shell and Tube Heat Exchanger. In Proceedings of the ASME International Mechanical Engineering Congress and Exposition, Proceedings (IMECE), San Diego, CA, USA, 15–21 November 2013; Volume 6. [CrossRef]
4. Anthony, F.M.; Stephen, R.T.; Dominic, A.H. *Ship Resistance and Propulsion: Practical Estimation of Ship Propulsive Power*; Cambridge University Press: Cambridge, UK, 2011.
5. Shi, W.; Grimmelius, H.; Stapersma, D. Analysis of ship propulsion system behavior and the impact on fuel consumption. *Int. Shipbuild. Prog.* **2010**, *57*, 35–64. [CrossRef]
6. Bilgin, B.; Magne, P.; Malysz, P.; Yang, Y.; Pantelic, V.; Preindl, M.; Korobkine, A.; Jiang, W.; Lawford, M.; Emadi, A. Making the Case for Electrified Transportation. *IEEE Trans. Transp. Electrif.* **2015**, *1*, 4–17. [CrossRef]
7. Sun, X.; Liang, X.; Shu, G.; Wang, Y.; Wang, Y.; Yu, H. Effect of different combustion models and alternative fuels on two-stroke marine diesel engine performance. *Appl. Therm. Eng.* **2017**, *115*, 597–606. [CrossRef]
8. Deniz, C.; Zincir, B. Environmental and economical assessment of alternative marine fuels. *J. Clean. Prod.* **2016**, *113*, 438–449. [CrossRef]
9. Woud, H.K.; Stapersma, D. *Design of Propulsion and Electric Power Generation Systems*; IMarEST, The Institute of Marine Engineering, Science and Technology: London, UK, 2002.
10. International Maritime Organization. *Resolution mepc.212(63), Guidelines on the Method of Calculation of the Attained Energy Efficiency Design Index (Eedi) for New Ships*; International Maritime Organization: London, UK, 2012.
11. International Maritime Organization. *Mepc.1/circ.681, Interim Guidelines on the Method of Calculation Of the Energy Efficiency Design Index for New Ships*; International Maritime Organization: London, UK, 2012.
12. Prousalidis, J.; Patsios, C.; Kanellos, F.; Sarigiannidis, A.; Tsekouras, N.; Antonopoulos, G. Exploiting shaft generators to improve ship efficiency. In Proceedings of the 2012 Electrical Systems for Aircraft, Railway and Ship Propulsion, Bologna, Italy, 16–18 October 2012; pp. 1–6. [CrossRef]
13. Sarigiannidis, A.; Kladas, A.; Chatzinikolaou, E.; Patsios, C. High efficiency Shaft Generator drive system design for Ro-Ro trailer-passenger ship application. In Proceedings of the 2015 International Conference on Electrical Systems for Aircraft, Railway, Ship Propulsion and Road Vehicles (ESARS), Aachen, Germany, 3–5 March 2015; pp. 1–6. [CrossRef]
14. Sarigiannidis, A.G.; Chatzinikolaou, E.; Patsios, C.; Kladas, A.G. Shaft Generator System Design and Ship Operation Improvement Involving SFOC Minimization, Electric Grid Conditioning, and Auxiliary Propulsion. *IEEE Trans. Transp. Electrif.* **2016**, *2*, 558–569. [CrossRef]
15. Denis Griffiths. *Marine Low Speed Diesel Engines*; Institute of Marine Engineers: London, UK, 2001.
16. Kouro, S.; Malinowski, M.; Gopakumar, K.; Pou, J.; Franquelo, L.G.; Wu, B.; Rodriguez, J.; Pérez, M.A.; Leon, J.I. Recent Advances and Industrial Applications of Multilevel Converters. *IEEE Trans. Ind. Electron.* **2010**, *57*, 2553–2580. [CrossRef]
17. Pena, R.; Cardenas, R.; Asher, G.M.; Clare, J.C.; Rodriguez, J.; Cortes, P. Vector control of a diesel-driven doubly fed induction machine for a stand-alone variable speed energy system. In Proceedings of the IEEE 2002 28th Annual Conference of the Industrial Electronics Society, IECON 02, Sevilla, Spain, 5–8 November 2002; Volume 2, pp. 985–990. [CrossRef]
18. Roy, S.; Malik, O.P.; Hope, G.S. A k-step predictive scheme for speed control of diesel driven power plants. *IEEE Trans. Ind. Appl.* **1993**, *29*, 389–396. [CrossRef]
19. Haddad, S.; Watson, N. *Principles and Performance in Diesel Engineering*; Ellis Horwood Series Engineering Science: New York, NY, USA, 1984.
20. Yin, J.; Su, T.; Guan, Z.; Chu, Q.; Meng, C.; Jia, L.; Wang, J.; Zhang, Y. Modeling and Validation of a Diesel Engine with Turbocharger for Hardware-in-the-Loop Applications. *Energies* **2017**, *10*, 685. [CrossRef]
21. Jiang, J. Optimal gain scheduling controller for a diesel engine. *IEEE Control. Syst. Mag.* **1994**, *14*, 42–48. [CrossRef]

22. Holtz, J. The representation of AC machine dynamics by complex signal flow graphs. *IEEE Trans. Ind. Electron.* **1995**, *42*, 263–271. [CrossRef]
23. Novotny, D.W.; Lipo, T.A. *Vector Control and Dynamics of AC Drives*; Oxford University Press: Oxford, UK, 1996; Volume 41.
24. Rahoui, A.; Bechouche, A.; Seddiki, H.; Abdeslam, D.O. Grid Voltages Estimation for Three-Phase PWM Rectifiers Control Without AC Voltage Sensors. *IEEE Trans. Power Electron.* **2018**, *33*, 859–875. [CrossRef]
25. Baggu, M.M.; Chowdhury, B.H.; Kimball, J.W. Comparison of Advanced Control Techniques for Grid Side Converter of Doubly-Fed Induction Generator Back-to-Back Converters to Improve Power Quality Performance During Unbalanced Voltage Dips. *IEEE J. Emerg. Sel. Top. Power Electron.* **2015**, *3*, 516–524. [CrossRef]
26. Clarksons. Clarksons World Fleet Register. Available online: http://www.clarksons.net (accessed on 20 December 2019).
27. MAN. MAN Diesel & Turbo Engine Builder. Available online: http://www.mandieselturbo.com (accessed on 20 December 2019).

Journal of
Marine Science and Engineering

Article

Effects of Marine Exhaust Gas Scrubbers on Gas and Particle Emissions

Hulda Winnes *, Erik Fridell and Jana Moldanová

IVL Swedish Environmental Research Institute, 400 14 Gothenburg, Sweden; erik.fridell@ivl.se (E.F.);
jana.moldanova@ivl.se (J.M.)
* Correspondence: hulda.winnes@ivl.se; Tel.: +46-10-788-6760

Received: 24 March 2020; Accepted: 21 April 2020; Published: 24 April 2020

Abstract: There is an increase in installations of exhaust gas scrubbers on ships following international regulations on sulphur content in marine fuel from 2020. We have conducted emission measurements on a four-stroke marine engine using low sulphur fuel oil (LSFO) and heavy fuel oil (HFO) at different steady state engine loads. For the HFO the exhaust was probed upstream and downstream of an exhaust gas scrubber. While sulphur dioxide was removed with high efficiency in the scrubber, the measurements of particle emissions indicate lower emissions at the use of LSFO than downstream of the scrubber. The scrubber removes between 32% and 43% of the particle mass from the exhaust at the HFO tests upstream and downstream of the scrubber, but levels equivalent to those in LSFO exhaust are not reached. Decreases in the emissions of polycyclic aromatic hydrocarbons (PAH-16) and particulate matter as black carbon, organic carbon and elemental carbon, over the scrubber were observed for a majority of the trials, although emissions at LSFO use were consistently lower at comparable engine power.

Keywords: scrubber; EGCS; emissions; particles; PM; BC; exhaust gases; on board measurements

---

1. Introduction

Regulations ban the use of marine residual fuels with sulphur mass content above 0.5% from 2020, in all ocean areas. The sulphur limits in emission control areas (SECAs) at 0.1% m/m will remain. The regulation further permits the use of exhaust gas cleaning systems (EGCS) that reduce the SO_2 concentration in the exhaust to levels equivalent to those from the regulated fuel sulphur limits [1]. The same regulation that limits the marine fuel sulphur content also aims at reducing particle emissions. A fundamental difference compared to the sulphur limit is that the permitted particle emission levels are not quantified but relies on the decrease of particle emissions with fuel sulphur content [2]. The emissions of particles after the exhaust gas scrubbing are accordingly neither regulated to a standard level nor is a specific measurement standard decided. The installation and operation of exhaust gas scrubbers on ships is often an attractive choice compared to operating a ship on refined fuel such as marine gasoil (MGO), from an economic perspective, see for example reference [3]. The classification society DNV GL estimates that approximately 4100 ships will be equipped with one or more EGCSs in 2021, based on statistics from a number of scrubber manufacturers [4]. Dry and liquid bulk carriers, container ships, cruise ships and roll-on/roll-off (RoRo) ships are the ship types most represented in the statistics. The engine power and fuel consumption of these ships are more important aspects than number of ships from an environmental perspective. A complete inventory on these parameters is not available. It seems though that large ships and high power engines are fitted with scrubbers to a larger extent than is the case for smaller ships.

Different scrubber designs exist. Open sea water scrubbers utilize the natural alkalinity of sea water and keep a high flow of process water in order to reduce SO_2 in the exhaust. The wash water is discharged to sea, most often without substantial treatment. In closed-loop scrubber designs, the process water is fresh- or seawater with an added alkaline chemical, often sodium hydroxide. The water is recirculated in the system and the sodium hydroxide neutralizes the sulphuric acid, see [5] for chemical reactions. Small portions of the process water are extracted from the process and passed through a water treatment unit before being discharged to the sea. This is often referred to as bleed-off water. The treatment produces a residue sludge that is brought ashore to a port reception facility. Open systems can use over 100 times more scrubber fluid per kWh output of the engine than a closed loop system. Washwater discharge criteria set minimum pH, maximum polycyclic aromatic hydrocarbons (PAH) level, maximum nitrate level, and maximum turbidity level with guidelines given in Resolution 259(68) of the Marine Environmental Protection Committee of the International Maritime Organization (IMO). The same criteria are valid for effluents from closed-loop systems as for the open systems. An important aspect for all types of exhaust gas scrubbers is the relocation of substances in the exhaust gas from air to the marine environment when scrubber systems are employed [6].

Quantification and characterization of particle emissions from marine engines fitted with exhaust gas scrubbers are interesting from environmental and health perspectives. The effects of an extensive use of scrubbers could significantly impact particulate air emissions from shipping. A limited number of previous studies have been published with the purpose to quantify and communicate the potential to reduce harmful exhaust gases and particles by using marine exhaust gas scrubbers; in total, specific emissions from six marine engines equipped with scrubbers have been found in scientific literature [7–11]. The ability of scrubbers to efficiently abate SO_2 is unchallenged while the effect of the exhaust gas scrubbing process on particle mass concentration varies significantly between the studies. Some measurements have indicated 75% reduction and more [7,9] and other studies conclude on more moderate particle reductions or even increases in particles over the scrubber [8,11]. A measured increase in particle mass over a wet scrubber has been explained as an increase in salt particles originating from salt in the process water and as an increase in sulphate particles due to low temperatures and high humidity after the scrubber [8]. Two peer reviewed studies comprise results from four engines in total and indicate between 7% and 75% particulate matter mass (PM) reduction over the scrubber of which one study presents the higher extreme, and the other find reductions below 45% from all tests [7,11]. The engine types, fuel oil, exhaust gas systems, and scrubber design differ between the two studies and could explain the differences in results.

Factors that influence particle removal in the scrubber include the design of the scrubber and scrubber process, and the composition of the exhaust gas particles [10]. The occurrence of semi-volatile species in the exhaust can be an important parameter since the temperatures in the scrubber process and during sampling will influence whether these species are solid or in gas phase. The sampling system setup could potentially be an important contributor to the large differences in results in different studies.

Only few studies have reported on the effects of scrubbers on specific particle contents and the variations are high. Elemental carbon (EC) concentration reductions range from 50% to 81% and reductions of condensable organic carbon (OC) concentrations of 13%–41% and 73%, in [7] and [8] respectively, and the reductions presented in [11] appear to be within this range however not quantified. Particulate sulphate increased over the scrubber in [8]. PAH contents were analysed by [7] who observed a significant reduction of the tested PAHs over the scrubber. There is a need for more measurement data on particle emissions in order to assess the potential environmental and health impacts of the increased use of scrubbers in shipping.

This paper presents the emission reduction potential of a closed-loop scrubber and compares emissions with those from combustion of low sulphur fuel oil (LSFO). LSFO is sometimes referred to as ultra-low sulphur fuel oil (ULSFO) or hybrid oil. There is no standard specifying density or viscosity for the LSFO fuels, and the name can be collectively used for marine fuels that do not fulfil the marine distillate specifications for e.g. viscosity and density in the ISO 8217 standard, but that are below the regulated limit for sulphur content in fuel. In this case the limit is 0.1%, and the LSFOs became available on the market shortly after the sulphur SECA regulations became effective. It costs less than MGO and is, therefore, the preferable choice among some ship operators. With the 0.5% global sulphur limit it is likely that hybrid fuels will be common also outside SECAs.

Our measurements cover both gases and particulate matter. Particle contents have been analysed for organic and elemental carbon, elements, and sulphur. The PAH content in particle and gas phase was analysed, and the SO_3/H_2SO_4 content in gas phase was sampled. Furthermore, online instruments were used to measure gas concentrations, black carbon concentrations and particle number concentrations (PN). Analyses of the scrubber water and ecotoxicology tests were part of the same campaign but are reported separately [6].

2. Materials and Methods

The test ship is a RoPax ferry taking rolling goods and passengers, on a scheduled timetable between the ports and is one of the largest in its category. The engine used for the trials is of model 8LMAN48/60. It is a 9600 kW, four-stroke engine with common rail assisted fuel injection. The scrubber system tested uses seawater, with added sodium hydroxide for increased alkalinity, as process water. Tests on LSFO were conducted in February 2017 and tests on HFO in September 2017.

All sampling is performed through sampling holes cut in the exhaust pipe for the probes. One set of holes are cut on deck 11, upstream of the scrubber and another set on deck 15, downstream of the scrubber. During the measurements at LSFO operations, the set of holes on deck 11 are used for sampling, while both decks are used during the tests with HFO.

Tests with similar scopes are made at multiple steady state engine loads close to these load points:

85%, 75%, 50%, and 34% of maximum continuous rating (MCR), for measurements on LSFO.

76%, 49%, and 32% MCR, for measurements on heavy fuel oil upstream of the scrubber.

76%, 48%, and 41% MCR, for measurements on heavy fuel oil downstream of the scrubber.

The engine loads are equally relevant to everyday operations, although long periods on loads below 35% MCR are avoided.

A number of gases and metrics of particles are included in the measurement scheme. The gas phase pollutants are sampled without dilution, extracting the hot exhaust sample through a heated probe directly to online instruments and adsorbent tubes. A Horiba PG 350 measures sulphur dioxide (SO_2), carbon monoxide (CO) and carbon dioxide (CO_2) by non-dispersive infrared (NDIR), and nitrogen oxide (NO_X) by chemiluminiscence. Raw gas is prior to the instrument conducted through a heated tube with Teflon lining via a ceramic filter to a preparation unit. The tube is heated to 190°C. The gas preparation unit cools the gas to 4°C and removes particles by filtration. The dry and particle free gas is used for continuous concentration measurements in the instruments with the interval 1 second. Hydrocarbons are measured with a flame ionisation detector instrument of model Graphite 52M-D. The instrument monitors total hydrocarbon, non-methane hydrocarbon and methane simultaneously. Hydrocarbons are measured as total carbon in unit ppb(V) of CH_4 equivalents.

Gaseous SO_3/H_2SO_4 concentrations are sampled in adsorbent glass tubes containing NaCl placed directly inside the exhaust channel. A flow is pumped through the NaCl tube and gas-phase H_2SO_4 reacts to form sodium sulphate and sodium bisulphate. Any gas-phase SO_3 is converted to H_2SO_4 in the flue gas at temperatures below 500°C, and is not present at temperatures below 200°C due to the presence of water vapour in the exhaust [12]. Subsequent analysis of the salt columns by ion chromatography for sulphates is made at the IVL Swedish Environmental Research Institute laboratory in Gothenburg.

Two different devices for exhaust gas dilution are used prior to particle sampling and measurements. The first system is a dilution tunnel designed to comply with ISO standard 8178:2/8178:1 for partial dilution systems. Filtered ambient air is used as dilution gas. The dilution tunnel dilutes the exhaust gas sample to a ratio of ~1:5-20 and allows for minor adjustments of flow through the system. The exact dilution rate is checked by CO_2 measurements using a sensor from Servomex in the diluted sample and comparing with the readings from the Horiba PG 350 in the raw exhaust gas. The second system is a Dekati Fine Particle Sampler (FPS) model 4000, using pressurized and optionally heated air in a two-stage dilution, with an adjustable dilution to a ratio of up to 1:500. A heated probe is used in all measurements with the FPS. The primary diffusion dilution stage dilutes the extracted exhaust sample with preheated air (set to ~250°C). The secondary ejector dilution step uses pressurized air of ambient temperature. The dilution ratios are determined from the ratio of CO_2 in raw and diluted exhaust gas using a CO_2 analyser from LiCor in the diluted sample.

The dilution tunnel is used prior to filter sampling for determination of PM mass and composition at LSFO operations and at HFO operations upstream of the scrubber while filter sampling downstream of the scrubber is conducted using dilution with the FPS. The choice not to use the dilution tunnel downstream of the scrubber is mainly due to space restrictions and inaccessibility of that measurement point; the dimensions of the dilution tunnel are approximately $1.5 \times 1 \times 0.3$ m and the sampling point downstream of the scrubber is located on deck 15 and accessible only via ladders from six decks below. In order to assure comparability of the two dilution alternatives, filter sampling was conducted at similar engine loads both with the dilution tunnel and FPS at operations on LSFO and on HFO upstream of the scrubber. For particle measurements performed with the online particle instruments, the FPS is always used for the dilution.

Sampling and dilution procedures bring uncertainties to the particle measurements. The first group of uncertainties relates to the representativeness of the extracted partial flow exhaust sample. Both dilution devices use isokinetic probes with adjustable inlet nozzle sizes to achieve near-isokinetic sampling. However, deviations can occur especially for the FPS device as the sample flux changes with the dilution ratio and the set-up does not allow for changes of the nozzle during the experiments. For particles in the typical size-range of the diesel exhaust particles, i.e., with a large part of the particle mass in sizes with particle diameter below 100 nm [13,14], deviation from the isokinetic sampling has only a small impact, and the isokinetic conditions are not required by ISO 8178 standard. Hence, potential influence of parameters that relates to particles' kinetics is not investigated further. Secondly, uncertainties are related to the dilution process affecting both condensation and nucleation of semi-volatile species and hence the measured PM mass and number. Achieving ISO 8178 standard sampling downstream of the scrubber is not possible as the exhaust temperature at this point is too low. Our filter sampling experienced deviations from the standard, including a stack temperature downstream of the scrubber that was too low and often lower filter sampling temperatures than prescribed. The FPS device is not designed to fulfil the ISO 8178. We see that sampling with FPS at LSFO combustion results in higher particle emission estimates than when using the dilution tunnel. Contrary to this, the particle emissions at HFO combustion upstream of the scrubber are often indicated to be lower for measurements with the FPS than for measurements with the dilution tunnel. A large variability of the measured emission factors, also when only samples taken with the dilution tunnel are considered, reveals large uncertainties, especially at low engine loads. An analysis of all individual filter samples still indicates agreement between the sampling systems at the tests with high engine loads. For the low engine load, the individual samples vary more and there is less agreement between the two sampling systems, see Table A1 in the Appendix A and Figure 1.

At the tests with LSFO and the dilution tunnel, the temperatures at the filters were between 29 °C and 33 °C, which is close to the dilution air temperature. This is lower than prescribed by ISO 8178. Filter temperatures are not recorded for the other tests. The raw gas transfer line is a 5 meter long tube with Teflon lining, heated to 190 °C. Dilution ratios were between 15.9 and 23.8. The ambient temperature was approximately 35 °C at tests on HFO at the location upstream of the scrubber and approximately 50 °C at the location downstream of the scrubber. We can assume similar temperatures also at the filters. The FPS gives a record of the temperature of the sample leaving the device, and the PM sample temperature was indicated to be between 32 and 50 °C during the tests (see the table in the Appendix A for details). The automatic logging of temperature data during the tests were performed in high surrounding temperature and it is probable that the high temperature cause problems with the dilution system signals. The dilution ratios were however confirmed, also when logging failed, by the CO_2 instruments. The transfer lines include approximately 1 meter insulated metal tubing prior to the FPS and 0.5 meter conductive tubing for the diluted sample. Table A1 in the Appendix A includes details on temperatures during sampling.

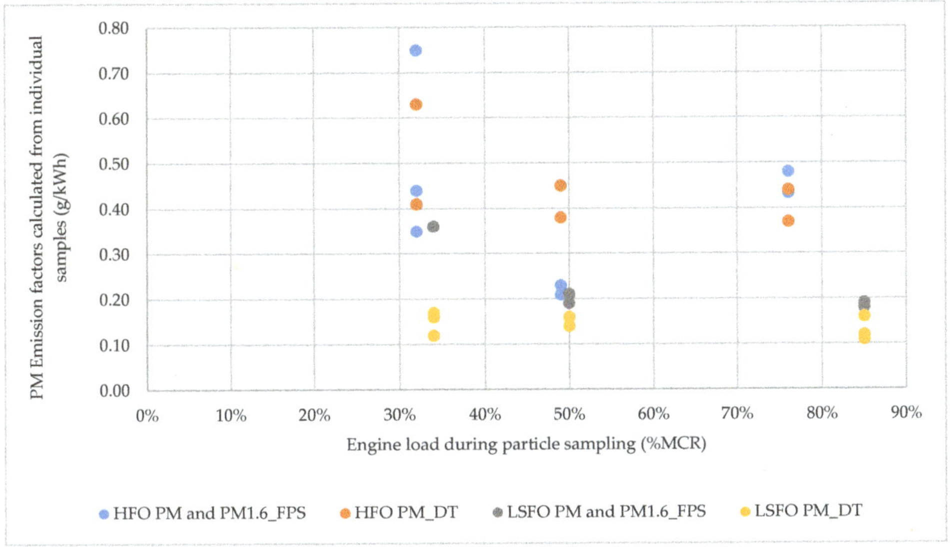

Figure 1. Particulate matter (PM) emission factors calculated from individual filter samples in a comparison between the two sampling systems used for exhaust gas dilution.

Particulate matter mass is sampled as PM_{tot} and $PM_{1.6}$ on polytetrafluoroethylene (PTFE) filters. Cyclones are used for the $PM_{1.6}$ sampling, assuring that large particles are removed from the sampling stream before the filter. The cyclones used are primarily designed for a cutoff at particle diameter of 2.5 µm, however, the flow through the cyclone during the tests was higher than the design flow, resulting in a lower cutoff diameter. The size cutoff at particle diameter 1.6 µm is calculated from the actual flow through the system. The gravimetrical analysis of the filter samples was performed at the certified laboratory of IVL. The filters are weighed in a controlled environment before and after sampling. A Mettler Toledo model MT5 balance is used. The balance is calibrated to an uncertainty limit of ±3 µg in the range 0–10 000 µg and ±7 µg in the range 10 000–100 000 µg, our sample weights are in the range 15 000–100 000 µg (Table A1 in the Appendix A).

Elemental and organic carbon (EC/OC) content on particles are determined by sampling using filter holders with pre-heated double quartz filters (Pall, Tissuequarz). Backup filters are used to correct for the positive sampling artefact from condensation of volatile organics on the filter. A filter section with a total area of 2.01 cm² is cut out of the filter for the analysis of OC and EC by a thermal–optical method (EC/OC analyser Model 4, Sunset Laboratory, USA) using the EUSAAR_2 temperature protocol [15]. The analyses of filters are conducted by Laboratory of Aerosols Chemistry and Physics; Institute of Chemical Process Fundamentals, in Prague. Reported uncertainties by the laboratory for their analyses are 10% for OC and 20% for EC.

An on-line electromobility-based instrument (TSI EEPS 3090) measures the number concentrations of particles between 5.6 nm and 560 nm in diameter in 32 size channels at 10 Hz. The instrument is used at HFO operations, upstream and downstream of the scrubber. There are no EEPS results for the LSFO fuel due to EEPS instrument failure during this fuel testing. A thermodenuder of model Dekati ELA 423 is used to vaporize volatile particles from the sample. We can thereby identify total number concentrations and size distributions of solid particles. The thermodenuder is heated to

300 °C in order to vaporize the volatile fraction of particles including many organics and sulphate. The use of a thermodenuder has been shown to also cause a loss of solid particles in the denuder through thermophoresis and diffusion depends on particle size, temperature, and velocity through the thermodenuder. Size dependent particle losses in the thermodenuder are calculated according to the manufacturer's instructions.

Magee Scientific's Aethalometer AE33, with continuous measurement of the attenuation of transmitted light at eight wavelengths is used for the detection of black carbon (BC) content. Measurement of absorption at 880 nm is interpreted as concentration of BC; in this context the BC refer to equivalent black carbon (eBC) as recommended in [16].

A schematic of the scrubber system and the instrument and sampling setup is presented in Figure 2.

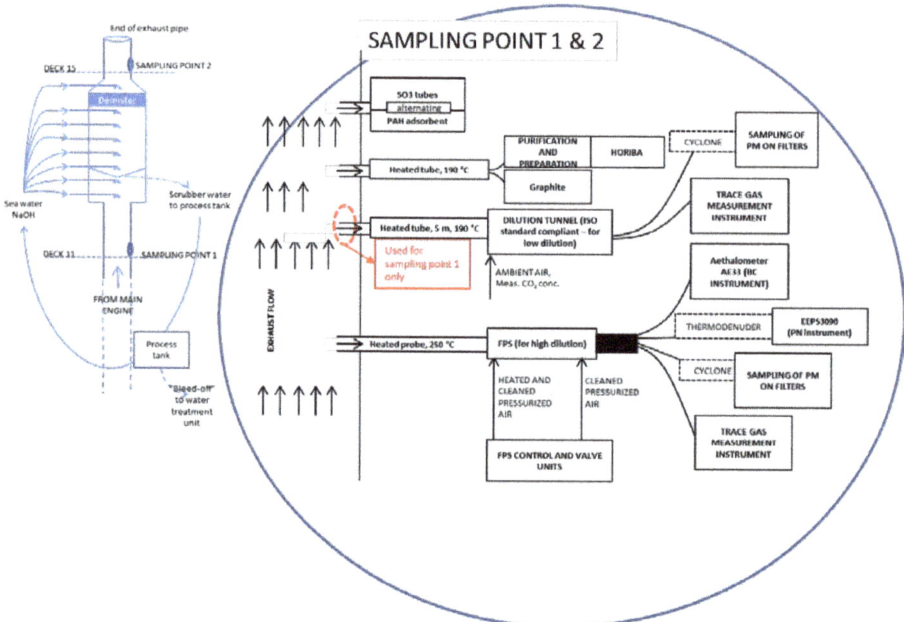

Figure 2. Schematic of the scrubber system and the instrument and sampling setup.

Uncertainties in the on-line instruments are expressed as coefficients of variance of average values. The variations in dilution ratios are assumed to add more uncertainty to these results.

PAHs are sampled from the undiluted exhaust by an adsorbent for subsequent Soxhlet extraction and analysis by high-performance liquid chromatography (HPLC) in the IVL laboratory in Gothenburg. Particle-bound PAHs are collected on filters coupled in series to glass columns with XAD7 and polyurethane foam (PUF) absorbent for capture of gas phase PAH. This method gives an analysis of combined content of PAHs in gaseous and particulate form. Contents of USEPA's PAH-16 priority pollutants are analysed. These include naphthalene, acenaphthylene, acenaphthene, fluorene, phenantrene, anthracene, fluoranthene, pyrene, benz[a]anthracene, chrysene, benzo[b]fluoranthene, benzo[k]fluoranthene, benzo[a]pyrene, dibenz[a, h]anthracene, benzo[ghi]perylene, and indeno[1,2,3-cd]pyrene.

Exhaust gas flow and emission factors of gases are calculated based on the carbon balance method as specified in ISO 8178:4. Fuel oil samples were sent for analysis of density, viscosity, calorific value and elemental content to the Saybolt Laboratory in Gothenburg. Analyses of the lubrication oil were also made. The density and viscosity of the HFO were significantly higher than those of the LSFO. The LSFO was still too viscous and dense to qualify as marine distillate oil (DMX-DFB) according to ISO 8217:2017 standard for marine fuels. The results from the analyses of the LSFO, the HFO, and the lube oil are presented in Table 1. Sulphur dioxide emissions are calculated from the sulphur content of fuel except for measurement downstream of the scrubber, since this presents a more reliable value than SO_2 measurements with the gas analyser.

Table 1. Fuel and lube oil analyses.

	Unit	Low Sulphur Fuel Oil (LSFO)	Lube Oil (at LSFO Tests)	Heavy Fuel Oil (HFO)	Lube Oil (at HFO Tests)
Date of Analysis		2017-03-15	2017-03-15	2017-10-13	2017-10-13
Density (at 15°C)	kg/m^3	908.5	909.1	989.5	919.9
Viscosity (at 50° C)	mm^2/s	81.49	n.a.	420.0	n.a.
Gross heat of combustion	MJ/kg	44.25	n.a.	42.19	n.a.
Net heat of combustion	MJ/kg	41.52	n.a.	39.97	n.a.
Sulphur	Mass %	0.10	0.43	2.77	0.55
Carbon	Mass %	86.8	84.3	85.4	83.4
Hydrogen	Mass %	12.9	13.2	10.5	12.7
Nitrogen	mg/kg	1800	520	4300	1200
Oxygen (calculated)	Mass %	<0.1	0.4	0.9	<0.1
Vanadium	mg/kg	<1	3	122	72
Nickel	mg/kg	21	7	32	33
Iron	mg/kg	11	7	22	40
Ash content	Mass %	0.041	1.636	0.035	>0.180
Total aromatic hydrocarbons	Mass %	15.5	25.5	23.5	24.9
Asphaltenes	Mass %	<0.50	n.a.	8.9	n.a.

3. Results

We present results from two viewpoints. One aspect is the reduction of different pollutants over the scrubber at tests on HFO. This is interesting from the viewpoint of which substances in the exhaust gas that can be expected to be removed in the scrubbing process, and to what extent. We also compare emissions downstream of the scrubber with emissions from LSFO combustion. This is an interesting comparison to make from an environmental point of view since both solutions are alternatives to comply with existing regulation on sulphur in marine fuel and corresponding emissions. In addition, the composition of particles is presented and discussed separately.

The summary of results for specific fuel oil consumption and emissions are presented in Table 2, and further elaborated upon in the following text.

Uncertainties in the results cannot be described by the coefficient of variation alone. In addition, uncertainty is added from difficulties in determining the fuel flow. We estimate a load-dependent uncertainty of the specific fuel oil consumption (SFOC) which of approximately ±10% at 50% engine load and lower, and ±5% at more than 50% engine load for this engine and measurement. Determination of the dilution ratios is another source for uncertainties as well as the handling of samples in the rough environment in the funnel. Uncertainties in the PM sampling system setup further includes a loss of particles in the system that increases with the length of the transfer line from the exhaust gas channel to the dilutor.

3.1. Scrubber emission reduction efficiency

The scrubber efficiently reduces emissions of sulphur dioxide in the exhaust gases. The SO_2 emission factor at HFO tests is reduced >99% over the scrubber, at all engine loads.

Table 2. Specific emissions at the different tests. Coefficients of variation (cov) at continuous measurements are presented in italics. MCR = Maximum continuous rating; SFOC = specific fuel oil consumption.

Parameter	Unit	Low Sulphur Fuel Oil (LSFO)				Heavy Fuel Oil (HFO) Upstr. of Scrubber		Heavy Fuel Oil (HFO) Downstr. of Scrubber			
		Engine Load (%MCR)				Engine Load (%MCR)		Engine Load (%MCR)			
		85%	75%	50%	34%	76%	49%	32%	76%	48%	41%
SFOC	g/kWh	181	180	198	239	187	208	253	187	211	239
CO_2	g/kWh	601	598	659	793	617	687	739	618	690	847
	cov	*2.4*	*3.0*	*4.0*	*8.7*	*3.7*	*9.6*	*11.8*	*4.9*	*9.0*	*17.8*
SO_2	g/kWh	0.36 [1]	0.36 [1]	0.40 [1]	0.48 [1]	10.4 [1]	11.6 [1]	14.1 [1]	0.06	0.03	0.02
	cov	*n.d.*	*n.d.*	*n.d.*	*n.d.*	*n.d.*	*n.d.*	*n.d.*	*92.1%*	*64.1%*	*41.0%*
SO_3	g/kWh	b.d.l.	b.d.l.	b.d.l.	b.d.l.	0.37	0.13	0.16	0.08	0.05	0.06
NO_X	g/kWh	11.84	9.73	11.85	15.38	11.0	12.6	16.3	10.9	12.4	14.6
	cov	*5.2%*	*0.5%*	*0.5%*	*1.8%*	*0.7%*	*4.2%*	*4.3%*	*1.4%*	*5.9%*	*5.1%*
nmHC	g/kWh	0.24	n.d.	0.30	0.45	0.36	0.30	0.40	0.16	0.24	n.d.
	cov	*6.9%*	*n.d.*	*4.1%*	*4.2%*	*12.6%*	*2.7%*	*4.1%*	*1.5%*	*1.6%*	*n.d*
CH_4	g/kWh	0.002	n.d.	0.002	0.004	0.001	0.004	0.005	0.004	0.002	n.d.
	cov	*9.6%*	*n.d.*	*7.4%*	*17.4%*	*29.8%*	*4.9%*	*4.1%*	*4.3%*	*6.3%*	*n.d*
CO	g/kWh	0.42	0.53	0.88	0.96	0.93	1.72	1.87	0.79	1.40	1.50
	cov	*2.6%*	*4.2%*	*5.0%*	*5.8%*	*6.9%*	*8.6%*	*8.5%*	*6.5%*	*16.6%*	*12.3%*
PM_{tot}	g/kWh	0.17	0.12	0.16	0.17	0.44	0.30	0.41	0.27	0.19	0.25
	Number of filter samples	4	2	3	3	4	4	4	3	3	2
$PM_{1.6}$	g/kWh	0.14	0.11	0.16	0.14	0.32	0.36	0.60	0.24	0.25	0.24
	Number of filter samples	2	2	3	2	2	2	2	3	3	2
PM_{all} [2]	g/kWh	0.16	0.11	0.16	0.16	0.40	0.36	0.44	0.27	0.23	0.25
PN	#/kWh	n.d.	n.d.	n.d.	n.d.	6×10^{15}	5×10^{15}	7×10^{15}	1×10^{15}	9×10^{14}	2×10^{15}
	cov	*n.d.*	*n.d.*	*n.d.*	*n.d.*	*9.7%*	*24.9%*	*15%*	*8.9%*	*4.0%*	*25.0%*
PN (non-volatile)	#/kWh	n.d.	n.d.	n.d.	n.d.	9×10^{14}	2×10^{15}	9×10^{14}	1×10^{15}	9×10^{14}	8×10^{14}
	cov	*n.d.*	*n.d.*	*n.d.*	*n.d.*	*0.7%*	*7.7%*	*8.8%*	*2.2%*	*1.4%*	*1.1%*
BC	g/kWh	0.006	0.004	0.01	0.012	0.022	0.035	0.065	0.022	0.022	0.028
	cov	*7.3%*	*7.5%*	*6.0%*	*4.5%*	*4.9%*	*8.5%*	*4.9%*	*5.1%*	*3.4%*	*5.3%*
EC	g/kWh	0.0030	0.0091	0.0043	0.0052	0.022	0.052	0.096	0.015	0.017	0.018
	Number of filter samples	3	2	2	3	1	2	1	2	2	1
OC	g/kWh	0.066	0.076	0.092	0.16	0.084	0.13	0.26	0.084	0.11	0.086
	Number of filter samples	3	2	2	3	1	2	1	2	2	1
PAH-16 [3]	mg/kWh	0.43	0.40	0.68	0.85	1.4	1.5	1.9	0.77	1.1	0.92

[1] Calculated from fuel sulphur content; [2] Median values from all samples of PM_{tot} and $PM_{1.6}$; [3] Analysis includes both particulate phase and gas phase polycyclic aromatic hydrocarbons (PAHs); n.d. = no data; b.d.l. = below detection limit.

Around 1%–8% of the sulphur oxides that form during the combustion in a diesel engine are sulphur trioxide SO_3 [17]. SO_3 will react rapidly with water in the exhaust gas to form gaseous H_2SO_4. Condensed H_2SO_4 can cause corrosive damage in locations with low temperatures in the cylinder and the exhaust channel and levels are preferably kept at a minimum. The temperature at which the H_2SO_4 condenses depends on the concentration of SO_3 and H_2O in the gas. In a wet scrubber, the exhaust is rapidly cooled to below the acid dew point; the hot exhaust in our measurements were approximately 250 °C and reduced to 20 °C, over the scrubber. Since the rate of cooling exceeds that of H_2SO_4 gas absorption in the scrubber fluid, it has been suggested that sub-micron H_2SO_4 particles are formed [18]. A reduction of these particles in the scrubber is dependent on mass-transfer through Brownian diffusion. As a removal mechanism, it is not efficient enough to remove all H_2SO_4 from the exhaust gas [18].

The concentrations of gas phase sulphuric acid and sulphur trioxide (H_2SO_4/SO_3) in our tests were significantly reduced over the scrubber. At 76% engine load there is a removal of 78% of the H_2SO_4/SO_3. At the two lower engine loads the reduction is less, 61%–63%. The observed reduction efficiency for SO_2 of over 99% for all loads indicates that the scrubber is less efficient in removing H_2SO_4/SO_3 than in removing SO_2. The measured H_2SO_4/SO_3 concentrations are higher than the SO_2 concentrations in the tests downstream of the scrubber. Metals such as V and Ni, present in the exhaust, can act as catalysts for the oxidation of SO_2 to SO_3, which can be a reason for the higher SO_3 levels at combustion of HFO [19].

There is also a reduction of the CO concentration over the scrubber. At lower engine loads this is more pronounced. We saw minor differences in the emissions of NO_X upstream and downstream of the scrubber at low engine loads. These differences are, however, most likely more related to the different engine loads than to scrubbing of NO_X. The engine load is around 32% at tests upstream of the scrubber and 41% at tests downstream of the scrubber.

Furthermore, specific PM emissions at HFO combustion are reduced over the scrubber. Reductions are approximately 34% at 76% engine load, and 42% at 48%–49% engine load. These reductions are in line with a central value of PM reductions in a joint analysis of previous studies, although values to compare with are few and results vary significantly.

As presented in Table 2, both BC and EC measurement indicate that there are reductions over the scrubber. BC results are not as consistent as EC results, although also these indicate a removal rate that increase at low engine loads.

Particle number (PN) concentrations were measured only at tests on HFO. Measurements were conducted using a thermodenuder (TD) to remove volatile species. By removing a majority of the volatiles with the TD (some organic matter can still remain at 300 °C), the large uncertainties coupled to the nucleation of volatile particles during the sampling dilution process is eliminated. The fraction of the solid particles that remains downstream of the exhaust gas scrubber can therefore be argued to give a more reliable value on cleaning efficiency than total particles. No clear reducing effect on solid PN concentration by the scrubber is seen, see Table 2, although the size distribution is changed. Size distributions of solid particle number concentration are presented in Figure 3.

Recalculating particle numbers in the different size channels to particle mass indicates higher specific emissions downstream than upstream of the scrubber at 75-76% engine load. This can possibly be explained as salt formation during the scrubber process. The increase in mass despite the loss in EC over the scrubber further emphasizes that there might be an addition of solids during the process. There is no indication of increases in solid particle mass at tests at the lower engine loads, see Table 3.

While upstream of the scrubber the thermodenuder tests indicate large part of particles being volatile (50–85% by number), measurements downstream of the scrubber give comparable number concentrations of particles with and without the thermodenuder. This could be due to a high hydrophilic content of the particles, which would cause them to react with, and be removed by, the scrubber liquid during the scrubbing process to a large extent. Additional explanations include that as the exhaust temperature decreases in the scrubber stack, the volatile compounds condensate on any available surface, such as water droplets and solid particles, and the effect of the reheating of the cold sample in the dilution system.

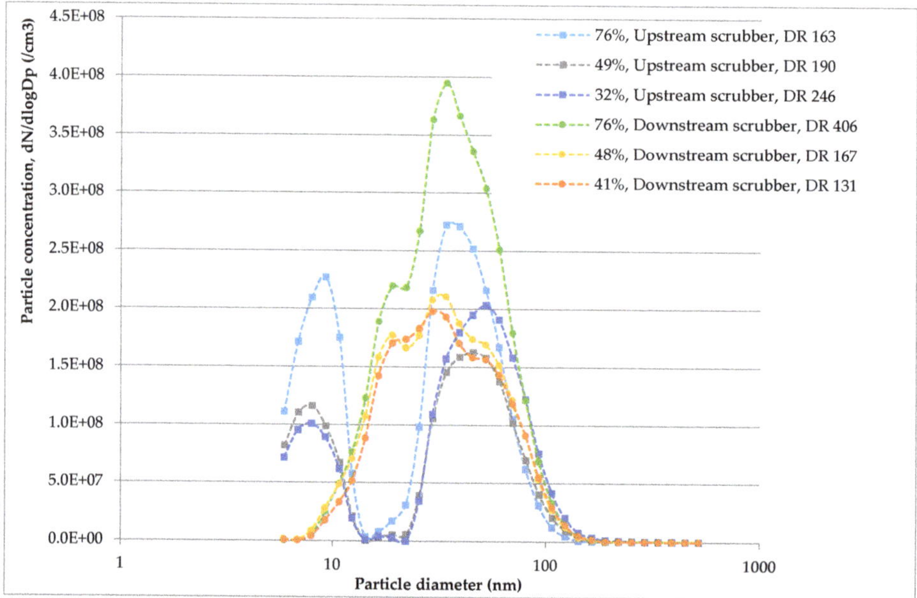

Figure 3. Particle number concentrations of solid particles measured in diluted exhaust gas upstream and downstream of the scrubber at different engine loads. A thermodenuder is used to remove volatile particles; DR = dilution ratio.

Table 3. Particle mass calculated from particle number size distribution for the size range 5.6–532 nm. Engine loads during trials are indicated. Cov of emission factors (EF) in parenthesis.

Trial		EF PM [1] (g/kWh)
Upstream scrubber; 76%		0.050 (8%)
Downstream scrubber; 76%		0.084 (3%)
Upstream scrubber; 49%		0.071 (7%)
Downstream scrubber; 48%	TD ON	0.067 (3%)
Upstream scrubber; 32%		0.11 (7%)
Downstream scrubber; 41%		0.066 (4%)
Upstream scrubber; 76%		0.18 (6%)
Downstream scrubber; 76%		0.084 (7%)
Upstream scrubber; 49%		0.15 (7%)
Downstream scrubber; 48%	TD OFF	0.075 (5%)
Upstream scrubber; 32%		0.22 (7%)
Downstream scrubber; 41%		0.10 (5%)

[1] Calculated from measured PN concentration in different size channels of the EEPS.

There was a significant decrease of total particle number over the scrubber (Table 2). At the highest engine loads tested (75% and 76% MCR) the reduction was 79% and at the medium engine loads tested (48% and 49% MCR) the reduction was 82%. The lowest engine loads are not fully comparable since one test is run on 32% MCR and the other on 41% MCR, but the measurements indicate a significant reduction also at low loads.

There is also a reduction of the total PAH-16 concentrations over the scrubber at all engine loads, Table 2 and Figure 4a,b. The specific emissions of total PAH-16 are in most cases higher at lower engine loads.

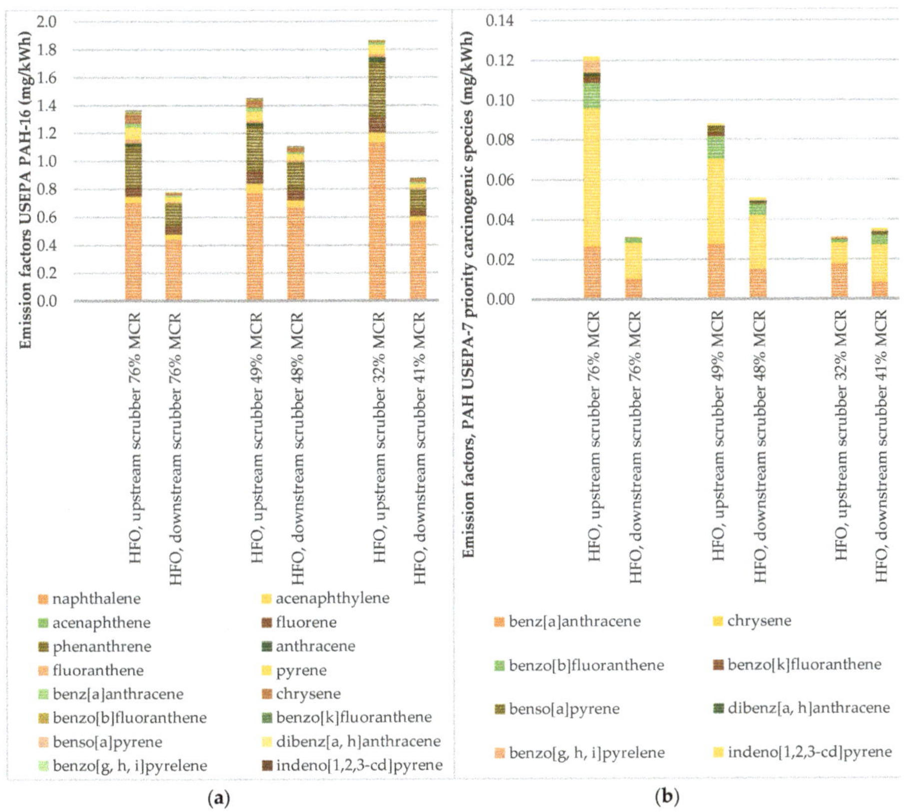

(a) (b)

Figure 4. Specific emissions of US Environmental Protection Agency's (EPA) PAH-16 (**a**) and PAH-7 (**b**) at HFO combustion upstream and downstream of the scrubber.

3.2. From an Environmental Perspective

The measured SO_2 emission factors with a scrubber and HFO fuel are lower than when the engine use LSFO, 84% to 96% lower from high to low engine load. Significant differences in emissions between the use of HFO combined with a scrubber, and LSFO, are observed for a number of gases (Table 2). There is a significant reduction of the total hydrocarbon (THC) concentration over the scrubber and specific emissions downstream of the scrubber are lower than at LSFO combustion for all engine loads tested. Emissions of CO and NO_X are however lower for LSFO combustion than for HFO combustion combined with the scrubber.

Average specific PM mass emissions at combustion of LSFO with 0.1% sulphur are lower than PM emissions downstream of the scrubber when using HFO. Comparisons at 75% (LSFO) and 76% (downstream) engine loads indicate emissions are 59% lower at LSFO combustion. At engine loads of 50% (LSFO) and 48% (downstream) the specific PM mass emission is 36% lower.

Compared to previous studies on particle emissions from low sulphur fuels, our results are in the lower end. PM emissions increase with sulphur content of fuel due to the sulphate content of particles. However, the correlation between particle emissions and sulphur content in fuel is weak for fuels with sulphur content below <0.5%. Emission factors derived for fuels with sulphur contents in the interval 0–0.5% give a mean value for the PM_{tot} emission factor of 0.2 g/kWh [2,8,11,14,20–27], compared to 0.12–0.17 g/kWh in our study. Other characteristics of fuel that impact particle formations include density, since a high density will increase combustion thermal efficiency, and viscosity, due to its impact on atomization during combustion. The low sulphur fuel oils will have higher density, increasing combustion efficiencies and oxidation of particles, and higher viscosity, which cause higher particle formation compared to distillate oils (Gysel et al). It is further difficult to relate how differences between engines and measurement setups influence the results. In two published studies, emissions from low-sulphur fuel oil and distillate oils have been tested on the same engine; Gysel et al., tested an MGO and a low sulphur (≤0.5%) fuel oil in the same engine and observed three times higher PM emissions from the latter [27]. Sulphur contents were 0.005% in the MGO and 0.009% in the LSFO. Winnes et al. observed similar differences in particle emissions when testing an MGO with 0.1% fuel sulphur content and a LSFO with 0.5% sulphur in two engines [20].

Specific emissions of both BC and EC downstream of the scrubber seem little influenced by engine load. Although some BC is removed by the scrubber, the specific emissions are higher than those from LSFO combustion by approximately 1.5 to 4 times.

PAH emissions are in our tests lower at LSFO combustion than at HFO combustion downstream of the scrubber, at each comparable engine load. The share of different PAH species is similar for all loads. Naphthalene is the most abundant and accounts for around 50% or more at all trials. The two- and three ring PAHs constitute between 78% and 91% of total PAHs, which is close to, but higher than, the 76% concluded as a typical value for diesel engine exhausts reported by [28]. The emissions of heavier PAHs, which can be represented by the EPA PAH-7 priority carcinogenic species, are at all occasions higher for LSFO than downstream of the scrubber. For the PAH-7 species an increase in emissions with decreasing engine load is observed at the LSFO tests. Specific emissions of US Environmental Protection Agency's PAH-16 and PAH-7 for tests at LSFO combustion and HFO combustion downstream of the scrubber are compared in Figure 5a,b, respectively.

3.3. Particle Composition

Filter samples of particles from two of the tested engine load points are analysed for composition. Our results indicate that also solid particles are removed during the scrubber process. An efficient removal of EC is seen at the high engine load. The removal of EC at the tests at low engine loads appear to be higher still, although tests upstream were made at 32% and tests downstream at 41% engine load, which could possibly exaggerate the effect seen. At LSFO combustion, the organic carbon dominates particle mass at both engine loads. Sulphates are not significantly contributing to particle mass and constitute 1.2% and 2.7% of total mass at 34% and 75% engine load, respectively. The particles from HFO downstream of the scrubber are similar in composition at the two loads tested. Upstream of the scrubber both organic and elemental carbon contribute significantly to particle mass, especially at the lower engine load. Sulphur is more abundant in particles at the higher load in the upstream tests. The absolute and relative contributions of components are presented in Table 4 and Figure 6. The undefined mass of particles at the respective test conditions varies between 1% and 25%.

3.4. Effects of Sampling System Parameters

Sampled particles (PM) include all solid and condensed material present in the diluted and cooled diesel exhaust. The composition can be expected to change throughout their passage in the exhaust duct and during the sampling due to changing conditions. The PM, both in terms of mass and number concentration, are sensitive to variations in temperature and humidity in the exhaust and during the sample conditioning (dilution). In the scrubber, the gases are cooled from around 300 °C to around

20 °C using water that is added to the exhaust as part of the scrubber design. Water that remains in the exhaust is removed with a demister system as part of the scrubber design. In the hot exhaust tests on HFO, upstream of the scrubber, we followed the ISO 8178 for dilution and sampling with a few exceptions. The standard is not applicable for cold exhaust gas sampling and an alternative sampling system was used downstream of the scrubber. Downstream of the scrubber, the sample line is heated, and the dilution is first made with hot and then ambient-temperature air. The temperature influences the volatile content of particles to a great extent. Detailed analysis of emission factors for PM mass sampled with the dilution tunnel and with FPS shows a great deal of variability, especially for the low engine load emission factors. The variability is, however, not only between the two dilution devices, but also within the samples taken with the same device. The variability is larger for HFO tests upstream the scrubber where a large influence of condensable sulphate is expected. Development of a robust sampling methodology for testing of emissions of particulate matter in the exhaust of widely varying physical and chemical properties is an important subject for further research.

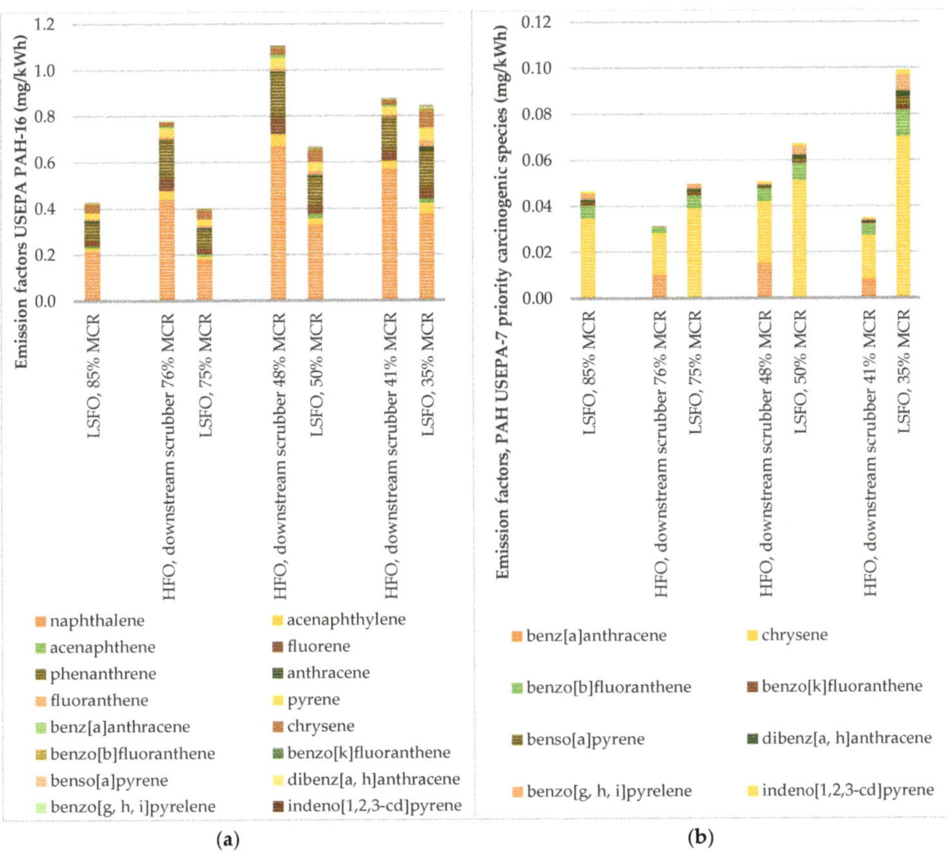

Figure 5. Specific emissions of US EPA's PAH-16 (**a**) and PAH-7 (**b**) at HFO and downstream of the scrubber and at LSFO combustion.

Table 4. Absolute and relative contributions of particulate components to particle mass at different tests. Relative values in brackets.

	Engine Load (%MCR)	OC mg/kWh	EC mg/kWh	Sulphate + Associated Water [1] mg/kWh	Elements mg/kWh	Undefined mg/kWh	PM [2] mg/kWh
Upstream scrubber	76%	77 (20%)	41 (11%)	150 (39%)	21 (5.3%)	95 (25%)	380
Downstream scrubber	76%	84 (34%)	15 (5.9%)	110 (46%)	24 (9.5%)	11 (5%)	250
LSFO	75%	78 (67%)	9.4 (8.0%)	3.2 (2.7%)	7.6 (6.4%)	19 (16%)	120
Upstream scrubber	32%	250 (49%)	94 (18%)	53 (10%)	36 (7.0%)	80 (16%)	510
Downstream scrubber	41%	87 (38%)	18 (8%)	89 (39%)	30 (13%)	2.4 (1.0%)	230
LSFO	34%	130 (91%)	4.9 (3.4%)	1.7 (1.2%)	2.3 (1.7%)	3.8 (2.7%)	140

[1] sulphate calculated from analyses of S on filters, H_2O calculated as SO_4^{2-}*0.8, see [14]; [2] average at test conditions (considering dilution ratio) from gravimetric analysis on filter.

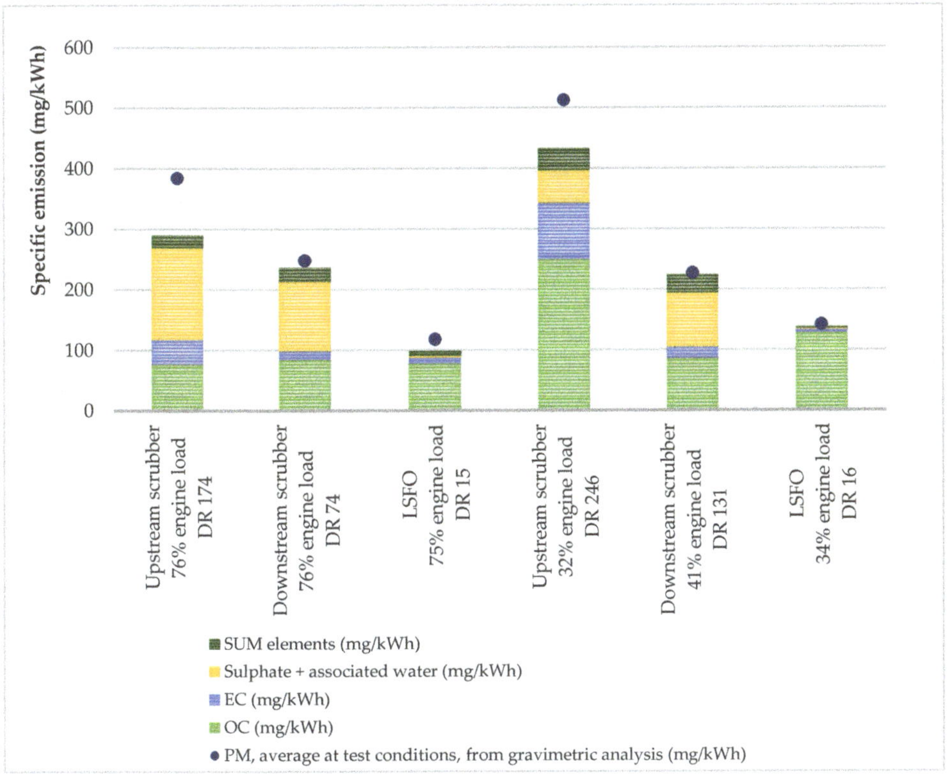

Figure 6. Absolute and relative contributions of particulate components to particle mass at different tests. N.B. analysis of filters from similar sampling and dilution conditions are considered, and total PM values may differ from the average values given in Table 2.

4. Conclusions

This study suggests that the reduction of particle mass over an exhaust gas treatment system is not efficient enough to accomplish particle emission levels equal to those from combustion of low sulphur fuel oil. With a traditional marine gasoil or other low sulphur fuel oil as marine fuel, the particle emissions can be expected to be significantly lower than at the use of HFO and a scrubber. The results from our measurement and previous studies suggest particle mass reductions of on average around 40% over exhaust gas scrubbers. Results of measurement studies on particle emission spread largely. This likely reflects the influence of sampling systems on the results. Particularly challenging are measurements in the cold exhaust gas downstream of the exhaust gas scrubber which are not covered by any standard.

The measurements also showed lower specific emissions of BC, EC, PAH-16 at operations on low sulphur fuel oil than downstream of the scrubber at heavy fuel oil operations. The scrubber efficiently reduced concentrations of gaseous SO_2, SO_3, total hydrocarbons and the heavier fraction of the measured PAHs.

The internationally agreed sulphur regulations for ships aim at reducing emissions to air of particles as well as sulphur oxides. In a wider context, the results from this study indicate that an extensive use of scrubbers may partly cancel the aimed for effect of reduced particle emissions to air that would be achieved by only using fuels with low sulphur content.

Further research on the amount of fuel used in the engines on ships equipped with scrubbers is important to draw conclusions on environmental and health effects of the technology.

Author Contributions: Conceptualization, H.W, E.F. and J.M.; investigation, H.W, E.F. and J.M.; writing—original draft preparation, H.W.; writing—review and editing, E.F and J.M. All authors have read and agreed to the published version of the manuscript. Authorship must be limited to those who have contributed substantially to the work reported.

Funding: This research was funded by the EU via Connecting Europe Facility and the IVL foundation.

Acknowledgments: The captains, chief engineers and crew are gratefully acknowledged for their support during emission measurements. The efforts on board by Håkan Salberg and Kjell Peterson were essential for the work.

Conflicts of Interest: The authors declare no conflict of interest.

Appendix A

Table A1 presents information on individual filters. Backup filters, background filters and blanks are excluded from the table. Fuel and engine load identify the test. EGCS indicates that samples are taken downstream of the scrubber. Filter type is indicated, the quartz filters are used for OC/EC analysis as indicated in the table. The OC/EC filters are also gravimetrically analysed although it is not recommended to use these results in further analyses. We have not included these results for calculating specific emissions. The dilution systems used are either a dilution tunnel (DT) or a fine particulate sampler from Dekati (FPS). The FPS gives temperatures at different locations in the dilutor. T1 is measured after primary dilution with heated air, T2 is measured upstream of the second dilution step and T4 is the temperature of the sample leaving the dilutor. The flows of the two dilution steps is also indicated. At measurements downstream of the scrubber (EGCS) the surrounding temperature was reaching above 50 (°C). It is likely that this was a cause of logging failure of the FPS. DR was continuously checked with trace gas measurements and the Horiba to make sure the dilutor worked though communication and logging suffered. Plausible ranges of sample temperatures are estimated when no log was retrieved from the FPS. These are based on logged values during tests that are adjacent in time. Filter temperatures during dilution tunnel experiments are given as a range based on dilution air temperature when measurements were not logged.

It is indicated in the table which filters that were used for analysis of PM_{tot}, $PM_{1.6}$, elements, sulphur and OC/EC. The table also presents PM mass concentrations and PM EF calculated from individual filters if applicable.

Table A1. Details on temperatures, flows dilution ratios, analyses and results from individual filter samples at different test settings.

Fuel	Filter Type	Engine Load	Dilution System (DT = Dilution Tunnel)	Dilution T, Primary Dilutor (°C) (T1)	T Up-Stream Ejector Dilutor (°C) (T3)	T Down-Stream Ejector Dilutor = Sample T (°C) (T4)	Flow Primary Dilutor (l/min)	Flow Ejector Dilutor (l/min)	Dilution Ratio	Filter T (°C)	PM_{tot}	$PM_{1.6}$	Elemetnts	S	OC/EC	PM Mass Conc (g/m3)	PM EF (g/kWh)
LSFO	Teflon	85%	DT	n.a.	n.a.	n.a.	n.a.	n.a.	15.9	33		X				0.026	0.16
LSFO	Teflon	85%	DT	n.a.	n.a.	n.a.	n.a.	n.a.	15.9	33	X			X		0.026	0.16
LSFO	Teflon	85%	DT	n.a.	n.a.	n.a.	n.a.	n.a.	16.4	32.5		X				0.021	0.12
LSFO	Teflon	85%	DT	n.a.	n.a.	n.a.	n.a.	n.a.	16.4	32.5	X					0.018	0.11
LSFO	Teflon	75%	DT	n.a.	n.a.	n.a.	n.a.	n.a.	15.9	29		X				0.018	0.10
LSFO	Teflon	75%	DT	n.a.	n.a.	n.a.	n.a.	n.a.	15.9	29	X					0.019	0.11
LSFO	Teflon	75%	DT	n.a.	n.a.	n.a.	n.a.	n.a.	16.3	25–35 *		X	X			0.021	0.12
LSFO	Teflon	75%	DT	n.a.	n.a.	n.a.	n.a.	n.a.	16.3	25–35 *	X					0.022	0.13
LSFO	Teflon	50%	DT	n.a.	n.a.	n.a.	n.a.	n.a.	16.7	33	X			X		0.027	0.16
LSFO	Teflon	50%	DT	n.a.	n.a.	n.a.	n.a.	n.a.	16.7	33		X				0.026	0.16
LSFO	Teflon	50%	DT	n.a.	n.a.	n.a.	n.a.	n.a.	16.8	33		X				0.024	0.14
LSFO	Teflon	50%	DT	n.a.	n.a.	n.a.	n.a.	n.a.	16.8	33	X					0.023	0.14
LSFO	Teflon	34%	DT	n.a.	n.a.	n.a.	n.a.	n.a.	20.2	29		X				0.017	0.12
LSFO	Teflon	34%	DT	n.a.	n.a.	n.a.	n.a.	n.a.	20.2	29	X					0.016	0.12
LSFO	Teflon	34%	DT	n.a.	n.a.	n.a.	n.a.	n.a.	23.8	31		X	X			0.021	0.16
LSFO	Teflon	34%	DT	n.a.	n.a.	n.a.	n.a.	n.a.	23.8	31	X			X		0.023	0.17
LSFO	Teflon	50%	FPS	34.7 ± 0.6	39.1 ± 0.5	36.3 ± 0.6	9.3 ± 0.2	78.2 ± 0.7	40	n.d.		X				0.034	0.21
LSFO	Teflon	50%	FPS	34.7 ± 0.6	39.1 ± 0.5	36.3 ± 0.6	9.3 ± 0.2	78.2 ± 0.7	40	n.d.	X					0.031	0.19
LSFO	Teflon	34%	FPS	95.7 ± 1.3	71.7 ± 0.7	40.8 ± 0.4	9.6 ± 0.2	80.8 ± 1.0	80	n.d.	X					0.049	0.36
LSFO	Teflon	85%	FPS	147.9 ± 1.8	106.2 ± 1.1	49.8 ± 0.8	9.5 ± 0.1	81.3 ± 1.0	92	n.d.	X					0.033	0.18
LSFO	Teflon	85%	FPS	147.9 ± 1.7	105.9 ± 0.9	50.3 ± 0.4	9.4 ± 0.1	81.3 ± 1.1	82	n.d.	X					0.032	0.19
LSFO	Quartz	75%	DT	n.a.	n.a.	n.a.	n.a.	n.a.	15.7	25–35 *	(X)				X	0.025	n.a.
LSFO	Quartz	75%	DT	n.a.	n.a.	n.a.	n.a.	n.a.	15.7	25–35 *		(X)			X	0.026	n.a.
LSFO	Quartz	50%	DT	n.a.	n.a.	n.a.	n.a.	n.a.	16.6	25–35 *	(X)				X	0.024	n.a.
LSFO	Quartz	50%	DT	n.a.	n.a.	n.a.	n.a.	n.a.	16.6	25–35 *		(X)			X	0.024	n.a.
LSFO	Quartz	34%	DT	n.a.	n.a.	n.a.	n.a.	n.a.	20.1	25–35 *		(X)			X	0.032	n.a.
LSFO	Quartz	34%	DT	n.a.	n.a.	n.a.	n.a.	n.a.	20.1	25–35 *	(X)				X	0.041	n.a.

Table A1. *Cont.*

Fuel	Filter Type	Engine Load	Dilution System (DT = Dilution Tunnel)	Dilution T, Primary Dilutor (°C) (T1)	T Up-Stream Ejector Dilutor (°C) (T3)	T Down-Stream Ejector Dilutor = Sample T (°C) (T4)	Flow Primary Dilutor (l/min)	Flow Ejector Dilutor (l/min)	Dilution Ratio	Filter T (°C)	PM$_{tot}$	PM$_{1.6}$	Elemetnts	S	OC/EC	PM Mass Conc (g/m3)	PM EF (g/kWh)
LSFO	Quartz	34%	FPS	95.6 ± 1.3	71.7 ± 0.7	40.1 ± 0.5	9.6 ± 0.2	80.8 ± 1.0	71.9	n.d.	(X)				X	0.059	n.a.
LSFO	Quartz	85%	DT	n.a.	n.a.	n.a.	n.a.	n.a.	22.8	25–35 *	(X)				X	0.027	n.a.
LSFO	Quartz	85%	DT	n.a.	n.a.	n.a.	n.a.	n.a.	22.8	25–35 *		(X)			X	0.021	n.a.
LSFO	Quartz	85%	FPS	149 ± 1.7	106.5 ± 0.9	49.6 ± 0.6	9.4 ± 0.3	80.5 ± 1.1	83.9	n.d.	(X)				X	0.034	n.a.
EGCS	Teflon	41%	FPS	148.7 ± 14.1	n.d.	49.2 ± 0.9	5.1 ± 0.3	101.1 ± 4.6	131	n.d.	X		X			0.0367	0.23
EGCS	Teflon	41%	FPS	148.7 ± 14.1	n.d.	49.2 ± 0.9	5.1 ± 0.3	101.1 ± 4.6	131	n.d.		X				0.0342	0.22
EGCS	Teflon	41%	FPS	163.8 ± 0.9 *	n.d.	53.2 ± 0.9	3.6 ± 1.1	40.9 ± 8.0	55	n.d.		X				0.0416	0.26
EGCS	Teflon	41%	FPS	163.8 ± 0.9 *	n.d.	53.2 ± 0.9	3.6 ± 1.1	40.9 ± 8.0	55	n.d.	X			X		0.0428	0.27
EGCS	Teflon	48%	FPS	n.d.	n.d.	40-55 *	n.d.	n.d.	55	n.d.	X					0.0444	0.28
EGCS	Teflon	48%	FPS	n.d.	n.d.	40-55 *	n.d.	n.d.	55	n.d.		X				0.0399	0.25
EGCS	Teflon	48%	FPS	165.5 ± 0.7	n.d.	46.8 ± 1.1	5.1 ± 0.03	98.9 ± 1.2	165	n.d.	X					0.031	0.19
EGCS	Teflon	48%	FPS	165.5 ± 0.7	n.d.	46.8 ± 1.1	5.1 ± 0.03	98.9 ± 1.2	165	n.d.		X				0.040	0.25
EGCS	Teflon	48%	FPS	n.d.	n.d.	40-55 *	n.d.	n.d.	53	n.d.	X					0.0401	0.16
EGCS	Teflon	48%	FPS	n.d.	n.d.	40-55 *	n.d.	n.d.	53	n.d.		X				0.0341	0.21
EGCS	Teflon	76%	FPS	104.8 ± 19.0	n.d.	33.7 ± 0.8	5.2 ± 0.1	99.2 ± 1.0	348	n.d.	X					0.0595	0.54
EGCS	Teflon	76%	FPS	104.8 ± 19.0	n.d.	33.7 ± 0.8	5.2 ± 0.1	99.2 ± 1.0	348	n.d.		X				0.0467	0.27
EGCS	Teflon	76%	FPS	n.d.	n.d.	30-45 *	n.d.	n.d.	69	n.d.		X				0.0412	0.24
EGCS	Teflon	76%	FPS	n.d.	n.d.	30-45 *	n.d.	n.d.	69	n.d.	X			X		0.0469	0.24
EGCS	Teflon	76%	FPS	n.d.	n.d.	30-45 *	n.d.	n.d.	74	n.d.		X				0.0388	0.22
EGCS	Teflon	76%	FPS	n.d.	n.d.	30-45 *	n.d.	n.d.	74	n.d.	X		X			0.0465	0.27
HFO	Teflon	32%	FPS	89.5 ± 2.0	78.9 ± 1.9	34.5 ± 1.1	13.1 ± 0.4	99.6 ± 0.9	151	n.d.		X				0.1011	0.75
HFO	Teflon	32%	DT	n.a.	n.a.	n.a.	n.a.	n.a.	14	30–45 *	X		X			0.0816	0.63
HFO	Teflon	32%	FPS	93.4 ± 1.5	82.3 ± 1.1	36.9 ± 0.7	13.4 ± 0.2	99.7 ± 1.0	246	n.d.		X	X			0.0596	0.44
HFO	Teflon	32%	FPS	93.4 ± 1.5	82.3 ± 1.1	36.9 ± 0.7	13.4 ± 0.2	99.7 ± 1.0	246	n.d.	X			X		0.0465	0.35
HFO	Teflon	32%	DT	n.a.	n.a.	n.a.	n.a.	n.a.	14	30–45 *	X					0.0713	0.41
HFO	Teflon	49%	DT	n.a.	n.a.	n.a.	n.a.	n.a.	16	30–45 *	X					0.0708	0.45
HFO	Teflon	49%	DT	n.a.	n.a.	n.a.	n.a.	n.a.	16	30–45 *	X					0.0597	0.38
HFO	Teflon	49%	FPS	88.0 ± 1.2	75.7 ± 1.0	35.9 ± 0.5	11.0 ± 0.0.5	98.8 ± 1.0	190	n.d.	X					0.0581	0.23
HFO	Teflon	49%	FPS	88.0 ± 1.2	75.7 ± 1.0	35.9 ± 0.5	11.0 ± 0.0.5	98.8 ± 1.0	190	n.d.		X				0.0598	0.36
HFO	Teflon	49%	FPS	86.6 ± 3.3	75.0 ± 3.2	35.1 ± 1.3	10.9 ± 0.2	98.2 ± 1.0	162	n.d.	X					0.0534	0.21
HFO	Teflon	49%	FPS	86.6 ± 3.3	75.0 ± 3.2	35.1 ± 1.3	10.9 ± 0.2	98.2 ± 1.0	162	n.d.		X				0.0588	0.36

Table A1. *Cont.*

Fuel	Filter Type	Engine Load	Dilution System (DT = Dilution Tunnel)	Dilution T, Primary Dilutor (°C) (T1)	T Up-Stream Ejector Dilutor (°C) (T3)	T Down-Stream Ejector Dilutor = Sample T (°C) (T4)	Flow Primary Dilutor (l/min)	Flow Ejector Dilutor (l/min)	Dilution Ratio	Filter T (°C)	PM_{tot}	$PM_{1.6}$	Elemetns	S	OC/EC	PM Mass Conc (g/m3)	PM EF (g/kWh)
HFO	Teflon	76%	FPS	96.3 ± 1.7	75.9 ± 1.4	37.7 ± 0.7	13.4 ± 0.09	99.0 ± 1.1	174	n.d.	X		X			0.0755	0.43
HFO	Teflon	76%	FPS	96.3 ± 1.7	75.9 ± 1.4	37.7 ± 0.7	13.4 ± 0.09	99.0 ± 1.1	174	n.d.		X				0.048	0.28
HFO	Teflon	76%	DT	n.a.	n.a.	n.a.	n.a.	n.a.	14	30–40 *	X		X			0.0645	0.37
HFO	Teflon	76%	DT	n.a.	n.a.	n.a.	n.a.	n.a.	14	30–40 *	X			X		0.077	0.44
HFO	Teflon	76%	FPS	97.5 ± 2.3	77.6 ± 2.3	41.4 ± 2.6	13.3 ± 0.2	99.0 ± 1.0	154	n.d.	X					0.084	0.48
HFO	Teflon	76%	FPS	97.5 ± 2.3	77.6 ± 2.3	41.4 ± 2.6	13.3 ± 0.2	99.0 ± 1.0	154	n.d.		X		X		0.0623	0.36
EGCS	Quartz	41%	FPS	n.d.	n.d.	40-55 *	n.d.	n.d.	44	n.d.	(X)				X	0.0209	n.a
EGCS	Quartz	48%	FPS	n.d.	n.d.	40-55 *	n.d.	n.d.	54	n.d.		(X)			X	0.055	n.a
EGCS	Quartz	48%	FPS	n.d.	n.d.	40-55 *	n.d.	n.d.	54	n.d.	(X)				X	0.0357	n.a
EGCS	Quartz	76%	FPS	n.d.	n.d.	30-45 *	n.d.	n.d.	71	n.d.		(X)			X	0.0477	n.a
EGCS	Quartz	76%	FPS	n.d.	n.d.	30-45 *	n.d.	n.d.	71	n.d.		(X)			X	0.0491	n.a
HFO	Quartz	32%	FPS	92.9 ± 1.4	82.3 ± 1.2	37.4 ± 1.0	13.4 ± 0.06	99.6 ± 1.0	230	n.d.	(X)				X	0.0795	n.a
HFO	Quartz	49%	FPS	88.9 ± 1.3	75.0 ± 3.2	36.4 ± 0.9	11.2 ± 0.05	99.8 ± 1.0	255	n.d.	(X)				X	0.0864	n.a
HFO	Quartz	49%	FPS	88.9 ± 1.3	75.0 ± 3.2	36.4 ± 0.9	11.2 ± 0.05	99.8 ± 1.0	255	n.d.		(X)			X	0.1189	n.a
HFO	Quartz	76%	FPS	42.0 ± 1.0	44.2 ± 1.0	40.3 ± 1.8	12.3 ± 0.3	96.1 ± 1.0	60	n.d.	(X)				X	0.0769	n.a
HFO	Quartz	76%	FPS	42.0 ± 1.0	44.2 ± 1.0	40.3 ± 1.8	12.3 ± 0.3	96.1 ± 1.0	60	n.d.	(X)				X	n.d.	n.a

* Missing temperature readings in sample gas and filter. The range is an estimate based on temperatures of dilution air. n.a. = not applicable, n.d. = no data.

References

1. Marine Environmental Protection Committee, Resolution MEPC. *Guidelines for Exhaust Gas Cleaning Systems*; IMO: London, UK, 2015.
2. Winnes, H.; Moldanová, J.; Anderson, M.; Fridell, E. On-board measurements of particle emissions from marine engines using fuels with different sulphur content. *Proc. Inst. Mech. Eng. Part M J. Eng. Marit. Environ.* **2014**, *230*, 45–54. [CrossRef]
3. Jiang, L.; Kronbak, J.; Christensen, L.P. The costs and benefits of sulphur reduction measures: Sulphur scrubbers versus marine gas oil. *Transp. Res. Part D Transp. Environ.* **2014**, *28*, 19–27. [CrossRef]
4. DNVGL, Alternative Fuel Insight Scrubber Statistics. Available online: www.https://afi.dnvgl.com/Statistics?repId=2 (accessed on 13 April 2020).
5. Caiazzo, G.; Di Nardo, A.; Langella, G.; Scala, F. Seawater scrubbing desulfurization: A model for SO2 absorption in fall-down droplets. *Environ. Prog. Sustain. Energy* **2011**, *31*, 277–287. [CrossRef]
6. Magnusson, K.; Thor, P.; Granberg, M. Scrubbers: Closing the loop Activity 3: Task 2. In *Risk Assessment of Marine Exhaust Gas Scrubber Water*; 2018; Available online: https://www.researchgate.net/publication/333973881_Scrubbers_Closing_the_loop_Activity_3_Task_2_Risk_Assessment_of_marine_exhaust_gas_scrubber_water (accessed on 20 April 2020).
7. Fridell, E.; Salo, K. Measurements of abatement of particles and exhaust gases in a marine gas scrubber. *Proc. Inst. Mech. Eng. Part M J. Eng. Marit. Environ.* **2014**, *230*, 154–162. [CrossRef]
8. Johnson, K.; Miller, W.; Durbin, T.; Jiang, Y.; Yang, J.; Karavelakis, G.; Cocker, D. Black Carbon Measurement Methods and Emission Factors from Ships. International Council on Clean Transportation (ICCT), 2016. Available online: https://www.ccacoalition.org/en/resources/black-carbon-measurement-methods-and-emission-factors-ships (accessed on 20 April 2020).
9. Wärtsilä, Exhaust Gas Scrubber Installed Onboard MT "Suula". Public Test Report. 2010. Available online: http://203.187.160.133:9011/www.annualreport2010.wartsila.com/c3pr90ntc0td/files/wartsila_2010/Docs/Scrubber_Test_Report_onboard_Suula.pdf (accessed on 20 April 2020).
10. Hansen, J.-P. *Exhaust Gas Scrubber Installed Onboard MV Ficaria Seaways—Public Test Report*; Miljöstyrelsen DK, 2012; Available online: https://www2.mst.dk/Udgiv/publications/2012/06/978-87-92903-28-0.pdf (accessed on 20 April 2020).
11. Lehtoranta, K.; Aakko-Saksa, P.T.; Murtonen, T.; Vesala, H.; Ntziachristos, L.; Rönkkö, T.; Karjalainen, P.; Kuittinen, N.; Timonen, H. Particulate Mass and Nonvolatile Particle Number Emissions from Marine Engines Using Low-Sulfur Fuels, Natural Gas, or Scrubbers. *Environ. Sci. Technol.* **2019**, *53*, 3315–3322. [CrossRef] [PubMed]
12. Cooper, D.; Ferm, M. Jämförelse av mätmetoder för bestämning av SO3- koncentrationer i rökgaser. In *IVL for "Stiftelsen för Värmeteknisk Forskning"*; 1994. (in Swedish)
13. Jonsson, Å.M.; Westerlund, J.; Hallquist, M. Size-resolved particle emission factors for individual ships. *Geophys. Res. Lett.* **2011**, *38*, 13809. [CrossRef]
14. Moldanová, J.; Fridell, E.; Winnes, H.; Holmin-Fridell Boman, J.; Jedynska, A.; Tishkova, V.; Demirdijan, B.; Joulie, S.; Bladt, H.; Ivleva, N.P.; et al. Physical and chemical characterisation of PM emissions from two ships operating in European Emission Control Areas. *Atmos. Meas. Tech. Discuss.* **2013**, *6*, 3931–3982. [CrossRef]
15. Cavalli, F.; Viana, M.; Yttri, K.E.; Genberg, J.; Putaud, J.-P. Toward a standardised thermal-optical protocol for measuring atmospheric organic and elemental carbon: The EUSAAR protocol. *Atmos. Meas. Tech.* **2010**, *3*, 79–89. [CrossRef]
16. Petzold, A.; Ogren, J.; Fiebig, M.; Laj, P.; Li, S.-M.; Baltensperger, U.; Popp, T.; Kinne, S.; Pappalardo, G.; Sugimoto, N.; et al. Recommendations for reporting "black carbon" measurements. *Atmospheric Chem. Phys. Discuss.* **2013**, *13*, 8365–8379. [CrossRef]
17. Cordtz, R.; Schramm, J.; Andreasen, A.; Eskildsen, S.; Mayer, S. Modeling the distribution of sulphur compounds in a large two stroke diesel engine. *Energy Fuels* **2013**, *27*, 1652–1660. [CrossRef]
18. Srivastava, R.; Miller, C.; Erickson, C.; Jambhekar, R. Emissions of sulfur trioxide from coal-fired power plants. *J. Air Waste Manag. Assoc.* **2004**, *54*, 750–762. [CrossRef] [PubMed]
19. Popovicheva, O.; Kireeva, E.; Shonija, N.; Zubareva, N.; Persiantseva, N.; Tishkova, V.; Demirdjian, B.; Moldanová, J.; Mogilnikov, V. Ship particulate pollutants: Characterization in terms of environmental implication. *J. Environ. Monit.* **2009**, *11*, 2077. [CrossRef] [PubMed]

20. Winnes, H.; Fridell, E. Particle emissions from ships: Dependence on fuel type. *J. Air Waste Manag. Assoc.* **2009**, *59*, 1391–1398. [CrossRef] [PubMed]
21. Agrawal, H.; Welch, W.A.; Miller, J.W.; Cocker, D.R. Emission Measurements from a Crude Oil Tanker at Sea. *Environ. Sci. Technol.* **2008**, *42*, 7098–7103. [CrossRef]
22. Cooper, D. Exhaust emissions from high speed passenger ferries. *Atmos. Environ.* **2001**, *35*, 4189–4200. [CrossRef]
23. Cooper, D. Exhaust emissions from ships at berth. *Atmos. Environ.* **2003**, *37*, 3817–3830. [CrossRef]
24. Zetterdahl, M.; Moldanová, J.; Pei, X.; Pathak, R.K.; Demirdjian, B. Impact of the 0.1% fuel sulfur content limit in SECA on particle and gaseous emissions from marine vessels. *Atmos. Environ.* **2016**, *145*, 338–345. [CrossRef]
25. Kasper, A.; Aufdenblatten, S.; Forss, A.; Mohr, M.; Burtscher, H. Particulate Emissions from a Low-Speed Marine Diesel Engine. *Aerosol Sci. Technol.* **2007**, *41*, 24–32. [CrossRef]
26. Fridell, E.; Steen, E.; Peterson, K. Primary particles in ship emissions. *Atmos. Environ.* **2008**, *42*, 1160–1168. [CrossRef]
27. Gysel, N.R.; Welch, W.A.; Johnson, K.; Miller, W.; Cocker, D.R. Detailed Analysis of Criteria and Particle Emissions from a Very Large Crude Carrier Using a Novel ECA Fuel. *Environ. Sci. Technol.* **2017**, *51*, 1868–1875. [CrossRef]
28. Khalili, N.R.; Scheff, P.A.; Holsen, T.M. PAH source fingerprints for coke ovens, diesel and, gasoline engines, highway tunnels, and wood combustion emissions. *Atmos. Environ.* **1995**, *29*, 533–542. [CrossRef]

Journal of
Marine Science and Engineering

Article

One-Dimensional Gas Flow Analysis of the Intake and Exhaust System of a Single Cylinder Diesel Engine

Kyong-Hyon Kim [1] and Kyeong-Ju Kong [2],*

[1] Training Ship Management Center, Pukyong National University, Busan 48513, Korea; bluefishkk@hanmail.net
[2] Department of Mechanical System Engineering, Graduate School, Pukyong National University, Busan 48513, Korea
* Correspondence: kjkong@pknu.ac.kr; Tel.: +82-51-629-6188

Received: 1 December 2020; Accepted: 18 December 2020; Published: 20 December 2020

Abstract: In order to design a diesel engine system and to predict its performance, it is necessary to analyze the gas flow of the intake and exhaust system. Gas flow analysis in a three-dimensional (3D) format needs a high-resolution workstation and an enormous amount of time for analysis. Calculation using the method of characteristics (MOC), which is a gas flow analysis in a one-dimensional (1D) format, has a fast calculation time and can be analyzed with a low-resolution workstation. However, there is a problem with poor accuracy in certain areas. It was assumed that the reason was that 1D could not implement the shape. The error that occurs in the shape of the bent pipe used in the intake and exhaust ports of the diesel engine was analyzed and to find a solution to the low accuracy, the results of the experiment and 1D analysis were compared. The discharge coefficient was calculated using the average mass flow rate, and as a result of applying it, the accuracy was improved for the maximum negative pressure by 0.56–1.93% and the maximum pressure by 3.11–7.86% among the intake pipe pressure results. The difference in phase of the exhaust pipe pressure did not improve. It is considered as a limitation of 1D analysis that does not improve even by applying the discharge coefficient. In the future, we intend to implement a bent pipe that cannot be realized in 1D using a 3D format and to prepare a method to supplement the reliability by using 1D–3D coupling.

Keywords: method of characteristics; one-dimensional numerical analysis; single cylinder diesel engine; mass flow rate; intake and exhaust system

1. Introduction

The intake and exhaust system of a diesel engine is related to the performance of the engine [1]. In a four-stroke engine, the gas exchange process releases combustion gases at the end of the power stroke, bringing more air into the next intake stroke [2]. The volumetric efficiency of an engine appears in a form similar to torque and is affected by the mass flow rate. It may be affected by various variables such as engine speed, air–fuel ratio, compression ratio, intake and exhaust valve geometries, and intake and exhaust pipe length [3]. Diesel engines are designed to maximize engine performance at the commercial speed; if the engine was operating outside of the commercial speed, the volumetric efficiency decreases and a lot of environmental pollutants are discharged. However, research is required to improve the performance and emission of environmental pollutants during operation, and not just at the commercial speed [4–6].

Regulations on the emission of environmental pollutants from marine diesel engines have been strengthened. Greenhouse gases, sulfur oxides, and nitrogen oxides are regulated and more regulation will be tightened in the future [7]. Scrubber, exhaust gas recirculation (EGR), selective catalytic reduction (SCR), etc., are used as devices to reduce the emission of environmental pollutants [8]. Gas flow analysis of the intake and exhaust system is required for the design of a diesel engine, such as when installing environmental pollution reduction devices and calculating the kinetic energy of exhaust gas acting on a turbine for turbocharger matching [9]. If only the components of the diesel engine are separated and the gas flow is analyzed, it is difficult to determine the effect of the components on a diesel engine [10].

Gas flow analysis of the entire diesel engine system in three-dimensional (3D) format is inefficient because it requires a high-resolution workstation and an enormous amount of time for the analysis [11–13]. It is indicative the fact that to carry out 3D gas flow analysis of the intake and exhaust gas exchange process of a single cylinder diesel engine without a combustion reaction using Ansys Fluent R15.0, a commercial flow analysis program, takes approximately 61 hours (based on a 16-core central processing unit) [14]. Therefore, the method of characteristics (MOC) approach has been used for one-dimensional (1D) format gas flow analysis with a fast calculation time and a low-resolution workstation [15].

The MOC is a method of calculating a pressure wave using Riemann variables called characteristic curves. It is a method with a fast calculation time and a high accuracy in a straight pipe, nozzle, and orifice. However, there is a disadvantage of low accuracy in complex geometries such as branches and bent pipes [16].

Diesel engines are designed according to the effects of reflected waves in the flow areas of intake and exhaust pipes, underscoring the need for numerical analysis based on reflected waves [17]. For 1D flow analysis using the MOC, the calculation includes the influence of reflected waves [18]. In this study, MOC was used to perform unsteady 1D gas flow analysis on a single cylinder diesel engine.

This 1D gas flow analysis was performed under the same boundary conditions as the experiment and the results were compared. The object of comparison was the average mass flow rate of the intake air and the intake and exhaust pipe pressure. The discharge coefficient was calculated using the average mass flow rate of the intake air and was applied to the 1D gas flow analysis. The results of the intake and exhaust pipe pressure of the gas flow analysis applying the discharge coefficient were compared with those of the experiment.

Previous studies using 1D gas flow analysis have focused on merits and verification. The purpose of the study is to analyze the errors in the 1D gas flow analysis caused by not realizing the shape as it is. By evaluating the validity of the results, we tried to determine the cause of the low accuracy of the 1D gas flow analysis, and to find a way to solve the problem in diesel engine intake and exhaust system.

2. Experimental Apparatus

Figure 1 representing the experimental apparatus. The engine was a 35 kW three-cylinder direct injection four-stroke diesel engine. Intake and exhaust pipes were installed only in the No.1 cylinder to measure the intake and exhaust pipe pressure of a single cylinder. To observe the intake and exhaust gas flow in a cold flow state where no combustion reaction occurred, an electric motor was connected to a fly wheel in order to rotate. The mass flow rate was measured by installing a laminar flowmeter (LFE-100B, Tsukasa Sokken, Tokyo, Japan) at the starting point of the intake pipe, and the differential pressure of the laminar flowmeter was measured using a differential pressure transmitter (FCO332, Furness Controls, Bexhill-on-Sea, England) [19].

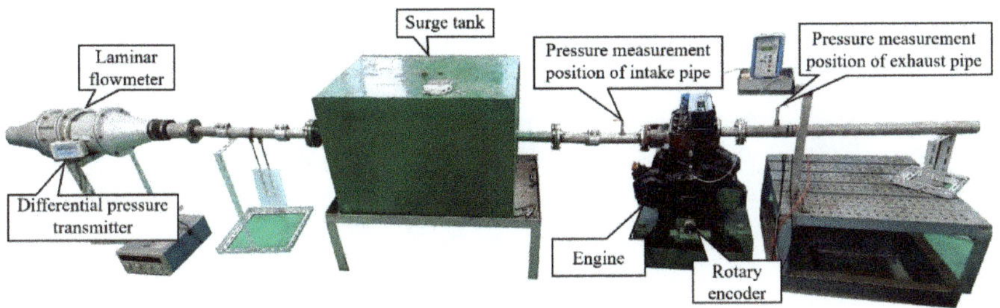

Figure 1. Photograph of the experimental apparatus.

Table 1 provides the specifications of the experimental apparatus. The intake air method was in a naturally aspirated state, and the experimental data were measured at intervals of 200 rpm in the range of 700–1500 rpm of engine speed. The intake pipe from the surge tank to the cylinder head was a straight pipe with a diameter of 0.04 m and a length of 0.5 m. The exhaust pipe was a straight pipe with a diameter of 0.04 m and a length of 1.0 m, and the end of the exhaust pipe was an open-end through which the exhaust gas was released into the atmosphere. CA° refers to the crank angle degrees.

Table 1. Specifications of the experimental apparatus.

Parameter	Value	Unit
Engine speed	700, 900, 1100, 1300, 1500	rpm
Cylinder bore × stroke	0.10 × 0.11	m
Compression ratio	17.6	-
Intake air method	Naturally aspirated	-
Volume of intake port	94.25	cm^3
Length of exhaust pipe	0.5	m
Diameter of exhaust pipe	0.04	m
Air–intake valve opens (AVO)	342	CA°
Air–intake valve closes (AVC)	580	CA°
Volume of exhaust port	32.96	cm^3
Length of exhaust pipe	1.0	m
Diameter of exhaust pipe	0.04	m
Exhaust valve opens (EVO)	130	CA°
Exhaust valve closes (EVC)	378	CA°

Figure 2 shows a block diagram of the experimental system. The intake and exhaust pipe pressure was measured using a data acquisition (DAQ) system consisting of a piezoresistive amplifier and a rotary encoder (E6C2-CWZ3E, Omron, Kyoto, Japan) [20]. The pressure sensors were piezoresistive (4045A5, Kistler, Winterthur, Switzerland) and the measurement point was the intake and exhaust pipe 0.15 m away from the cylinder head. The data were acquired at intervals of 0.5 CA° during 100 revolutions while the engine was stable. In order to obtain an accurate measurement, the ensemble average was applied to measured data. The ensemble average was used as a method of calculating the average of remaining values excluding the top 10% and the bottom 10% of the measured data [21]. The pressure results of the intake and exhaust pipe obtained by the ensemble average were used to verify the results of the 1D gas flow analysis.

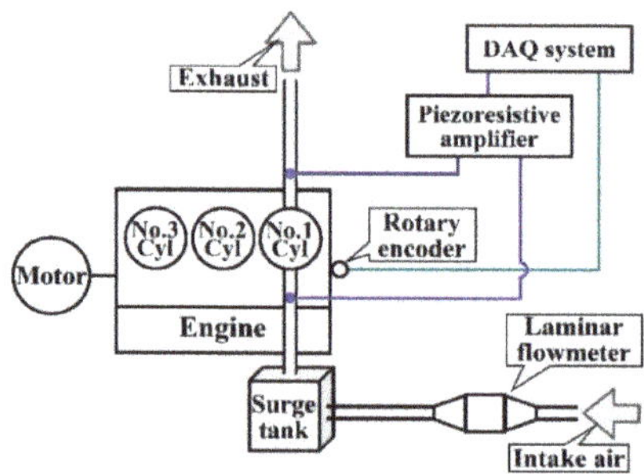

Figure 2. Block diagram of the experimental system. DAQ: data acquisition.

3. Theoretical Interpretation

3.1. Average Mass Flow Rate

The equations for calculating the experimental mass flow rate to obtain the discharge coefficient is as follows. The average mass flow rate of the intake air $\dot{m}_{exp}$ was measured with a laminar flowmeter using Equation (1) [22].

$$\dot{m}_{exp} = Q_t \times \rho_h,\tag{1}$$

where Q_t is the volumetric flow rate and calculated using Equation (2).

$$Q_t = K_{20} \times \left(\frac{\mu_{t(20°C)}}{\mu_{t(intake)}}\right) \times P_x,\tag{2}$$

where K_{20} is the laminar coefficient under standard conditions, μ_t is the viscosity depending on the air temperature, and P_x is the differential pressure of the laminar flowmeter measured using a differential pressure meter. The average value measured during 100 revolutions for each engine speed was used, and the values are shown in Table 2.

Table 2. Differential pressure according to the engine speed.

Engine Speed (rpm)	P_x (mmH$_2$O)
700	2.0
900	2.6
1100	3.1
1300	3.6
1500	4.1

ρ_h is the density of moist air and was calculated using Equation (3).

$$\rho_h = \rho_d \times \frac{1+x}{1+1.609 \times x},\tag{3}$$

where ρ_d is the density of dry air, x is the absolute humidity, and the values of each coefficient are shown in the Table 3.

Table 3. Values of coefficients.

Coefficient	Value
K_{20}	$1.7323 \times \left(1 - 2.823 \times 10^{-4} \times P_x\right)$
μ_t (kg/m·s)	$2.791 \times 10^{-7} \times (air\ temperature)^{0.7355}$
ρ_d (kg/m^3)	1.2250
x (kg/kg)	0.01062

3.2. Method of Charateristics

The equations for calculating the mass flow rate in 1D model to obtain the discharge coefficient is as follows.

The mass flow rate of the intake air $\dot{m}_{moc}$ was calculated using the MOC, as follows [18]:

$$\dot{m}_{moc} = \rho_2 \times u_2 \times F_2,\tag{4}$$

where the subscript 2 refers to the downstream conditions.

The air density can be calculated with equation related to pressure, specific heat ratio and speed of sound. ρ_2 is the air density under downstream conditions and was calculated using Equation (5).

$$\rho_2 = \left(\frac{p_2}{p_{01}}\right)^{\frac{1}{\kappa}} \times \frac{\kappa \times p_{01}}{a_{01}{}^2},\tag{5}$$

where subscript 01 refers to the upstream stagnation conditions.

The speed of sound $a_{01}{}^2$ was calculated using the energy equation.

$$a_{01}{}^2 = a_2{}^2 + \frac{\kappa - 1}{2} \times u_2{}^2.\tag{6}$$

Using Equations (5) and (6), Equation (4) can be expressed as follows:

$$\dot{m}_{moc} = \frac{p_{01} \times F_2}{a_{01}} \times \left[\left(\frac{2 \times \kappa^2}{\kappa - 1}\right) \times \left(\frac{p_2}{p_{01}}\right)^{\frac{2}{\kappa}} \times \left\{1 - \left(\frac{p_2}{p_{01}}\right)^{\frac{\kappa-1}{\kappa}}\right\}\right]^{\frac{1}{2}},\tag{7}$$

to obtain the characteristics, nondimensional form is required:

$$\xi_f = \frac{\dot{m}_{moc} \times a_{01}}{p_{01} \times F_2} \times \left[\left(\frac{2 \times \kappa^2}{\kappa - 1}\right) \times \left(\frac{p_2}{p_{01}}\right)^{\frac{2}{\kappa}} \times \left\{1 - \left(\frac{p_2}{p_{01}}\right)^{\frac{\kappa-1}{\kappa}}\right\}\right]^{\frac{1}{2}}.\tag{8}$$

Equation (8) can be computed directly from the pressure ratio $\left(\frac{p_2}{p_{01}}\right)$ across the valve or port depending on the direction of flow. When the flow is choked in the throat, Equation (8) becomes:

$$\xi_f = \frac{\dot{m}_{moc} \times a_0}{p_0 \times F_2} = \kappa \times \left(\frac{2}{\kappa + 1}\right)^{\frac{\kappa+1}{2(\kappa-1)}},\tag{9}$$

where subscript 0 refers to the stagnation conditions.

The actual gas flow has to take into account the discharge coefficient C_d, which can be calculated by multiplying Equations (8) and (9) by C_d. If the value of ρ_h is the same, C_d can be expressed using Equation (10) [23,24].

$$C_d = \frac{\dot{m}_{exp}}{\dot{m}_{moc}}.\tag{10}$$

4. One-Dimensional Gas Flow Analysis

Figure 3 shows the 1D modeling of the gas flow analysis of a single cylinder diesel engine. The intake and exhaust ports are bent pipes and cannot be modeled in 1D [18,25]. Therefore, the geometry of the intake and exhaust ports was excluded from the modeling, and the length of the intake and exhaust ports was included in the intake and exhaust pipes. The only difference between the experimental and 1D modeling is the intake and exhaust ports. The purpose is to observe the errors caused by such differences, and all other boundary conditions were modeled in the same as in the experiment.

The number of meshes was 50 meshes in the intake pipe and 100 meshes in the exhaust pipe. Benson et al. verified the mesh independence in more than 12 meshes of the exhaust pipe [18], and this study confirmed the mesh independence in more than 15 meshes.

The 1D gas flow analysis of a single cylinder diesel engine was programmed using the C language. The structure of the program consisted of a main program for applying the initial conditions and calculating the pressure, velocity, and time, and a subroutine function for calculating the cylinder and intake and the exhaust gas flow [18]. Figure 4 shows the algorithm for calculating the intake air mass flow rate among the subroutine functions.

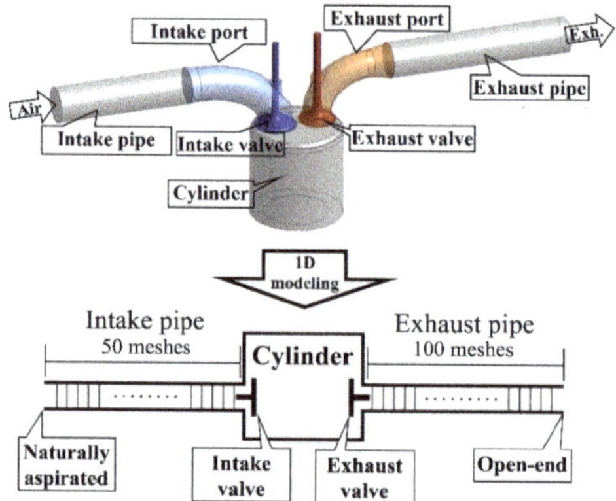

Figure 3. One-dimensional (1D) modeling of the gas flow analysis of a single cylinder diesel engine.

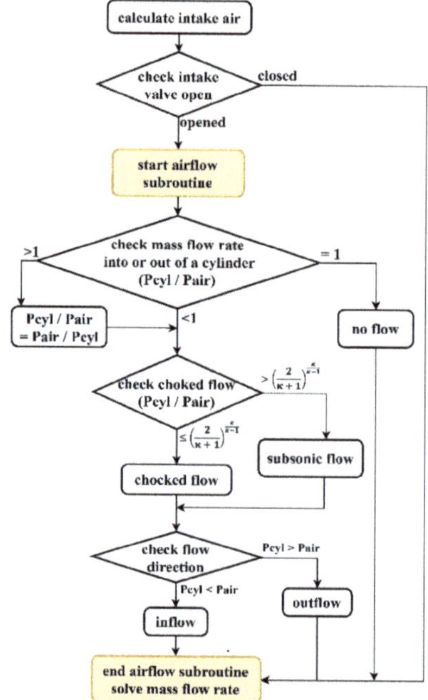

Figure 4. Algorithm for calculating the intake mass flow rate of the one-dimensional gas flow analysis program.

The calculation of the intake subroutine function came after the cylinder subroutine function, which calculated the pressure and mass flow rate of the cylinder. If the intake valve was open, the intake subroutine function was started. Through the cylinder pressure and intake pressure ratio, it was possible to determine the choked flow, and to calculate whether the type of flow from the cylinder was inflow or outflow. When the characteristics of the flow are determined through the calculation algorithm, the mass flow rate can be calculated using Equation (8) or Equation (9).

5. Results

5.1. Average Mass Flow Rate and Discharge Coefficient

The discharge coefficient was used to correct the difference between the 1D gas flow analysis and the measurement results. The reason is to use the gas flow analysis results to predict the performance of the experimental apparatus.

Figure 5 and Table 4 show the average mass flow rate and the discharge coefficient of the intake air according to the engine speed.

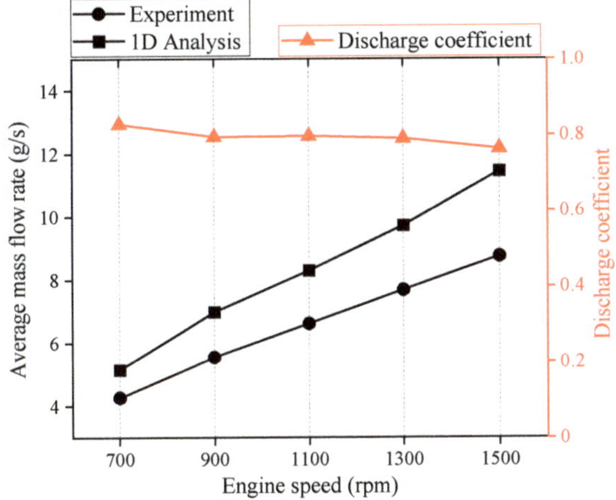

Figure 5. Average mass flow rate and discharge coefficient of intake air.

Table 4. Comparison of the average mass flow rate and the discharge coefficient according to the engine speed.

Engine Speed (rpm)	Average Mass Flow Rate (g/s)		Discharge Coefficient
	Experiment	1D Analysis	
700	4.2684	5.1533	0.8283
900	5.5480	6.9780	0.7951
1100	6.6140	8.2969	0.7972
1300	7.6797	9.7150	0.7905
1500	8.7451	11.4471	0.7640

The experimental results refer to the average mass flow rate measured using a laminar flowmeter. In order to compare the average mass flow rate, the mass flow rate calculation results of the 1D gas flow rate analysis were converted to the average mass flow rate [26].

The volumetric efficiency associated with the average mass flow rate did not continue to increase in proportion to the engine speed due to valve overlap, the influence of the reflected waves, etc. However, in terms of the engine of this study, it was confirmed through a previous study that the volumetric efficiency increases in proportion to the engine speed up to 1700 rpm [22]. As the engine speed increased to 1500 rpm, the average mass flow rate increased, and the difference between the experiment and the 1D flow analysis results also increased. The reason for the difference in the average mass flow rate is the difference between the actual physical phenomena and the theoretical calculations [27,28]. The higher engine speed is the larger flow rate into the cylinder, so the difference is thought to have occurred because of this. In order to correct the differences in the results of the 1D gas flow analysis, the discharge coefficient was calculated using Equation (10). The discharge coefficient obtained in this way is a correction method for the difference between the experiment and 1D gas flow analysis. Therefore, it can be used to predict the performance of the experimental apparatus through the results of 1D gas flow analysis in the future.

5.2. Mass Flow Rate in One-dimensional Gas Flow Analysis

One-dimensional gas flow analysis can calculate the mass flow rate over time [18]. Figure 6 shows the intake air mass flow rate during the intake period compared to the case where the discharge coefficient was applied.

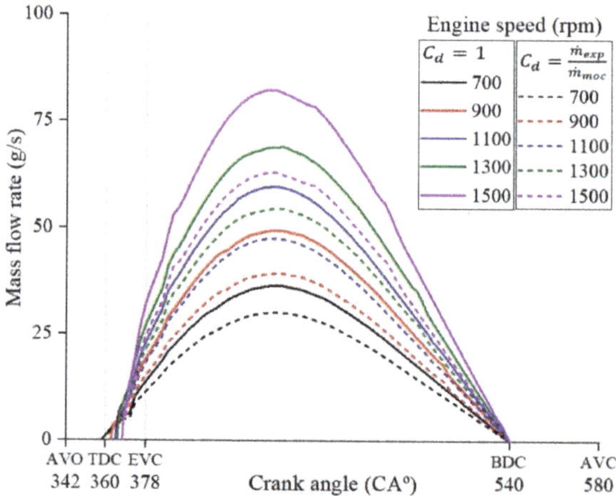

Figure 6. One-dimensional analysis results of the mass flow rate. AVC, air–intake valve closes; AVO, air–intake valve opens; BDC, bottom dead center; EVC, exhaust valve closes; TDC, top dead center.

The mass flow rate does not increase as soon as the intake valve was opened [29]. This is because, during the valve overlap period (342~378 CA°), the outflow from the cylinder exits through the exhaust port. The higher engine speed is the larger gas exchange flow rate, so the timing at which the mass flow rate increases is delayed. The mass flow rate increases after the top dead center (TDC; 360 CA°), and the intake air enters the cylinder until the bottom dead center (BDC; 540 CA°). If the mass flow rate is large, the corrected value due to the application of the discharge coefficient also increases. Therefore, the correction of the result through the application of the discharge coefficient has more influence on the results of the gas exchange process in the valve open state than in the valve closed state.

5.3. Intake Pipe Pressure

Figure 7 shows the intake pipe pressure results during the intake period. The higher engine speed is the greater negative pressure [30], and the time when the greatest negative pressure occurred was the same as in the experiment and in the 1D gas flow analysis. The phase difference occurred within 0.5 CA°.

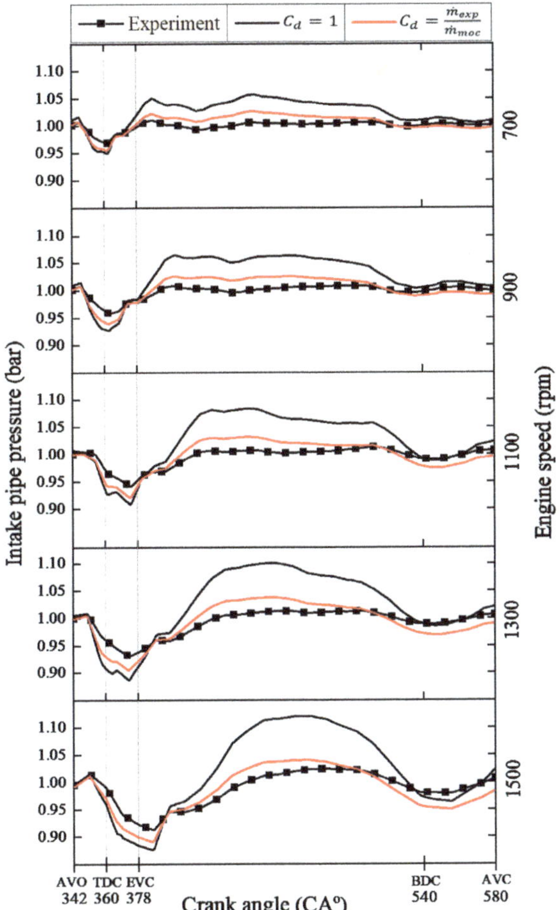

Figure 7. Comparison of the experiment and 1D gas flow analysis results of the intake pipe pressure during the intake period.

After the exhaust valve closes (EVC; 378 CA°), there is no flow from the cylinder to the exhaust, and the flow enters the cylinder through the intake system. The time it takes for the pressure to rise to atmospheric pressure appeared later as the engine speed increased, which is thought to be due to the large flow rate that entered the cylinder.

By applying a discharge coefficient, errors within the experimental results were reduced. Even when the discharge coefficient was applied, the negative pressure and the pressure change were larger in the 1D gas flow analysis compared to the experiment [31]. The error of the maximum negative pressure decreased by 0.56–1.93% after applying the discharge coefficient and the maximum pressure error decreased by 3.11–7.86%. This is expected to be an error caused by modeling the intake and exhaust ports as straight pipes. When the piston goes down from TDC to BDC, the amount of intake is more in a straight pipe and more gas exchanged in intake and exhaust because there is less loss compared to bent pipes due to an eddy through the pipe [32,33].

In straight pipes, reflected waves are generated at the end of the pipe. However, in bent pipes, reflected waves are generated passing through the bent area [34]. The error in the 1D gas flow analysis is thought to have been caused by not calculating the influence of the reflected wave in bent pipes, as in the experiment [35]. It is expected that errors can be reduced if the geometry of bent pipes, which cannot be modeled in 1D, such as intake ports, are analyzed in 3D and are coupled [36,37].

5.4. Exhaust Pipe Pressure

Figure 8 shows the exhaust pipe pressure results from the exhaust period to the BDC. Comparing the 1D gas flow analysis results at 700 rpm with those of the experiment, it can be seen that the amount of pressure change was small during the exhaust period (130~378 CA°). The phase difference occurred after the EVC. The reason for the phase difference is that the engine speed is slow and the reflected wave affects it many times [38]. In the case of 700 rpm, the fourth reflected wave was reached during EVC to BDC. At 900 rpm or more, the third or less reflected waves were reached. As the 1D gas flow analysis did not model the exhaust port of bent geometry, it is expected that more errors would have occurred as the influence of the reflected wave increased.

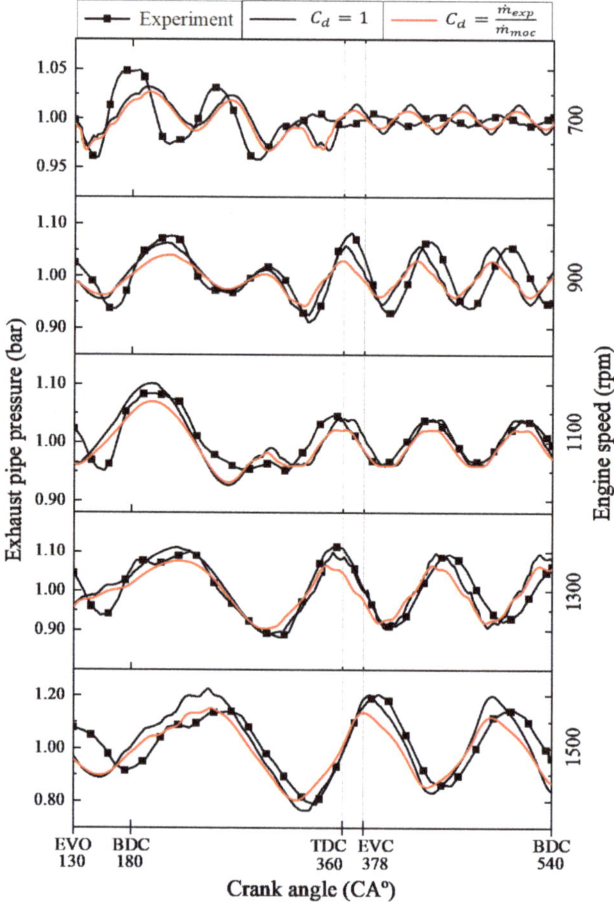

Figure 8. Comparison of the experiment and 1D gas flow analysis results of the exhaust pipe pressure during the exhaust period to the BDC.

In the results above 900 rpm, the amount of pressure change was similar or large during the exhaust period. The phase difference occurred after the peak of the pressure wave appeared, which was smaller than that of 700 rpm. The peak of the pressure wave occurred due to the influence of the reflected wave [39]. In all of the results, as the influence of the reflected wave increased, a larger error occurred.

As a result of the calculation by applying the discharge coefficient, the size of the pressure change was reduced, and the phase was not affected. The reason for the phase difference likely occurred because the effect of the reflected wave generated at the open end of the exhaust pipe reached the exhaust port and affected the calculation. For the same reason as the intake port, the exhaust port could not be modeled in the same geometry as in the experiment. It is necessary to reduce the error in the 1D gas flow analysis caused by bent pipes not only for the intake system, but also for the exhaust system.

When the discharge coefficient was applied, the error of the minimum pressure increased by 0.46–6.14%. In addition, the error of the maximum pressure increased by 0.53–5.03% at 900 rpm or higher. Applying the discharge coefficient was not effective in reducing the error of the exhaust pipe pressure.

Summarizing the intake and exhaust pipe pressure results, the error of the intake pipe pressure is improved. However, since it is a result of reducing the difference in the intake mass flow rate, it is not a fundamental improvement method for the error occurring in the 1D gas flow analysis. Also, the error of the exhaust pipe pressure was not improved even when the discharge coefficient was applied. It is expected that such errors will be improved only when a method to improve errors occurring in complex shapes, which is a disadvantage of 1D, is used.

6. Conclusions

A 1D gas flow analysis was performed using the MOC for a single cylinder diesel engine, and the results of comparing the average mass flow rate and the intake and exhaust pipe pressure with the results of the experiment are summarized as follows:

(1)　The average mass flow rate of the 1D gas flow analysis was larger than that of the experimental results, and the discharge coefficient was calculated using this.
(2)　The 1D gas flow analysis was performed by applying the discharge coefficient, and the accuracy of the intake pipe pressure results was improved.
(3)　In the exhaust pipe result of the 1D gas flow analysis, there was a difference in pressure and phase, and there was no improvement even when the discharge coefficient was applied.
(4)　Applying the discharge coefficient is not a fundamental method to improve the error of 1D gas flow analysis, and the error that occurs due to the complex shape must be improved.
(5)　If the shortcomings of the 1D gas flow analysis, which cannot model the bent pipe geometry of the intake and exhaust ports, are compensated, the error is expected to decrease.

As a result of the experiment and verification, there was a limit to increasing the accuracy, even when applying the discharge coefficient obtained using the average mass flow rate. This error is thought to be due to the disadvantage of 1D gas flow analysis, which cannot calculate complex geometries. In the future, it is expected that this problem can be solved by analyzing the complex geometries in 3D and using 1D–3D coupling gas flow analysis.

Author Contributions: Conceptualization and methodology, K.-J.K.; experimental system configuration and data curation, K.-H.K.; software, K.-J.K.; writing and validation, K.-J.K. and K.-H.K.; supervision, K.-J.K. and K.-H.K. All authors have read and agreed to the published version of the manuscript.

Funding: This research received no external funding.

Acknowledgments: The authors would like to acknowledgment the Turbo Engine Lab, Department of Mechanical System Engineering, Pukyong National University for purchasing and supporting the use of the experimental apparatus.

Conflicts of Interest: The authors declare no conflict of interest.

References

1. Abdullah, N.R.; Shahruddin, N.S.; Mamat, R.; Mohd, A.; Mamat, I.; Zulkifli, A. Effects of air intake pressure on the engine performance, fuel economy and exhaust emissions of a small gasoline engine. *J. Mech. Eng. Sci.* **2014**, *6*, 949–958. [CrossRef]
2. Schulten, P.J.M.; Stapersma, D. Mean value modelling of the gas exchange of a 4-stroke diesel engine for use in powertrain applications. *SAE Tech. Pap.* **2003**. [CrossRef]
3. Kocher, L.; Koeberlein, E.; Van Alstine, D.G.; Stricker, K.; Shaver, G. Physically based volumetric efficiency model for diesel engines utilizing variable intake valve actuation. *Int. J. Engine Res.* **2012**, *13*, 169–184. [CrossRef]
4. Lu, X.-C.; Yang, J.-G.; Zhang, W.-G.; Huang, Z. Effect of cetane number improver on heat release rate and emissions of high speed diesel engine fueled with ethanol–diesel blend fuel. *Fuel* **2004**, *83*, 2013–2020. [CrossRef]
5. Papagiannakis, R.G.; Rakopoulos, C.D.; Hountalas, D.T.; Rakopoulos, D.C. Emission characteristics of high speed, dual fuel, compression ignition engine operating in a wide range of natural gas/diesel fuel proportions. *Fuel* **2010**, *89*, 1397–1406. [CrossRef]
6. Kopac, M.; Kokturk, L. Determination of optimum speed of an internal combustion engine by exergy analysis. *Int. J. Energy* **2005**, *2*, 40–54. [CrossRef]
7. International Maritime Organization (IMO). Effective date of implementation of the fuel oil standard in regulation. *Marpol Annex VI Resolut. MEPC* **2016**, *280*, 1.
8. Johnson, B.T. Diesel Engine Emissions and Their Control. *Platin. Met. Rev.* **2008**, *52*, 23–37. [CrossRef]
9. Davies, P.O.A.L. Piston engine intake and exhaust system design. *J. Sound Vib.* **1996**, *190*, 677–712. [CrossRef]
10. Albarbar, A. Diesel engine fuel injection monitoring using acoustic measurements and independent component analysis. *Measurement* **2010**, *43*, 1376–1386. [CrossRef]
11. Onorati, A.; Montenegro, G.; D'Errico, G.; Piscaglia, F. Integrated 1d–3d fluid dynamic simulation of a turbocharged diesel engine with complete intake and exhaust systems. *SAE Tech. Pap.* **2010**. [CrossRef]
12. Wurzenberger, J.C.; Wanker, R. *Multi-Scale Scr Modeling, 1d Kinetic Analysis and 3d System Simulation*; SAE International: Warrendale, PA, USA, 2005. [CrossRef]
13. Zinner, C.; Jandl, S.; Schmidt, S. Comparison of different downsizing strategies for 2- and 3-cylinder engines by the use of 1d-cfd simulation. In Proceedings of the SAE/JSAE 2016 Small Engine Technology Conference & Exhibition, Charleston, SC, USA, 15–17 November 2016.
14. Kim, K.H.; Kong, K.J. Validation of diesel engine gas flow one-dimensional numerical analysis using the method of characteristics. *J. Korean Soc. Fish Ocean Technol.* **2020**, *56*, 230–237. [CrossRef]
15. Brown, G.L. Determination of two-stroke engine exhaust noise by the method of characteristics. *J. Sound Vib.* **1982**, *82*, 305–327.
16. Winterbone, D.E.; Pearson, R.J. *Theory of Engine Manifold Design: Wave Action Methods for IC Engines*; Wiley-Blackwell: Hoboken, NJ, USA, 2000; pp. 179–274.
17. Winterbone, D.E.; Pearson, R.J. *Design Techniques for Engine Manifolds: Wave Action Methods for IC Engines*; Wiley-Blackwell: Hoboken, NJ, USA, 1999; pp. 27–150.
18. Benson, R.S.; Horlock, J.H.; Winterbone, D.E. *The Thermodynamics and Gas Dynamics of Internal-Combustion Engines*; Oxford University Press: Oxford, UK, 1982; Volume 1, pp. 73–313.
19. Ohtsubo, H.; Yamauchi, K.; Nakazono, T.; Yamane, K.; Kawasaki, K. Influence of compression ratio on performance and variations in each cylinder of multi-cylinder natural gas engine with pcci combustion. *SAE Trans.* **2007**, *116*, 396–404.
20. Han, B.; Guan, X.; Ou, J. Electrode design, measuring method and data acquisition system of carbon fiber cement paste piezoresistive sensors. *Sens. Actuators A* **2007**, *135*, 360–369. [CrossRef]
21. Amann, C.A. Cylinder-pressure measurement and its use in engine research. *SAE Trans.* **1985**, *94*, 418–435.
22. Lee, S.D.; Kang, H.Y.; Koh, D.K.; Ahn, S.K. The effect of intake and exhaust pulsating flow on the volumetric efficiency in a diesel engine. *J. Korean Soc. Power Syst. Eng.* **2006**, *9*, 19–25.
23. Borghei, S.M.; Jalili, M.R.; Ghodsian, M. Discharge coefficient for sharp-crested side weir in subcritical flow. *J. Hydraul. Eng.* **1999**, *125*. [CrossRef]
24. Chu, C.R.; Chiu, Y.-H.; Chen, Y.-J.; Wang, Y.-W.; Chou, C.-P. Turbulence effects on the discharge coefficient and mean flow rate of wind-driven cross-ventilation. *Build. Environ.* **2009**, *44*, 2064–2072. [CrossRef]

25. Hufnagel, L.; Canton, J.; Örlü, R.; Marin, O.; Merzari, E.; Schlatter, P. The three-dimensional structure of swirl-switching in bent pipe flow. *J. Fluid Mech.* **2018**, *835*, 86–101. [CrossRef]

26. Tabaczynski, R.J.; Heywood, J.B.; Keck, J.C. Time-resolved measurements of hydrocarbon mass flowrate in the exhaust of a spark-ignition engine. *SAE Tech. Pap.* **1972**. [CrossRef]

27. Ramezanizadeh, M.; Alhuyi Nazari, M.; Ahmadi, M.H.; Chau, K. Experimental and numerical analysis of a nanofluidic thermosyphon heat exchanger. *Eng. Appl. Comput. Fluid Mech.* **2019**, *13*, 40–47. [CrossRef]

28. Kim, S.J.; Jang, S.P. Experimental and numerical analysis of heat transfer phenomena in a sensor tube of a mass flow controller. *Int. J. Heat Mass Transf.* **2001**, *44*, 1711–1724. [CrossRef]

29. Harrison, M.F.; Stanev, P.T. Measuring wave dynamics in ic engine intake systems. *J. Sound Vib.* **2004**, *269*, 389–408. [CrossRef]

30. Ceccarani, M.; Rebottini, C.; Bettini, R. Engine misfire monitoring for a v12 engine by exhaust pressure analysis. *SAE Tech. Pap.* **1998**. [CrossRef]

31. Takahashi, H.; Ogino, S.; Nishimura, T.; Okuno, Y. Experimental analysis for the improvement of radiator cooling air intake and discharge. *SAE Tech. Pap.* **1992**. [CrossRef]

32. Lu, J.-T.; Hsueh, Y.-C.; Huang, Y.-R.; Hwang, Y.-J.; Sun, C.-K. Bending loss of terahertz pipe waveguides. *Opt. Express* **2010**, *18*, 26332–26338. [CrossRef]

33. Simizu, Y.; Sugino, K.; Kuzuhara, S. Hydraulic losses and flow patterns in bent pipes. *Bull. JSME* **1982**, *25*, 24–31. [CrossRef]

34. Ohyagi, S.; Obara, T.; Nakata, F.; Hoshi, S. A numerical simulation of reflection processes of a detonation wave on a wedge. *Shock Waves* **2000**, *10*, 185–190. [CrossRef]

35. Parker, K.H.; Jones, C.J.H. Forward and Backward Running Waves in the Arteries: Analysis Using the Method of Characteristics. *J. Biomech. Eng.* **1990**, *112*, 322–326. [CrossRef]

36. Kong, K.-J.; Jung, S.-H.; Jeong, T.-Y.; Koh, D.-K. 1d–3d coupling algorithm for unsteady gas flow analysis in pipe systems. *J. Mech. Sci. Technol.* **2019**, *33*, 4521–4528. [CrossRef]

37. Coghe, A.; Brunello, G.; Tassi, E. Effects of Intake Ports on the In-Cylinder Air Motion under Steady Flow Conditions. *SAE Tech. Pap.* **1988**. [CrossRef]

38. HeyWood, J. *Internal Combustion Engine Fundamental*; McGraw-Hill Book Company: New York, NY, USA, 1988; pp. 748–8161.

39. Khir, A.W.; Parker, K.H. Measurements of wave speed and reflected waves in elastic tubes and bifurcations. *J. Biomech.* **2002**, *35*, 775–783. [CrossRef]

Publisher's Note: MDPI stays neutral with regard to jurisdictional claims in published maps and institutional affiliations.

Journal of
Marine Science and Engineering

Article

Development of a Marine Two-Stroke Diesel Engine MVEM with In-Cylinder Pressure Trace Predictive Capability and a Novel Compressor Model

Haosheng Shen [1], Jundong Zhang [1,*], Baicheng Yang [2] and Baozhu Jia [3]

[1] College of Marine Engineering, Dalian Maritime University, Dalian 116026, China; shen7231591@126.com
[2] College of Navigation, Dalian Maritime University, Dalian 116026, China; yangbaicheng@dlmu.edu.cn
[3] Marine College, Guangdong Ocean University, Zhanjiang 524088, China; skysky@dlmu.edu.cn
* Correspondence: zhjundong@dlmu.edu.cn; Tel.: +86-134-7894-8607

Received: 15 February 2020; Accepted: 13 March 2020; Published: 16 March 2020

Abstract: In this article, to meet the requirements of marine engine room simulator on both the simulation speed and simulation accuracy, a mean value engine model (MVEM) for the 7S80ME-C9.2 marine two-stroke diesel engine was developed and validated in the MATLAB/Simulink environment. In consideration of the significant influence of turbocharger compressor on both the engine steady state performance and transient response, a novel compressor model (mass flow rate and isentropic efficiency model) based on a previous study carried out by the first author was proposed with the aim of achieving satisfactory simulation accuracy within the whole engine operating envelope. The predictive and extrapolative capability of the proposed compressor model was validated by carrying out simulation experiments and analyzing the simulation results under steady state condition and during transient process. To make the traditional MVEM capable of predicting in-cylinder pressure trace, the cylinder pressure analytic model proposed by Eriksson and Andersson for the four-stroke SI (spark ignition) engine was adapted to the 7S80ME-C9.2 marine two-stroke diesel engine based on the characteristic of in-cylinder pressure trace of this type of engine and then coupled to the MVEM developed in this paper. Since there is no need to solve any differential equation as it is done in the 0-D model, the advantage of MVEM in running speed is not impaired. For achieving satisfactory simulation accuracy by using the analytic model, the model parameters were calibrated elaborately by using engine measured data and a 0-D model and the relevant tuning procedure was discussed in detail.

Keywords: marine two-stroke diesel engine; mean value engine model; compressor model; in-cylinder pressure trace; model calibration

---

1. Introduction

The marine large scale two-stroke diesel engine is widely adopted as the prime mover for the large merchant ships mainly because of its high thermal efficiency, reliability as well as the ability of burning low grade fuel, i.e., heavy fuel oil (HFO). For complying with the stringent international environmental legislations and obtaining improved fuel efficiency, engine manufacturers have developed new versions of marine engines, mainly including marine electronically controlled diesel engines and marine dual-fuel engines [1–3].

In consideration of the large size and weight of marine large scale two-stroke diesel engines as well as the substantial manpower and financial power required for carrying out experimental studies, various engine simulation techniques have been widely adopted for investigating engine performance, designing and testing the fault diagnosis algorithm, as well as developing the engine control system. Among these simulation models, 0-D models and mean value engine model (MVEM) are widely adopted by researchers mainly because of their fast running speed and satisfactory simulation accuracy. The difference between the 0-D model and MVEM lies in the modeling approach for the cylinder. For the 0-D model, the cylinder is assumed as an open thermodynamic system, where the working medium is uniformly distributed in it. By applying mass and energy conservation laws and incorporating several relevant sub-models, the in-cylinder pressure trace can be predicted. Furthermore, the 0-D model is also capable of predicting engine performance at varying settings (e.g., varying the start of injection timing, exhaust valve opening/closing timing, turbine area, etc.), which is a special advantage in comparison to the MVEM [1]. In the book published by Eriksson and Nielsen, the MVEM is defined as "Mean value engine models are models where the signals, parameters, and variables that are considered are averaged over one or several cycles" [4]. In this respect, the mass and energy flow through the cylinder is assumed continuous for the MVEM, and the engine average performance over one or several cycles can be obtained. Consequently, the MVEM is able to run much faster than the 0-D model, which is therefore very suitable for cases that require fast running speed, such as the simulation of engine transients for a long period. On the other hand, despite similar predictive accuracy can be achieved by the two models, the in-cylinder pressure trace cannot be predicted by the MVEM, which is its major limitation [1,5].

In the literature, researchers tried to improve the predictive ability of the MVEM and the running speed of the 0-D model by adopting a "hybrid" modeling approach, meaning that different modeling approaches can be adopted for different engine components or different phases of the engine cycle. This approach can effectively overcome the limitation caused by using only a single modeling approach. In the study carried out by Altosole et al., for meeting the requirement of real-time ship maneuvering simulation, the cylinder simulation was entirely based on a set of five-dimensional numerical matrices, each of which was generated by a 0-D model [6]. It was revealed from the simulation results that this modeling approach can achieve similar transient response but reduce the simulation time of about 99%; however, the in-cylinder pressure trace cannot be predicted. Nikzadfar and Shamekhi developed an extended MVEM for control-oriented modeling of diesel engines transient performance and emissions by replacing the cylinder model with two artificial neural networks (ANN). One is for predicting aspirated air mass flow, torque and exhaust gas temperature and the other for predicting soot and NO_x emission [7]. Despite the fact that satisfactory predictive accuracy and running speed can be achieved with this extended MVEM, it still cannot predict the in-cylinder pressure trace. Based on the modular MVEM developed by Theotokatos [8], Baldi et al. proposed a combined mean value-zero dimensional model for a large marine four-stroke diesel engine, where the closed part of the cycle was represented by the 0-D model and the open part by the MVEM. The combined model fully takes the respective advantage of the 0-D model (ability to predict in-cylinder pressure trace) and MVEM (fast running speed) [9]. Nevertheless, it was pointed out by Theotokatos et al. that the 0-D model still needed to be called at each calculation step for the combined mean value-zero dimensional model, which made its running speed still scant for cases where engine transient simulation for a long period was required [1]. In the study carried out by Tang et al., the hybrid modeling approach was further improved by simplifying the in-cylinder pressure calculation during the scavenging and exhausting phases with two linear functions and abandoning engine cycles at certain intervals [5]. It was revealed that the modified model was able to predict the in-cylinder pressure trace and run as fast as the MVEM at steady state condition; however, during the transient process, the improvement in running speed is at the expense of predictive precision.

For practical applications that have high requirement on both predictive accuracy and running speed, the MVEM seems to be the best choice. However, due to the absence of a detailed mathematical description of in-cylinder working process, the in-cylinder pressure trace cannot be predicted with MVEM, which limits its practical value to some extent. This is the reason why 0-D model was incorporated into the MVEM in several studies [5,9]. However, it should be noted that even limited 0-D modeling will increase the model complexity and thus affect the running speed of the MVEM obviously. To solve this problem, the cylinder pressure analytic model proposed by Eriksson and

Andersson for the four-stroke SI (spark ignition) engine was modified and adapted to the 7S80ME-C9.2 marine two-stroke diesel engine in this paper [10]. By coupling the cylinder pressure analytic model to the MVEM, the MVEM is able to estimate the in-cylinder pressure trace. Furthermore, as the running speed of the cylinder pressure analytic model is much faster than the 0-D model, the merit of the MVEM in running speed is not affected significantly.

In consideration of the significant influence of compressor model on the simulation accuracy of the whole engine model, a novel compressor mass flow rate and isentropic efficiency model was proposed in this paper based on the research results of a previous paper published by the first author, which compared and analyzed the predictive and extrapolative ability of several classical and recent proposed compressor mass flow rate and isentropic efficiency empirical models. The incorporation of the novel compressor model will be very helpful for the MVEM to achieve satisfactory predictive accuracy in the whole engine operating envelope under both steady and transient conditions.

The MVEM developed in this paper is very suitable for applications that require both fast running speed and in-cylinder pressure trace predictive capability, for example, this MVEM has been attempted to be used in a VLCC (Very Large Crude Carrier) marine engine room simulator [11].

2. Model Description

2.1. Engine Specificiation and MVEM Structure

The marine large scale two-stroke electronically controlled diesel engine MAN B&W 7S80ME-C9.2 is selected as the simulation object in the paper. This type of engine is widely adopted as the prime mover of merchant ships, especially the VLCC. The engine main technical parameters at the MCR (Maximum Continuous Rating) point are presented in Table 1. The MVEM is implemented in MATLAB/Simulink environment following a block oriented modeling approach, as it is depicted in Figure 1. Benefitting from the sub-system creation function of MATLAB/Simulink, the structure of the MVEM is very similar with that of the real engine, each component of which is represented by an individual block. The modeling approaches of these engine components will be introduced in the following sub-sections.

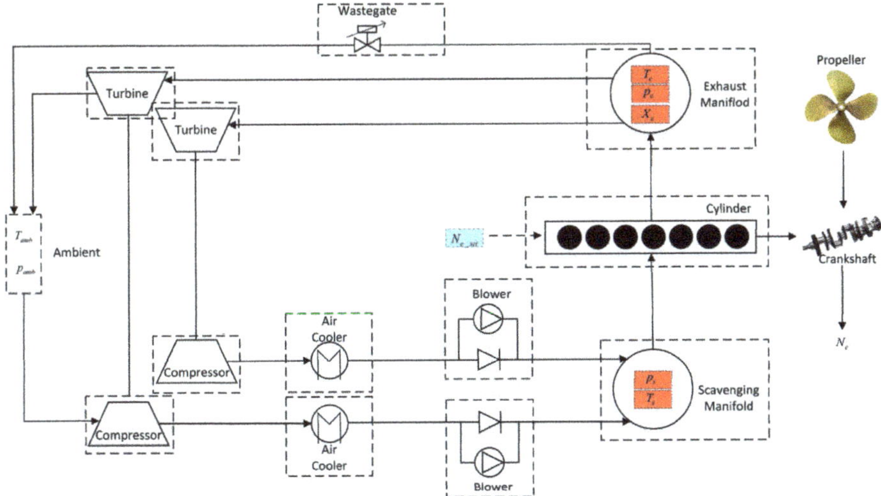

Figure 1. Mean value engine model (MVEM) structure.

Table 1. Main Technical parameters of 7S80ME-C9.2 diesel engine at MCR (Maximum Continuous Rating).

Parameters	Value
Number of cylinder (–)	7
Cylinder bore (mm)	800
Stroke (mm)	3450
Speed (rpm)	72
Power (kW)	25190
Maximum pressure (MPa)	17.1
Mean brake effective pressure (bar)	17.3
Firing order	1-7-2-5-4-3-6
Specific fuel oil consumption (g/kW·h)	165.9

2.2. Cylinder

For the MVEM, the actual intermittent gas flowing process through the scavenging ports and exhaust valve is simplified as a flowing process through an equivalent orifice with fixed area under sub-sonic flow consideration. Consequently, the cylinder inflowing air mass flow rate $\dot{m}_{sz}$ can be calculated with the following equation [5,8,9,12,13]:

$$\dot{m}_{sz} = C_z A_z \frac{p_s}{\sqrt{R_s T_s}} \sqrt{\frac{2k_s}{k_s - 1} \left[\left(\frac{p_e}{p_s}\right)^{\frac{2}{\gamma_s}} - \left(\frac{p_e}{p_s}\right)^{\frac{\gamma_s+1}{\gamma_s}} \right]} \tag{1}$$

where C_z and A_z are the flow coefficient and the equivalent orifice area, respectively; R_s, T_s, p_s and γ_s are the gas constant, temperature, pressure and specific heat ratio of the air in the scavenging manifold, respectively; p_e is the exhaust manifold pressure.

As the exhaust valve lifting curve is not provided by the engine manufacturer and the orifice flow coefficient is also unknown, therefore, the product of C_z and A_z($C_z A_z$), is treated as a calibration parameter in this paper. $C_z A_z$ can be approximated as a linear function of the brake power, that is $C_z A_z = k_{CA0} + k_{CA1}P_b$.

As similar with the engine air inflowing process, the MVEM also treats the fuel injection as a continuous process. According to the mass conservation law, the mass flow rate of exhaust gas exiting the cylinders $\dot{m}_{ze}$ can be calculated by adding $\dot{m}_{sz}$ and the fuel injection rate $\dot{m}_f$.

The energy released by the fuel burning cannot be fully exploited by the engine and converted to the mechanical energy directly, therefore, a portion of the thermal energy still remains in the exhaust gas. Consequently, based on the energy conservation law, the thermal energy of the exhaust gas exiting the cylinders can be written as [8,12,13]:

$$\dot{m}_{ze}h_{ze} = \dot{m}_{sz}c_{p,s}T_s + \zeta\eta_{comb}\dot{m}_f H_{LHV} \tag{2}$$

where h_{ze} is the specific enthalpy of exhaust gas; η_{comb} is the combustion efficiency, which is a function of the air-fuel ratio ($A/F = \dot{m}_{sz}/\dot{m}_f$); $c_{p,s}$ is the constant pressure specific heat of the scavenging air; H_{LHV} is the fuel lower heating value; ζ is the fuel chemical energy proportion in the exhaust gas, which can be fitted as a linear function of the brake power, that is $\zeta = k_{\zeta0} + k_{\zeta1}P_b$.

The mean indicated effective pressure $\bar{p}_i$ is fitted as a linear function of the fuel index. According to the engine shop trial report, the mean friction effective pressure $\bar{p}_f$ is always 1 bar for all the tested loading conditions, therefore, a constant value of $\bar{p}_f$ is assumed in this paper. Having obtained $\bar{p}_i$ and $\bar{p}_f$, the mean brake effective pressure $\bar{p}_b$ can be calculated as $\bar{p}_b = \bar{p}_i - \bar{p}_f$.

Finally, the engine brake torque Q_b, brake power P_b and brake specific fuel consumption (BSFC) can be derived as per the following equations:

$$Q_b = \frac{n_z V_d \bar{p}_b}{2\pi}, P_b = \frac{\pi N_e Q_b}{30}, \text{BSFC} = \frac{\dot{m}_f}{P_b} \tag{3}$$

where V_d is the engine displacement volume of a single cylinder.

2.3. Turbocharger

The power absorbed by the compressor can be written as:

$$P_c = \dot{m}_c c_{p,a}(T_{c,out} - T_{c,in}) \tag{4}$$

where $\dot{m}_c$ is the compressor mass flow rate; $c_{p,a}$ is the air constant pressure specific heat; $T_{c,in}$ and $T_{c,out}$ are the temperature of air entering and exiting the compressor, respectively.

The compressor mass flow rate depends on not only the turbocharger rotational speed but also the pressure ratio across it with the latter can be written as:

$$\Pi_c = \frac{p_s + \Delta p_{ac} - \Delta p_{bl}}{p_{amb} - \Delta p_{af}} \tag{5}$$

where p_s, Δp_{ac}, Δp_{bl}, p_{amb} and Δp_{af} are the scavenging manifold pressure, air cooler pressure drop, auxiliary blower pressure increase, ambient pressure and compressor air filter pressure drop, respectively. In this paper, Δp_{af} is fitted as a second-order polynomial of $\dot{m}_c$. Modeling approaches for Δp_{ac} and Δp_{bl} will be introduced in Section 2.5.

The temperature of air exiting the compressor can be calculated by the following equation, which is derived based on the definition of compressor isentropic efficiency:

$$T_{c\_out} = T_{c\_in}(1 + (\Pi_c^{(\gamma_a-1)/\gamma_a} - 1)/\eta_c) \tag{6}$$

where γ_a is the air specific heat ratio; η_c is the isentropic efficiency.

The power generated by the turbine can be written as:

$$P_t = \dot{m}_t c_{p,e}(T_{t,in} - T_{t,out}) \tag{7}$$

where $\dot{m}_t$ is the turbine mass flow rate; $c_{p,e}$ is the exhaust gas constant pressure specific heat; $T_{t,in}$ and $T_{t,out}$ are the temperature of exhaust gas entering and exiting the turbine.

In this paper, the turbine is simplified as a nozzle and the exhaust gas mass flow rate flowing through it can be calculated based on the assumption of one-dimensional isentropic adiabatic flow with the input data including the gas thermodynamic properties in the exhaust manifold, the pressure ratio across the turbine as well as the turbine equivalent flow area and flow coefficient.

The pressure ratio Π_t is calculated by the following equation:

$$\Pi_t = \frac{p_{amb} + p_{t,back}}{p_e} \tag{8}$$

where $p_{t,back}$ is the turbine back-pressure and it is fitted as an exponential function of the turbine mass flow rate in this paper.

The turbine equivalent flow area A_t can be computed as per the following equation:

$$A_t = \sqrt{\frac{(A_D \cdot A_S)^2}{A_D^2 + A_S^2}} \tag{9}$$

where A_D and A_S are the flow area of the turbine impeller and nozzle ring, respectively.

For the turbocharger turbine investigated in this paper, its flow coefficient C_t and isentropic efficiency η_t only depend on the expansion ratio Γ_t ($\Gamma_t = 1/\Pi_t$) and are not influenced by the turbocharger rotational speed. Therefore, C_t and η_t are modeled by using look-up table in this paper based on the turbine performance map.

The temperature of exhaust gas exiting the turbine can be calculated based on the definition of turbine isentropic efficiency as the following equation:

$$T_{t,out} = T_{t,in}(1 - \eta_t(1 - \Pi_t^{(\gamma_e-1)/\gamma_e}))$$ (10)

where γ_e is the specific heat ratio of exhaust gas.

Finally, the turbocharger shaft angular speed ω_{tc} can be calculated by integrating the following equation, which is derived based on the angular momentum conservation law:

$$\frac{d\omega_{tc}}{dt} = \frac{1}{J_{tc}}\left(\frac{P_t}{\omega_{tc}}\eta_{tc,m} - \frac{P_c}{\omega_{tc}}\right)$$ (11)

where J_{tc} is the moment of inertia of the turbocharger rotating part; $\eta_{tc,m}$ is the turbocharger mechanical efficiency.

2.4. Scavenging and Exhaust Manifolds

The scavenging and exhaust manifolds are treated as control volumes in the MVEM. By applying the mass conservation law, the mass changing rate in the manifold can be computed as:

$$\frac{dm}{dt} = \dot{m}_{in} - \dot{m}_{out}$$ (12)

where $k_{\zeta 0}$ and $\dot{m}_{out}$ are the mass flow rate entering and exiting the manifold, respectively.

The temperature changing rate in the manifold can be derived by applying the energy conservation law. The differential equation governing the temperature changing rate in the manifold can be written as:

$$\frac{dT}{dt} = \frac{\dot{m}_{in}c_v(T_{in} - T) + R(T_{in}\dot{m}_{in} - T\dot{m}_{out}) + \dot{Q}_{ht}}{mc_v}$$ (13)

where T_{in} is the gas temperature entering the manifold; $\dot{Q}_{ht}$ is the heat dissipation; c_v is the gas constant volume specific heat.

Due to the negligible temperature difference between the scavenging air and the surrounding, the heat dissipation in the scavenging manifold is neglected, whereas the heat dissipation in the exhaust manifold must be taken into account because the exhaust gas temperature is much higher than the surrounding. $\dot{Q}_{ht}$ is computed as it was done in Theotokatos and Tzelepis by using the overall heat transfer coefficient and heat transfer area [12].

Having obtained the stored mass and gas temperature by integrating Equations (12) and (13), the gas pressure in the manifold can be derived by using the ideal gas state equation.

2.5. Air cooler, Auxiliary Blower and Wastegate

The air temperature exiting the air cooler $T_{ac,out}$ can be written as:

$$T_{ac,out} = T_{ac,in} - \eta_{ac}(T_{ac,in} - T_{cw})$$ (14)

where $T_{ac,in}$ is the air temperature entering the air cooler; η_{ac} is the cooling efficiency; T_{cw} is the cooling water temperature, which is assumed to be constant and equals to 300 K in this paper.

In this paper, the air cooler cooling efficiency and the pressure drop is fitted as a second-order polynomial of the air mass flow rate flowing through it.

Auxiliary blower of the centrifugal type, which is driven by an induction motor running at fixed rotational speed, is commonly adopted for marine large scale two-stroke diesel engines. Consequently, the auxiliary blower can be modeled as a centrifugal compressor that runs at fixed rotational speed. Following this idea, the pressure increase across the blower Δp_{bl} and blower efficiency η_{bl} thus only depend on the air volume flow rate $\dot{V}_{bl}$. Therefore, Δp_{bl} and η_{bl} are fitted as a second and third order polynomial of $\dot{V}_{bl}$, respectively, based on the auxiliary blower performance map in this paper. Having obtained Δp_{bl} and η_{bl}, the air temperature exiting the blower can be computed by using Equation (6), which is originally used to calculate the air temperature existing the compressor.

To prevent the compressor from entering into the unstable surging area, many marine large scale two-stroke diesel engines are equipped with a wastegate in recent years, which is a bypass configuration in parallel with the turbine. In this paper, the wastegate is simplified as an ideal nozzle and the gas mass flow rate is calculated based on the assumption of one-dimensional isentropic adiabatic flow. It should be noted that the wastegate shares the same upstream and downstream condition with the turbine.

2.6. Engine Speed Governor

In this paper, the governor is modeled as a proportional-integral (PI) controller with the actual engine rotational speed as the feedback signal [1,5,8,9,12,13]. In addition, scavenging air and torque limiters are also incorporated in the governor model for protecting the engine integrity during fast transients.

As similar with the turbocharger shaft, the engine crankshaft rotational speed N_e can be calculated by integrating the following equation:

$$\frac{dN_e}{dt} = \frac{60}{2\pi} \cdot \frac{Q_b - Q_p}{J_e + J_{sh} + J_p} \tag{15}$$

where Q_p is the propeller resisting torque; J_e, J_{sh} and J_p are the moment of inertia of the engine, shaft system and propeller, respectively.

As the main focus of this paper is to investigate the engine steady-state performance and transient response by using MVEM, the extra propeller moment of inertia caused by the entrained water is neglected; in addition, the hydrodynamic characteristic of the propeller and ship hull is also neglected. For simplicity, the propeller resisting torque is calculated by using the propeller propulsive characteristic curve that passes through the engine MCR point [13]:

$$Q_p = K_p N_e^2, \; K_p = Q_{b,\text{MCR}} / N_{e,\text{MCR}}^2 \tag{16}$$

2.7. Compressor Model Improvement

2.7.1. Compressor Performance Map

The working characteristic of a compressor is usually represented by the performance map as shown in Figure 2 with the mass (or volume) flow rate and pressure ratio as the horizontal and vertical coordinate, respectively.

As shown in Figure 2, most of the compressor performance maps provided by the manufacturers only contain several discrete iso-speed and iso-efficiency curves in the design operating zone. However, for marine turbocharger compressor, its actual rotational speed may be lower than the lowest rotational speed presented in the performance map, or higher than the highest one; in addition, its pressure ratio may approach to unity under certain operating conditions, such as slow steaming, activation of the auxiliary blower and ship maneuvering [14]. Therefore, it is necessary for the developed compressor model to be capable of extrapolating to these off-design operating zones accurately and robustly. In addition, the developed compressor model is also required to accurately interpolate within the unknown areas between these discrete curves.

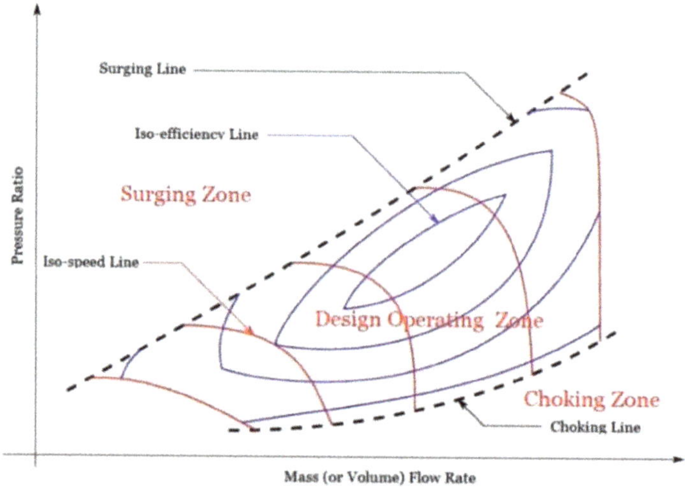

Figure 2. Compressor performance map.

2.7.2. Compressor Mass Flow Rate Model

In a previous study, the first author of this paper carried out an applicable and comparative research of compressor mass flow rate empirical models to two marine large-scale compressors (A270-L59 and TCA88-25070 marine compressor) [15]. The range of applicable and comparative analysis included both the predictive accuracy in the design operating area and the extrapolative ability to the off-design operating areas. These off-design operating areas include the area with rotational speed lower than the lowest speed available in the performance map, the area with rotational speed higher than the highest speed as well as the area to the right of the curve that connects the points with maximum flow rate at each iso-speed curve. These off-design operating areas are named as LS (Low Speed), HS (High Speed) and LPR (Low Pressure Ratio) area, respectively, for convenience of expression. By analyzing the applicable and comparative results, it can be found that none of these compressor empirical models was able to achieve satisfactory predictive and extrapolative accuracy in the whole operating area simultaneously. To solve this problem, a zonal compressor mass flow rate model is proposed in this paper, which selects the model with the best accuracy for each operating area.

Based on the applicable and comparative results presented in Shen et al. [15], it can be found that for the A270-L59 turbocharger compressor, GuanCong model achieves the best predictive accuracy in the design operating area, Leufvén and Llamas ellipse model is capable of capturing the compressor choking phenomenon accurately, whereas the Karlson-II exponential model is able to extrapolate to the LS and HS area robustly and reliably. The detail description of the three models can be found in Shen et al. [15], which will be not introduced in this paper.

For implementing the zonal compressor mass flow rate model, it is necessary to define the zone division standard firstly. As shown in Figure 3, the lowest and highest iso-speed curve available in the compressor performance map is used as the border between the LS and HS area and the design operating area, respectively, whereas the curve connecting the points with maximum flow rate at each iso-speed curve is used to divide the LPR area from the other areas.

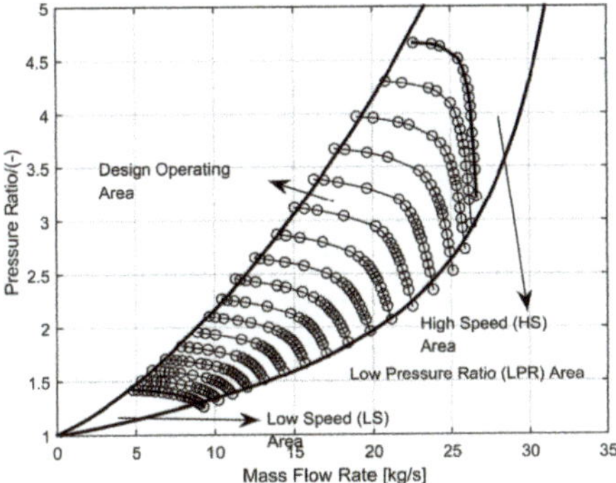

Figure 3. Zone division standard of the compressor performance map.

The changing trend of pressure ratio Π_{lb} and mass flow rate $\dot{m}_{lb}$ with rotational speed on this curve can be represented by using Equations (17) and (18), respectively. As shown in Figure 4, the exponential function is capable of capturing the changing trend of Π_{lb} with rotational speed accurately, which increases slowly under low rotational speed conditions and then rapidly under high rotational speed conditions; on the other hand, the changing trend of $\dot{m}_{lb}$ can be described satisfactorily with the arc-tangent function. It is also found that the changing trend of the pressure ratio Π_{sur} and mass flow rate $\dot{m}_{sur}$ with rotational speed on the surging line can be captured satisfactorily by using Equations (17) and (18) as shown in Figure 4.

$$\Pi_{lb} = a_1 e^{a_2 N_{tc}} + a_3 \tag{17}$$

$$\dot{m}_{lb} = b_1 + b_2 \arctan(b_3 N_{tc} + b_4) \tag{18}$$

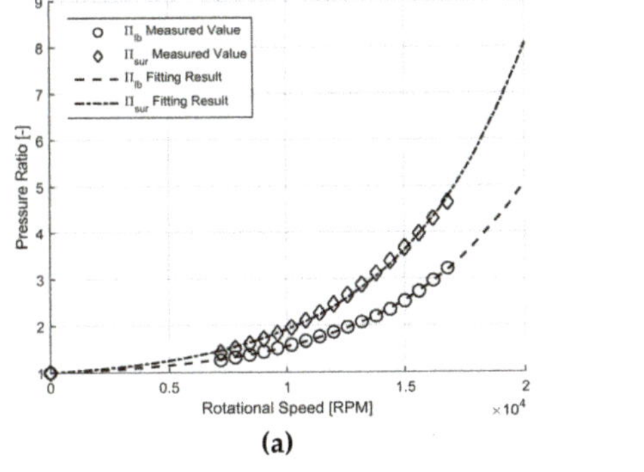
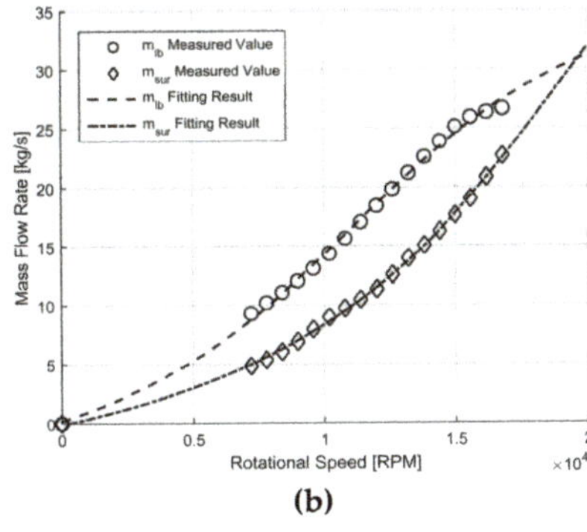

(a)　　　　　　　　　　(b)

Figure 4. Measured pressure ratio and mass flow rate at the boundary of surging zone and low pressure ratio zone and corresponding fitting result: (**a**) pressure ratio; (**b**) mass flow rate.

To increase the reliability of Equations (17) and (18) when extrapolating to the LS area, two additional data points are added when parameterizing the two equations, that is the value of Π_{lb}, Π_{sur}, $\dot{m}_{lb}$ and $\dot{m}_{sur}$ when the rotational speed is equal to zero. It is assumed that $\Pi_{lb} = \Pi_{sur} = 1$ and $\dot{m}_{lb} = \dot{m}_{sur} = 0$ in this paper.

For the area to the left of the surging line, it is assumed that the mass flow rate is equal to $\dot{m}_{sur}$ if the current pressure ratio is larger than the respective Π_{sur} under the current rotational speed condition.

Note that the mass flow rate predicted by the GuanCong model (or Karlson-II exponential model) when the pressure ratio is equal to Π_{lb} is usually different from the one predicted by the Leufvén and Llamas ellipse model. Therefore, to prevent the possible discontinuity phenomenon during the simulation process when the operating point enters into the LPR area, a simple curve blending method is proposed in this paper with the following steps:

(1) Estimate the mass flow rate by using the GuanCong model (or Karlson-II model) and the Leufvén and Llamas ellipse model, respectively, that is $\dot{m}_{GuanOrKarl}$ and $\dot{m}_{ell}$;

(2) Calculate the weighting coefficient z according to Equations (19) and (20):

$$z = 3q^2 - 2q^3 \tag{19}$$

$$q = \frac{\Pi - 1}{\Pi_{lb} - 1} \tag{20}$$

where the range of z and q is between 0 and 1.

(3) Blend $\dot{m}_{GuanOrKarl}$ and $\dot{m}_{ell}$ by using the weighting coefficient z according to Equation (21) to obtain the mass flow rate $\dot{m}_{LPR}$ under the current pressure ratio and rotational speed condition.

$$\dot{m}_{LPR} = z \cdot \dot{m}_{GuanOrKarl} + (1 - z) \cdot \dot{m}_{ell} \tag{21}$$

By applying this curve blending method, it can guarantee the smooth transition of the iso-speed curve when the operating point enters into the LPR area from the other areas; in addition, it can fully take the advantage of the GuanCong model (or Karlson-II exponential model) and the Leufvén and Llamas ellipse model, which presents satisfactory predictive capability at the operating point with pressure ratio equal to Π_{lb} and 1, respectively.

Figure 5 presents the blending results in the LPR area as well as two extrapolated iso-speed curves in the LS and HS area, respectively. As can be observed from this figure, the zonal compressor mass flow rate model is capable of not only predicting the available measured data points accurately but also extrapolating to the off-design operating area robustly and reasonably, which can effectively improve the steady and transient simulation accuracy of the MVEM developed in this paper. In addition, as also can be observed in Figure 5, under low and medium speed conditions, obvious difference exists between the blended curve and the original curve, indicating the deficiency of GuanCong model (or Karlson-II exponential model) in capturing the choking phenomenon; however, the difference gets less obvious with the increase in rotational speed and this is because that under high speed conditions, the measured data points available in the compressor map are very close to the choking point or have already choked, which can be estimated accurately by the GuanCong model (or Karlson-II exponential model), so the blending manipulation is unnecessary.

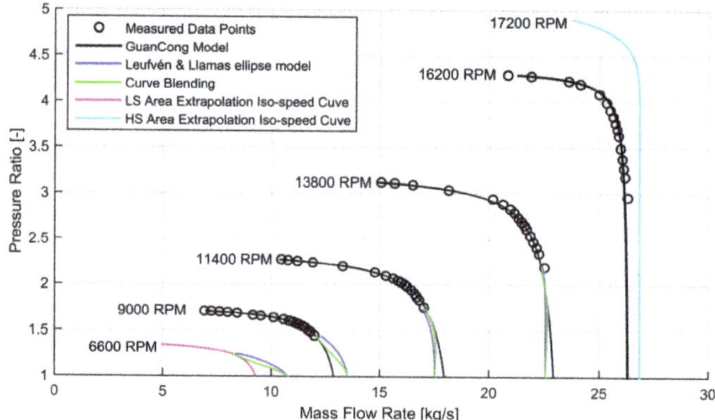

Figure 5. Prediction and extrapolation result of the zonal compressor mass flow rate model.

2.7.3. Compressor Isentropic Efficiency Model

The compressor isentropic efficiency model developed in this paper is based on the one originally proposed by Hadef et al., which is referred to as "Hadef model" herein [16]. The theoretical foundation of the Hadef model is the Euler equation for turbo-machinery, meanwhile, two assumptions are made: (1) air is only accelerated when flowing through the compressor impeller without being any compressed, thus, the air density at the impeller inlet and outlet can be assumed to be identical; (2) under given rotational speed condition, the airflow angle at the impeller outlet does not change with the mass flow rate. Based on the two assumptions, it can be concluded that under given rotational speed condition, the actual specific enthalpy change varies linearly with the mass flow rate as can be observed in Figure 6. Consequently, the Euler equation can be re-written as:

$$\Delta h_{act} = b(N_{tc}) - a(N_{tc})\dot{m}_c \tag{22}$$

$$a(N_{tc}) = 0 + a_1 N_{tc} + a_2 N_{tc}^2 \tag{23}$$

$$b(N_{tc}) = 0 + b_1 N_{tc} + b_2 N_{tc}^2 \tag{24}$$

where a and b are the slope and intercept, respectively.

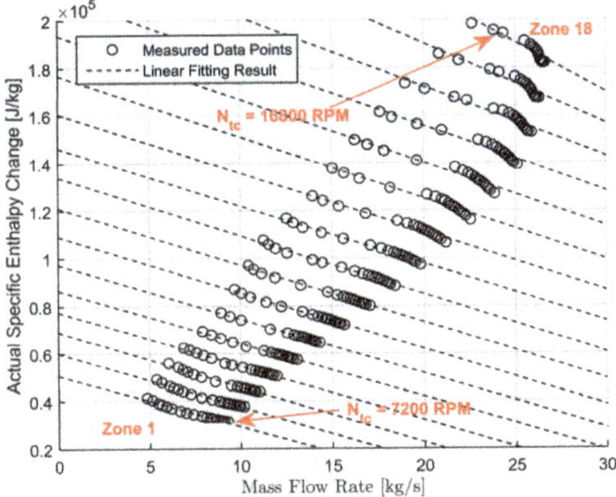

Figure 6. Relationship between actual specific enthalpy change and mass flow rate at each rotational speed condition (interval between each iso-speed curve is 600 RPM).

By applying Equations (22)–(24) with the model parameters calibrated by using Least Square Method (LSM), the actual specific enthalpy change can be calculated under the given rotational speed and mass flow rate condition. On the other hand, the specific enthalpy change under the ideal isentropic process can be derived by Equation (25). Subsequently, the compressor isentropic efficiency can be calculated according to its definition ($\eta_c = \Delta h_{is} / \Delta h_{act}$).

$$\Delta h_{is} = c_{p,a} T_{c,in} \left(\Pi_c^{\frac{\gamma_a - 1}{\gamma_a}} - 1 \right) \tag{25}$$

As can be observed from Figure 6, under high rotational speed conditions, distinguishable non-linear decreasing trend in the actual specific enthalpy change occurs as the mass flow rate gradually approaches to the choking point. As a result, the model's predictive accuracy will be polluted if the model parameters are calibrated by using all the available measured data points in the compressor map. To solve this problem, a zonal modeling approach based on the Hadef model is proposed, which is referred to as "zonal isentropic efficiency model" in this paper. The basic idea is to divide the whole $\Delta h_{act} - \dot{m}_c$ plane as shown in Figure 6 into several zones depending on the iso-speed curves available in the performance map, and then each zone is calibrated individually by using the measured data points at the current zone's upper and lower iso-speed curve.

To compare the predictive accuracy of the Hadef model and the zonal isentropic efficiency model proposed in this paper, five error evaluation criteria, including R_c^2, $RD_{\max}$, $MAPE$, $PEB_{\pm5\%}$ and $PEB_{\pm10\%}$, are adopted, the definitions of which can be found in Shen et al. [15].

Figures 7 and 8 presents the predictive results and the relative error distribution for the two isentropic efficiency models, respectively, whereas Table 2 presents respective error evaluation criteria results. Note that for depicting the predictive results clearly, only five iso-speed curves (7200, 9600, 12000, 14400, 16800 rpm) are presented in Figures 7a and 8a. As can be observed from Figures 7a and 8a, both models are capable of capturing the changing trend of the isentropic efficiency with mass flow rate. Nevertheless, it can be found from Figures 7b and 8b as well as Table 2 that the predictive accuracy of the zonal isentropic efficiency model is better than the Hadef model with lower $RD_{\max}$ and $MAPE$ and higher R_c^2; in addition, the relative errors of all the predictive results are within ±5%. The better predictive accuracy of the zonal isentropic efficiency model mainly benefits from the zonal modeling approach. With this approach, the working characteristic of the compressor within different speed range can be captured much more accurately.

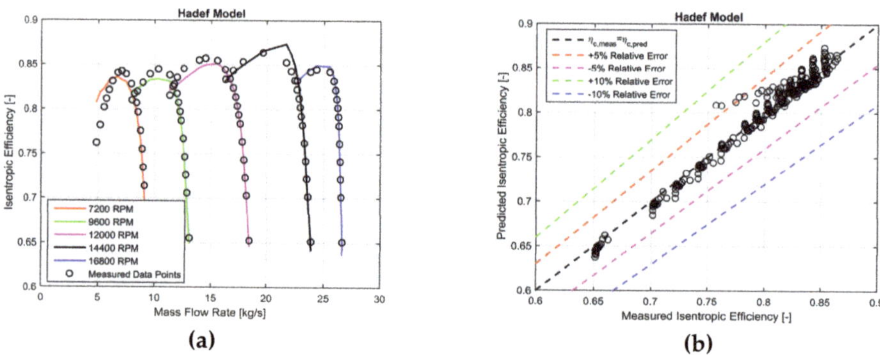

Figure 7. Predictive result and error distribution for Hadef isentropic efficiency model: (**a**) predictive result; (**b**) error distribution.

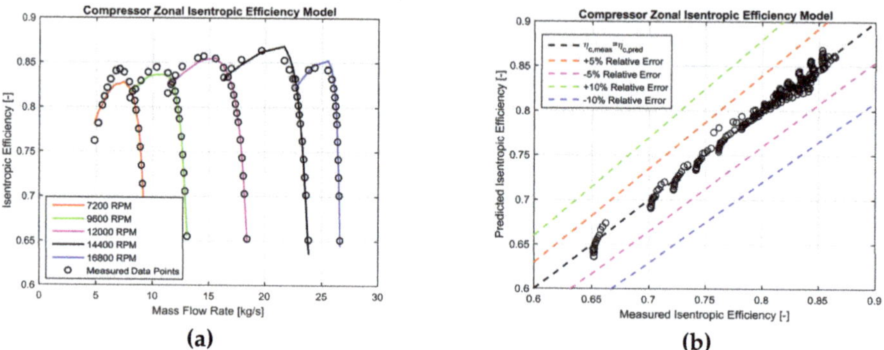

Figure 8. Prediction result and error distribution for compressor zonal isentropic efficiency model: (**a**) predictive result; (**b**) error distribution.

Table 2. Error evaluation criteria result for Hadef model and zonal compressor isentropic efficiency model.

Model	R_c^2	$RD_{\mathbf{max}}(\%)$	$MAPE(\%)$	$PEB_{\pm 5\%}(\%)$	$PEB_{\pm 10\%}(\%)$
Hadef model	0.9675	6.7967	0.9038	98.9619	100
Zonal isentropic efficiency model	0.9858	2.5535	0.6428	100	100

As similar with the compressor mass flow rate model, the isentropic efficiency model is also required to extrapolate to the off-design operating area robustly and accurately. When extrapolating to the LS and HS area, the model parameters belonging to the first and last zone is adopted, respectively. As the definition of isentropic efficiency is directly adopted, the isentropic efficiency value will necessarily be zero when the pressure ratio is equal to 1, which guarantees the model's LPR area extrapolative accuracy to a certain extent.

For the compressor isentropic efficiency model, besides the pressure ratio and rotational speed, the mass flow rate is also required as the input variable. By incorporating the compressor mass flow rate model developed in Section 2.7.2, which is capable of extrapolating to the off-design operating area robustly and accurately, the extrapolative ability of the zonal isentropic efficiency model can be investigated. Figure 9 presents the corresponding isentropic efficiency extrapolative results. As can be observed from this figure, when extrapolating to the LPR area, each iso-speed curve is capable of achieving a smooth transition to the operating point with pressure ratio and isentropic efficiency equal to 1 and 0, respectively; when extrapolating to the LS and HS area, the changing trend of the extrapolated iso-speed curve is similar with the other iso-speed curves available in the compressor performance map, respectively, which verifies the rationality of the extrapolative strategy adopted in this paper to a certain extent.

To further verify the extrapolative ability of the zonal isentropic efficiency model in the LS and HS area, the first and last iso-speed curves available in the performance map are removed firstly, and then the model is calibrated with the remaining measured data points and the removed iso-speed curves are extrapolated by using the model, finally the removed iso-speed curves are compared with the extrapolated results to evaluate the model's extrapolative ability. Figure 10 shows the extrapolative results in the LS and HS area, respectively. As can be observed from Figure 10b, satisfactory HS area extrapolative results are achieved with an *MAPE* of only 0.8422%; on the other hand, although it is slightly inferior with respective to that in the HS area, the model's LS area extrapolative accuracy is still satisfactory with an *MAPE* of 2.0318%.

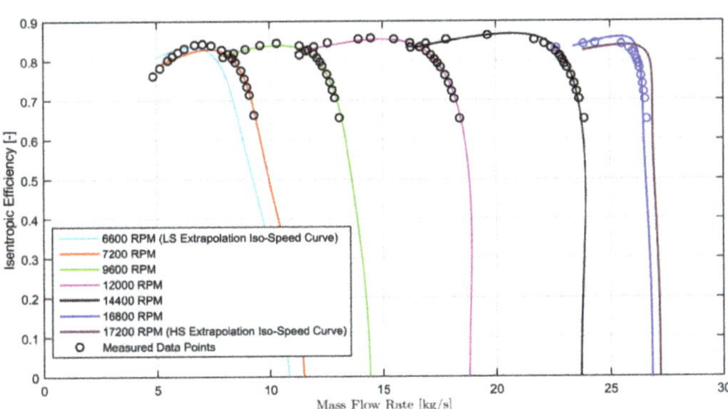

Figure 9. Prediction and extrapolation result of the zonal compressor isentropic efficiency model.

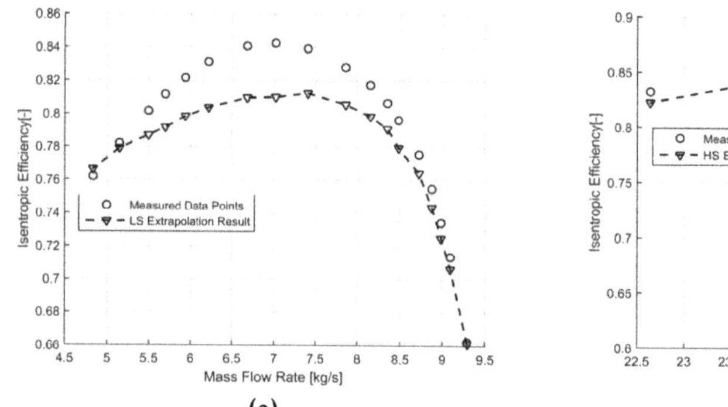

Figure 10. Extrapolation result of the zonal compressor isentropic efficiency model: (**a**) LS (Low Speed) area extrapolation result; (**b**) HS (High Speed) area extrapolation result.

3. Model Calibration and Results

3.1. Model Calibration

As can be found from Section 2, there are several model parameters to be calibrated. Engine geometrical parameters are extracted from the engine project guide. For the model parameters available in the turbocharger model, they are calibrated by using the performance map. Model parameters in several engine component sub-models, such as air cooler cooling efficiency and pressure drop, are estimated by using the engine shop trial report, which provides the measured engine performance parameters at several steady loading conditions. The last four model parameters remained to be determined include k_{CA0}, k_{CA1}, $k_{\zeta 0}$ and $k_{\zeta 1}$, which have significant influences on the MVEM's predictive accuracy. For estimating the four model parameters based on the measured data provided in engine shop trial report, the Parameter Estimation toolbox provided by MATLAB/Simulink is used, which converts the model parameter estimation problem to numerical optimization problem. In general, the model parameter calibration procedure is carried out in three steps as following:

(1) Initialization of k_{CA0} and k_{CA1}. In this step, it is assumed that $\dot{m}_{sz}$ is equal to the compressor mass flow rate $\dot{m}_c$, which, in turn, can be estimated by using the compressor model. Based on this assumption, the initial value of $C_z A_z$ at each engine loading condition can be estimated by using Equation (1), and then the initial value of k_{CA0} and k_{CA1} can be estimated;

(2) Initialization of $k_{\zeta 0}$ and $k_{\zeta 1}$. For estimating the initial value of ζ at each engine loading condition, the Parameter Estimation toolbox is adopted for the whole engine model. Non-linear least square method is selected as the optimization method. The initial value of $C_z A_z$ obtained in step 1 is regarded as known quantity in this step. The input variable required only includes the engine speed, whereas the output variables include pressure and temperature in the scavenging and exhaust manifolds as well as the turbocharger rotational speed. The sum of squares of the errors between the predicted and measured value of the selected output variables is used as the cost function. As a result, the initial value of ζ at each engine loading condition can be estimated. Consequently, $k_{\zeta 0}$ and $k_{\zeta 1}$ can be initialized;

(3) Final calibration of k_{CA0}, k_{CA1}, $k_{\zeta 0}$ and $k_{\zeta 1}$. The four model parameters are calibrated simultaneously in this step, the values of which obtained in step 2 are used as the initial values. Note that the calibration process in this step is carried out by using the measured data at all the engine loading conditions simultaneously with the cost function as shown in Equation (26).

$$V(\varphi) = \frac{1}{NS} \sum_{i=1}^{S} \sum_{n=1}^{N} \left(e^i[n] \right)^2 \tag{26}$$

where N is the number of measured loading points available in the engine shop trial report; S is number of selected output variables; φ represents the parameters to be estimated.

3.2. Engine Steady State Performance

In order to validate the MVEM developed in this paper, the engine steady state operation was simulated at 15%, 25%, 50%, 75%, 80% and 100% of the engine MCR point. The predicted engine performance parameters were compared to the respective measured values provided in the engine shop trial report, whereas their relative errors are presented in Table 3. To assess the prediction accuracy of the MVEM developed in this paper, the simulation results shown in the paper published by Tang et al. from the same research group is also presented in Table 3 for comparison [5]. The same engine was adopted as the simulation object in Tang et al. but the engine model was developed by using the 0-D modeling approach [5].

Table 3. Relative error of the engine performance parameters.

Engine Load (%)	15	25	50	75	80	100
	Error (%)					
Brake Power	0.74	−0.94	0.63	0.55	0.35	0.04
Brake Power [5]	null	1.61	2.21	0.38	0.46	0.56
BSFC	−0.18	0.58	−0.47	−0.50	−0.06	−0.74
BSFC [5]	null	3.02	1.70	0.38	0.46	0.56
Scavenging manifold pressure	1.43	1.80	3.45	1.94	2.14	1.16
Scavenging manifold pressure [5]	null	−3.23	−1.27	0	0.28	0
Exhaust manifold pressure	−3.20	−5.21	−3.19	−2.63	−1.26	−0.89
Exhaust manifold temperature	−0.02	−1.70	−2.38	−2.27	−1.92	−1.87
Exhaust manifold temperature [5]	null	0.73	1.75	1.31	−1	−0.22
Turbocharger speed	−0.043	−0.55	1.69	0.14	0.74	0.49
Turbocharger speed [5]	null	−2.43	−0.98	−1.14	−2.08	−1.34
Compressor outlet temperature	1.21	−0.42	2.68	0.23	0.60	−0.22
Turbine outlet temperature	3.59	2.85	3.42	2.31	2.23	3.09

As can be observed from Table 3, satisfactory predictive accuracy was obtained at each investigated engine loading condition with all the relative errors of brake power and BSFC (Brake Specific Fuel Consumption) less than 1%, whereas most of the relative errors of the other engine performance parameters are less than 4%. The measured turbocharger rotational speed is 4981 rpm at 15% engine load, which is lower than the lowest speed available in the compressor performance map. On the other hand, the predicted turbocharger rotational speed is 4979 rpm at this load condition with an extreme low relative error of only −0.043%, indicating the fidelity of the novel compressor model when extrapolating to the LS area. In addition, except for the 50% engine load, the relative errors of the compressor outlet temperature are all around 1%, indicating the satisfactory predictive accuracy of the zonal compressor isentropic efficiency model.

In the research carried out by Tang et al., only the relative errors of brake power, BSFC, scavenging manifold pressure, exhaust manifold temperature and turbocharger speed are presented; in addition, the relative errors at 15% engine load are not provided, perhaps it is because unsatisfactory simulation accuracy was achieved at this load condition [5]. For brake power, BSFC and turbocharger rotational speed, better simulation accuracy is generally obtained with the MVEM. The compressor model developed in this paper contributes to the better simulation accuracy of the MVEM in turbocharger rotational speed with respect to the 0-D model developed by Tang et al. where the compressor is modeled by using look-up table method [5]. Although better simulation accuracy is achieved with the 0-D model for scavenging manifold pressure and exhaust manifold temperature, the difference is not significant and they all meet the requirement on simulation accuracy; on the other hand, it should be noted that the MVEM runs much faster than the 0-D model.

As the model calibration procedure was carried out based on the measured engine performance parameters provided by the engine shop trial report and the model validation was also implemented by comparing the predicted results to these measured ones, therefore, in order to further validate the fidelity of the MVEM, the Computerized Engine Application System-Engine Room Dimensioning tool provided in the official website of MAN Diesel & Turbo is used to generate engine performance parameters within the engine load region from 15% to 100%. However, it should be noted that the engine performance parameters generated by using CEAS (Computerized Engine Application System) tool is for 7S80ME-C9.5 engine and the respective LHV (Lower Heating Value) of fuel is 42700 kJ/kg, whereas the investigated engine in this paper is 7S80ME-C9.2 and the LHV is 42151 kJ/kg, which will inevitably cause deviations on the engine performance parameters between the engine shop trial report and the CEAS. Despite these limitations, the results generated by CEAS is very valuable for reference in investigating the qualitative changing trend of engine performance parameters with engine load condition.

Figure 11 presents the engine performance parameters predicted by the MVEM and calculated by the CEAS within the load region from 15% to 100%. In addition, the measured values obtained from the engine shop trial report are also plotted in Figure 11 for reference. As can be observed from Figure 11, despite the existing of deviations, the simulation results for brake power, SFOC, scavenging manifold pressure and temperature and the turbine outlet temperature all agree well with the results obtained from the CEAS qualitatively. In addition, the other engine performance parameters also vary reasonably with the engine load.

At 65% engine load, the BSFC and turbine outlet temperature predicted by the MVEM reaches the minimum, indicating that the engine and turbocharger achieves the optimal operating state, which is consistent with the fitted BSFC curve based on the measured data as shown in Figure 11b. It should be noted that this optimal load is obviously lower than that of marine two-stroke diesel engines designed in earlier years. One reason of this relatively low optimal engine load is that low steaming is taken into account during the ship design phase in recent years. Noticeable discontinuity is observed at 35% engine load and this is because that below this load, the auxiliary blower is activated, which results in a larger amount of scavenging air supplied to the cylinder and thus lowers the fuel-air equivalence ratio as shown in Figure 11j. Consequently, with respect to 35% engine load, the fuel-air equivalence ratio at 30% load decreases from 0.3182 to 0.2872, the exhaust gas manifold temperature decreases from 573.3 K to 542.2 K and the turbine outlet temperature decreases form 515 K to 492.5 K; in addition, the scavenging manifold temperature increases from 299.2 K to 304.8 K mainly owing to the compression effect by the auxiliary blower. When the engine load decreases from 25% to 20%, the exhaust outlet temperature increases slightly from 486.9 K to 488.4 K, whereas it decreases from 488.4 K to 486 K when the engine load decreases from 20% to 15%. The opposite phenomenon in the two engine load interval (25%–20% and 20%–15%) is caused probably by the different changing rate of fuel injection rate and air mass flow rate with engine load, which is represented by the changing trend of fuel-air equivalence ratio as shown in Figure 11j.

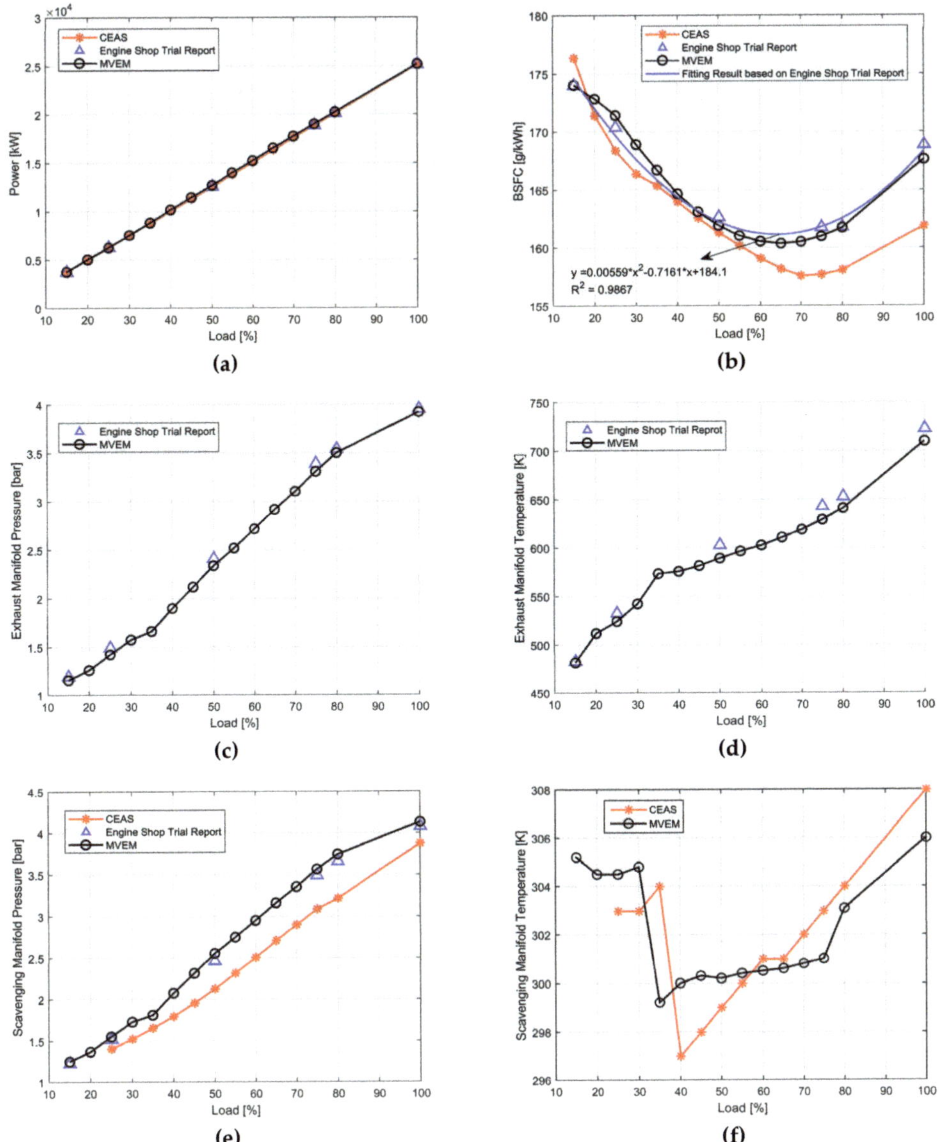

Figure 11. *Cont.*

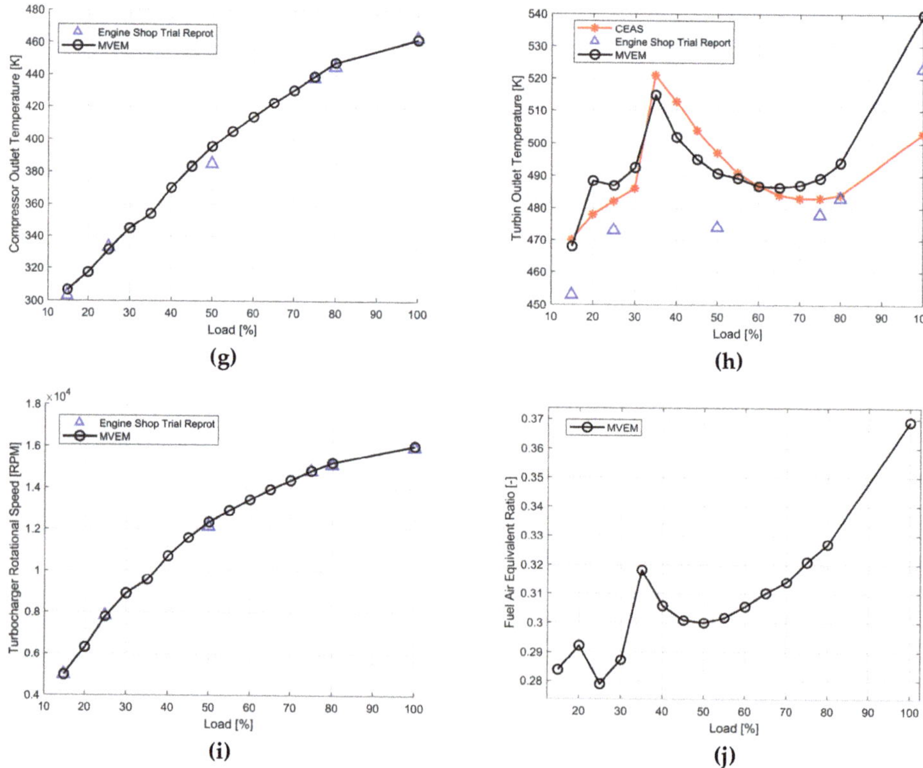

Figure 11. Steady state simulation results and comparison with engine shop trial report and CEAS (Computerized Engine Application System): (**a**) brake power; (**b**) brake specific fuel consumption (BSFC); (**c**) exhaust manifold pressure; (**d**) exhaust manifold temperature; (**e**) scavenging manifold pressure; (**f**) scavenging manifold temperature; (**g**) compressor outlet temperature; (**h**) turbine outlet temperature; (**i**) turbocharger rotational speed; (**j**) fuel-air equivalence ratio.

3.3. Engine Transient Response

For the purpose of investigating the transient response of the MVEM as well as the compressor model developed in this paper, two simulation experiments are carried out in this paper. In the first experiment, the engine setting speed steps down from 72 rpm to 66.8 rpm at 500 s, from 66.8 rpm to 60.7 rpm at 1000 s, from 60.7 rpm to 57.1 rpm at 1500 s, from 57.1 rpm to 50.7 rpm at 2000 s, from 57.1 rpm to 45.4 rpm at 3000 s, from 45.4 rpm to 38.3 rpm at 4000 s. These engine speeds correspond to 100%, 80%, 60%, 50%, 35%, 25% and 15% engine load, respectively, thus covering the whole engine operating envelope. As shown in Figure 12, the engine rotational speed, brake power and fuel index are able to stabilize in a short time, whereas it takes more time for the other engine performance parameters until stabilization. This phenomenon is caused mainly by the turbocharger inertia as well as the relatively large volume of the scavenging and exhaust manifolds. As shown in Figure 12b, with the decrease in engine setting speed, the turbocharger rotational speed gradually decreases. Starting from 4050 s, the turbocharger rotational speed becomes lower than 7200 rpm, which is the lowest speed available in the compressor performance map, and then enters into the LS off-design operating area, finally the turbocharger rotational speed stabilizes at 4990 rpm. As can be inferred from the trajectory of the compressor operating points as shown in Figure 12j, the compressor model developed in this paper is capable of interpolating within the design operating area and extrapolating to the LS off-design operating area reasonably and robustly.

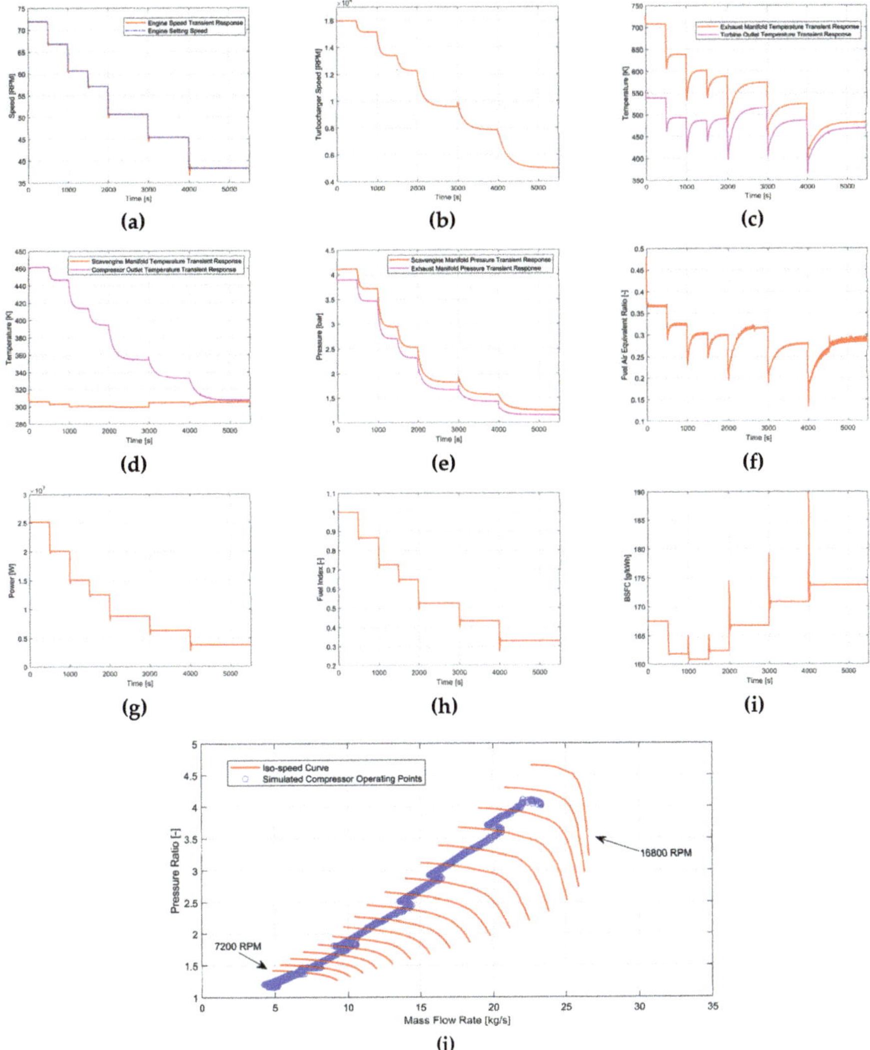

Figure 12. Simulation results for the engine transient operation with setting speed changes: (**a**) engine rotational speed; (**b**) turbocharger rotational speed; (**c**) exhaust manifold and turbine outlet temperature; (**d**) scavenging manifold and compressor outlet temperature; (**e**) scavenging and exhaust manifold pressure; (**f**) fuel-air equivalence ratio; (**g**) brake power; (**h**) fuel index; (**i**) BSFC; (**j**) compressor operating points trajectory.

Figure 13 presents the engine transient response with a setting speed of 72 RPM when the resistant torque steps up at the 100th s. As can be observed from this figure, the engine rotational speed drops down quickly when the resistant torque increases. Then, under the control of engine governor, more fuel will be injected into the cylinders to balance the increased resistant torque for stabilizing the engine speed. Consequently, more thermal energy will be stored in the exhaust gas, which drives the turbocharger to work with higher rotational speed. Finally, the scavenging manifold pressure increases and more air flows into the engine cylinders. In addition, as can be inferred from the trajectory of the compressor operating points as shown in Figure 13j, the compressor model is able to extrapolate to the HS off-design operating area reasonably and robustly.

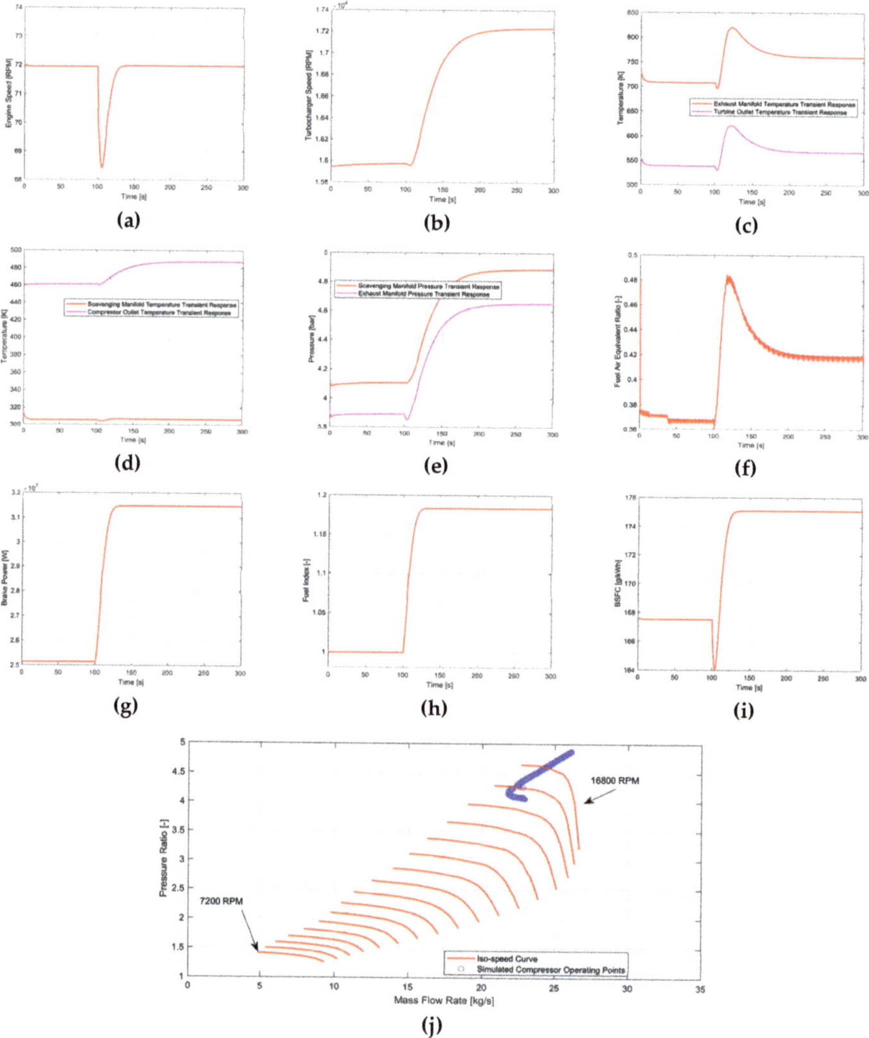

Figure 13. Simulation results for the engine transient operation with resistant torque changes: (**a**) engine rotational speed; (**b**) turbocharger rotational speed; (**c**) exhaust manifold and turbine outlet temperature; (**d**) scavenging manifold and compressor outlet temperature; (**e**) scavenging and exhaust manifold pressure; (**f**) fuel-air equivalence ratio; (**g**) brake power; (**h**) fuel index; (**i**) BSFC; (**j**) compressor operating points trajectory.

4. Coupling of MVEM with Cylinder Pressure Analytic Model

As discussed in the Introduction section, MVEM is unable to predict the in-cylinder pressure trace, which weakens MVEM's practical value to some extent. To solve this problem, the cylinder pressure analytic model proposed by Eriksson and Andersson for the four-stroke spark ignition (SI) engine was adapted to the 7S80ME-C9.2 marine two-stroke diesel engine and coupled to the MVEM developed in this paper [10]. The major merit of this analytic model is that the calculation of in-cylinder pressure is completely based on algebraic equations without needing to solve any differential equation as it is done in the 0-D model, thus greatly accelerating the model's simulating speed in predicting in-cylinder pressure trace.

In this section, the cylinder pressure analytic model is firstly adapted to the marine two-stroke diesel engine with its basic idea shown in Figure 14; then, the model parameter calibration procedure is discussed in detail; finally, the model is coupled to the MVEM and its in-cylinder pressure trace predictive ability is evaluated by comparing with the measured indicator diagram.

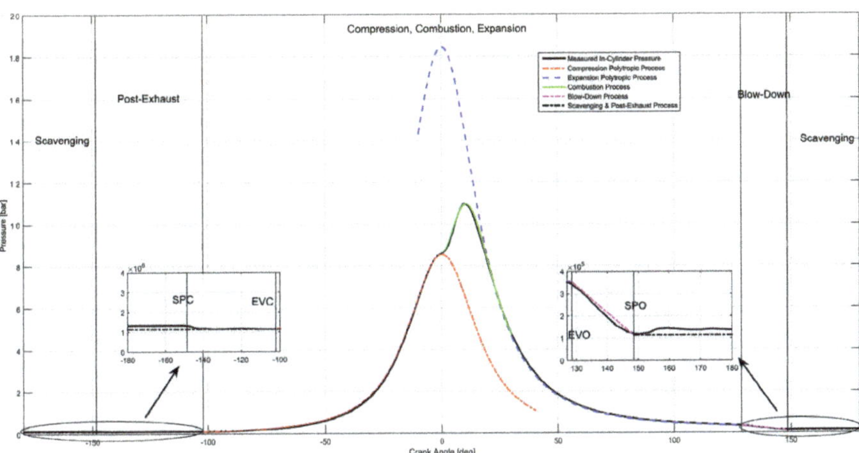

Figure 14. Basic idea of the cylinder pressure analytic model (SPC: Scavenging ports closing; SPO: Scavenging ports opening; EVC: Exhaust valve closing; EVO: Exhaust valve opening).

4.1. Cylinder Pressure Analytic Model

(1) Compression Phase (EVC to SOC)

The actual compression process can be approximated by a polytropic process with a polytropic index of n_{comp}. Based on this assumption, the instantaneous in-cylinder pressure p_{comp} and temperature T_{comp} during this phase can be calculated as:

$$p_{comp}(\theta) = p_{evc}\left(\frac{V_{evc}}{V(\theta)}\right)^{n_{comp}} \tag{27}$$

$$T_{comp}(\theta) = T_{evc}\left(\frac{V_{evc}}{V(\theta)}\right)^{n_{comp}-1} \tag{28}$$

where p_{evc}, V_{evc} and T_{evc} are the pressure, volume and temperature at EVC, respectively, and the EVC is treated as the reference point of the compression polytropic process.

As there is no mass exchange between the cylinder and surrounding during this phase, the in-cylinder trapped mass m_{trap} can be calculated by using ideal gas state equation.

(2) Expansion Phase (EOC to EVO)

The expansion phase is also treated as a polytropic process. The calculation method of instantaneous in-cylinder pressure p_{exp} and temperature T_{exp} during this phase is also as similar with the compression phase. The main difference lies in the determination of polytropic index and the pressure and temperature of the reference point.

The temperature at the reference point of expansion phase can be estimated by adding an additional temperature increment based on $T_{comp}(0°)$:

$$\Delta T = \frac{m_{f,cycle}H_{LHV}\eta_{comb}}{c_v(T_{comp}(0°),\phi_{trap})(m_{trap}+m_{f,cycle})} \tag{29}$$

$$T_{exp,ref} = T_{comp}(0°) + \Delta T \tag{30}$$

The theoretical foundation of Equations (29) and (30) is the constant volume heating process in the ideal Otto cycle. Once $T_{exp,ref}$ is determined, $p_{exp,ref}$ can be calculated by using ideal gas state equation.

(3) Combustion Phase

During the combustion phase, the in-cylinder instantaneous pressure p_{comb} can be calculated by interpolating between p_{comb} and p_{exp} with the Wiebe function f_{wiebe} selected as the interpolating function:

$$p_{comb} = (1 - f_{wiebe}) \cdot p_{comp} + f_{wiebe} \cdot p_{exp} \tag{31}$$

(4) Blow-down Phase (EVO to SPO)

As can be observed from Figure 14, the instantaneous in-cylinder pressure p_{exh} during this phase presents linear variation trend roughly, which, thus, can be approximated by a straight line crossing the points of (θ_{evo}, p_{evo}) and (θ_{spo}, p_{spo}) as the following equation:

$$p_{exh} = p_{evo} + \frac{p_{spo} - p_{evo}}{\theta_{spo} - \theta_{evo}}(\theta - \theta_{evo}) \tag{32}$$

(5) Scavenging and Post Exhaust Phase (SPO to EVC)

As can be observed from Figure 14, the instantaneous in-cylinder p_{scav_pe} during this phase does not fluctuate significantly, which is always between the scavenging and exhaust manifold pressure. Consequently, p_{scav_pe} can be approximated by a straight line with a constant pressure value.

4.2. Model Parameters Calibration

There are several model parameters in the cylinder pressure analytic model to be calibrated as following:

1. The compression and expansion polytropic index;
2. The pressure and temperature at the reference point of the compression polytropic process;
3. Wiebe function model parameters.

4.2.1. Calibration of Poyltropic Index

The polytropic process can be expressed as $pV^n = const$ and it can be transformed to $\ln p + n \ln V = const$ by taking the logarithm of each side. Consequently, by setting $\ln V$ and $\ln p$ as the horizontal and vertical coordinate, respectively, a straight line can be obtained with the negative of its slope equal to the polytropic index [17]. Figure 15 presents the estimation result of n_{comp} and n_{exp} for a certain operating condition. As can be observed from this figure, strong linear relationship exists between $\ln V$ and $\ln p$ with R^2 larger than 0.999 for both the compression and expansion phase, which verifies the rationality of the assumption that the actual compression and expansion process can be approximately by the polytropic process.

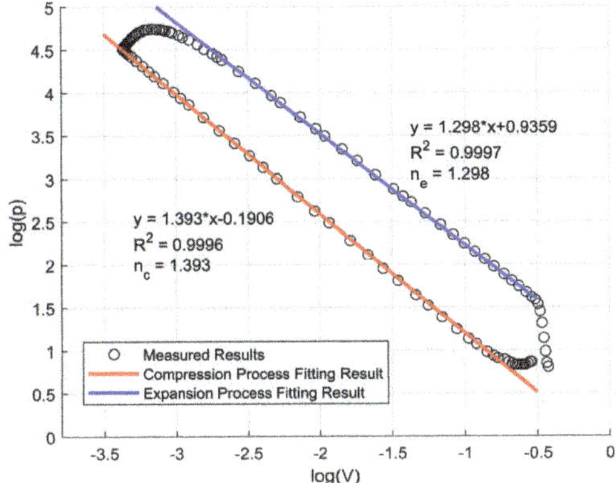

Figure 15. Relationship between log(V) and log(p) and linear fitting result for the compression and expansion process.

Figure 16 provides the compression and expansion polytropic index estimation results for 29 known engine operating points. These operating points are measured on-board a ship and cover a wide range of operating conditions, which are named as A1 to A29 in this paper. For n_{comp}, its maximum and minimum value is 1.416 and 1.391, respectively, with a standard deviation of 0.0058; for n_{exp}, its maximum and minimum value is 1.321 and 1.280, respectively, with a standard deviation of 0.0076. The results indicate that the polytropic index will not change significantly with the engine operating conditions. Actually, it was revealed by Brunt and Platts that the polytropic index was influenced by many factors including engine speed, working medium temperature, heat transfer, etc. [18]. However, due to lacking of experimental conditions especially for the marine large scale diesel engine investigated in this paper, it is difficult to derive correlations between the polytropic index and other engine performance parameters. Therefore, in this paper a 2-D look-up table is established to calculate the polytropic index with the engine speed and brake power as the input variables.

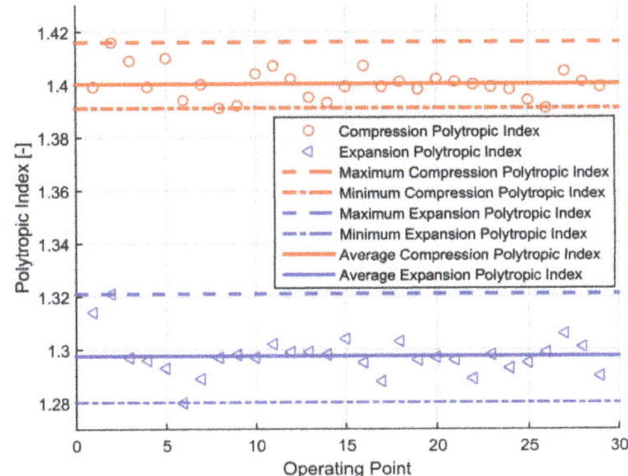

Figure 16. Compression and expansion polytropic index results for the 29 known operating conditions.

4.2.2. Calibration of p_{evc} and T_{evc}

In Figure 17, the measured value of p_s and p_{evc} for the 29 known engine operating points are presented and compared. It can be observed from this figure that p_s is always greater than p_{evc}. In order to evaluate the quantitative relationship between p_s and p_{evc}, they are set as the horizontal and vertical coordinate, respectively, as shown in Figure 18. It can be found from this figure that strong linear relationship exists between them with a R^2 greater than 0.99. Consequently, p_{evc} can be expressed as a linear function of p_{sm}, that is $p_{evc} = 0.9749p_s - 0.1029$ in this paper.

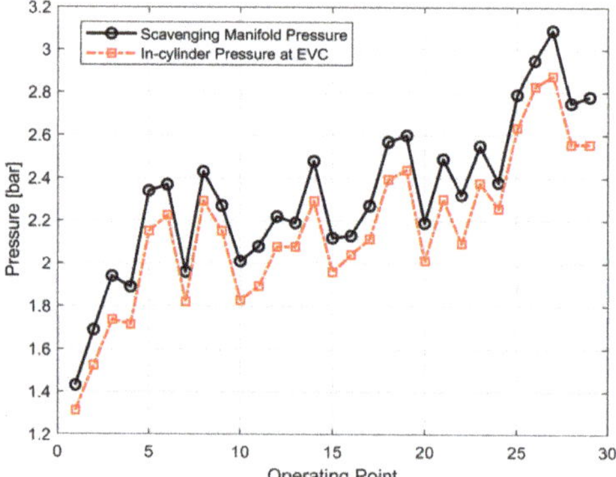

Figure 17. Comparison between scavenging manifold pressure and in-cylinder pressure at exhaust valve closing (EVC) based on measured data.

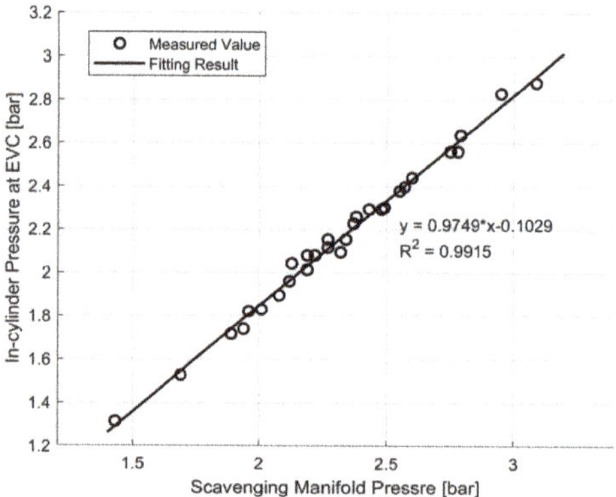

Figure 18. Fitting result between scavenging manifold pressure and in-cylinder pressure at EVC based on measured data.

To guarantee the continuity of the simulated in-cylinder pressure trace during the scavenging and post-exhaust phase and also to reduce the model complexity, it is assumed that the in-cylinder pressure during the scavenging and post-exhaust phase is equal to p_{evc}. Although it can be observed from Figure 14 that this assumption will make the predicted in-cylinder pressure during the two phases slightly lower than the measured pressure, this will not significantly influence the model's predictive accuracy because the pressure during the two phases is only about 0.5% of the in-cylinder maximum pressure. Based on this assumption, the slope and intercept of Equation (32) can be determined.

The research carried out by Tang et al [5] and Kharroubi and Söğüt [19] all indicated that relationship exists between the scavenging air temperature T_s and T_{evc}. As T_{evc} is not measured both during the engine shop trial and on-board the ship, a 0-D engine model is adopted to investigate the relationship between T_s and T_{evc} in this paper. In Figure 19, the simulated results of T_s and T_{evc} are set as the horizontal and vertical coordinate. It can be found from this figure that a second-order polynomial is sufficient to capture the changing trend of T_{evc} with T_s with a R^2 of 0.9958, indicating satisfactory fitting results.

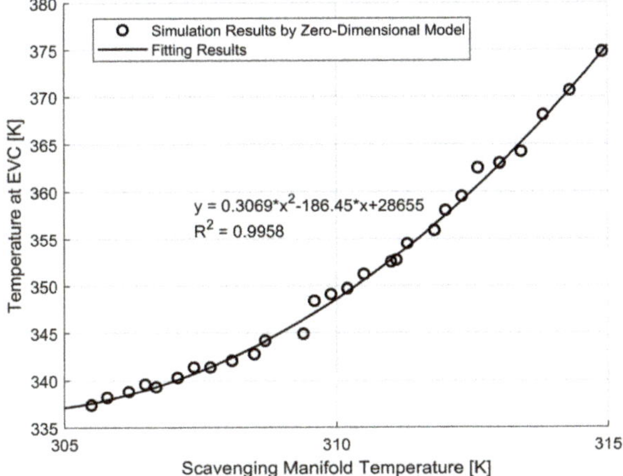

Figure 19. Fitting result between scavenging manifold temperature and in-cylinder temperature at EVC based on the 0-D model.

4.2.3. Calibration of Wiebe Function Model Parameters

In order to predict the in-cylinder pressure trace during the combustion phase, the Wiebe function is used to interpolate between the compression and expansion pressure. The general form of the single Wiebe function can be written as:

$$\chi = 1 - \exp[-a(\frac{\theta - \theta_{soc}}{\Delta\theta_{CD}})^{m+1}] \tag{33}$$

where χ is the mass fraction burned (MFB); φ_{soc} is the start of combustion crank angle; $\Delta\varphi_{CD}$ is the combustion duration; a is the efficiency factor; m is the form factor.

Step 1: Derive the MFB based on the measured in-cylinder pressure trace

Several classical MFB estimation methods can be found in the literature, such as the Apparent Heat Release model, the Gatowski model, the RW model, as well as the pressure ratio management (PRM) model [20]. In this paper, the PRM model is adopted to derive the MFB:

$$PR(\theta) = \frac{p_z(\theta)}{p_m(\theta)} - 1 \tag{34}$$

$$\chi(\theta) = \frac{PR(\theta)}{PR_{\max}(\theta)} \tag{35}$$

where PR is referred to as the pressure ratio; p_z is the measured in-cylinder pressure; p_m is the motored in-cylinder pressure; $PR_{\max}$ is the maximum value of PR, which is used to normalize PR to obtain the MFB.

In this paper, the motored in-cylinder pressure p_m is estimated by assuming the motored process as a polytropic process as it was done by many researchers [21–23]. In addition, it is assumed that the expansion polytropic index is equal to that of the compression process, the value of which can be estimated by following the calibration procedure as described in Section 4.2.1. Consequently, p_m can be determined by using the following equation:

$$p_m(\theta) = p_{ref}(\frac{V_{ref}}{V(\theta)})^{n_m} \tag{36}$$

where p_{ref} and V_{ref} are the pressure and volume at the reference point, which can be an arbitrary crank angle position between the EVC and SOC.

Step 2: Determine the number of single Wiebe function needed to be superposed

Ghojel pointed out that the form factor m in Wiebe function can reflect the distribution of combustion process, thus, the variation of m can be exploited to determine the number of single Wiebe function to be superposed [24]. Following this idea, the original single Wiebe function as shown in Equation (33) can be transformed to the following form by appropriate mathematical manipulation:

$$\ln[\ln(1-\chi)/\ln 0.5] = (m+1)\ln(\theta-\theta_{soc}) - (m+1)\ln(\theta_{50}-\theta_{soc}) \tag{37}$$

where θ_{50} is the crank angle corresponding to 50% MFB.

Set the $\ln(\theta-\theta_{soc})$ and $\ln[\ln(1-\chi)/\ln 0.5]$ as the horizontal and vertical coordinate, respectively. If a straight line is obtained, a single Wiebe function is sufficient to simulate the combustion process accurately, otherwise, two or more single Wiebe functions are needed, the number of which can be estimated by the number of broken lines in the plane of $\ln(\theta-\theta_{soc})\sim\ln[\ln(1-\chi)/\ln 0.5]$. For the 7S80ME-C9.2 marine engine, it is found that superposing two single Wiebe functions is sufficient in the whole engine operating envelope.

Step 3: Estimate the Wiebe function model parameters by using LSM

As mentioned in Step 2, double Wiebe function is sufficient to simulate the actual combustion process, which can be written as:

$$\chi = (1-\beta)\left\{1-\exp[-a_p(\frac{\theta-\theta_{soc,p}}{\Delta\theta_{CD,p}})^{m_p+1}]\right\} + \beta\left\{1-\exp[-a_d(\frac{\theta-\theta_{soc,d}}{\Delta\theta_{CD,d}})^{m_d+1}]\right\} \tag{38}$$

where subscript p and d represents premixed and diffusive combustion, respectively; β is the fraction of fuel burned during the diffusive combustion phase.

There are totally 9 model parameters in the double Wiebe function as shown in equation (38), making the model calibration process relatively laborious. Therefore, to reduce the number of Wiebe function model parameters to be calibrated, two assumptions are made herein:

(1) The premixed combustion duration is assumed to be equal to that of diffusive combustion. In addition, the crank angle range between 1% and 99% MFB is defined as the combustion duration by taking into account that the Signal to Noise Ratio (SNR) is relatively lower and the combustion is unstable at the start and end of the combustion;

(2) The crank angle at the start of premixed combustion is also assumed to be equal to that of diffusive combustion.

Based on the two assumptions, the number of Wiebe function model parameters to be calibrated reduces from 9 to 7; in addition, $\Delta\theta_{CD}$ and θ_{SOC} can be estimated from the MFB curve beforehand. As a result, only 5 Wiebe function model parameters remain to be estimated.

Table 4 presents the calibration results of the Wiebe function model parameters for four operating points as well as the relevant predictive errors in MFB. As can be observed from this table, PE is less than 1.07 %, whereas MAE is less than 0.31 %, indicating the sufficient accuracy of double Wiebe function in simulating the actual combustion process in CI engines as well as the satisfactory calibration results.

To obtain an overall accurate Wiebe function, a 2-D look-up table is established for the Wiebe function model parameters with the engine speed and brake power as the inputs. In addition, it is found that the combustion duration can be well approximated by a linear function of the fuel injection rate.

Table 4. Wiebe parameters and predictive error.

Operating Point	A2	A8	A14	A25
a_p	3.415	3.769	3.45	3.319
a_d	23.8	23.72	17.28	12.71
m_p	1.148	1.292	1.578	2.2
m_d	1.006	0.9729	0.9035	0.7247
β	0.6749	0.6481	0.6992	0.8023
PE (%) [1]	1.07	0.81	0.95	0.78
MAE (%) [2]	0.31	0.27	0.19	0.25

[1] *PE*: Peak Error; [2] *MAE*: Mean Absolute Error.

4.3. Results

In this section, the in-cylinder pressure trace simulated by the analytic model under four different operating conditions are compared with the measured ones to validate its correctness as shown in Figure 20. It can be observed from this figure that the simulation results agree well with the measured results and can capture the variation trend of in-cylinder pressure trace with crank angle during each working phase. During the gas exchange phase (blow-down, scavenging and post-exhaust), the relative errors of the simulation results are relatively higher than the other phases, which is mainly caused by the simplifications and assumptions made for these phases, i.e., two simple linear functions are adopted to approximate the in-cylinder pressure trace; however, it should be noted that the absolute errors during the gas exchange phase are less than 0.15 bar, which is still satisfactory with respect to the variation range of in-cylinder pressure during the whole working cycle. During the combustion phase, the relative errors of most of the simulation results are less than 5%, which is mainly benefiting from the effective calibration of Wiebe function model parameters as introduced in Section 4.2.3. It should be noted that by superposing more single Wiebe functions, the simulation accuracy during the combustion phase will improve, but this will make the Wiebe function model parameters calibration process laborious. Actually, the current simulation accuracy is already satisfactory to some extent. At the start of compression phase, the relative errors of part of the simulation results approach 20%, which is maybe attributed to two reasons: (1) the absolute value of in-cylinder pressure is relatively lower and small absolute errors will lead to obvious relative errors; (2) the actual compression process is approximated by a polytropic process. Note that as part of the input variables of the analytic model are from the MVEM, thus the predictive errors of MVEM will transform to the analytic model and lead to predictive errors.

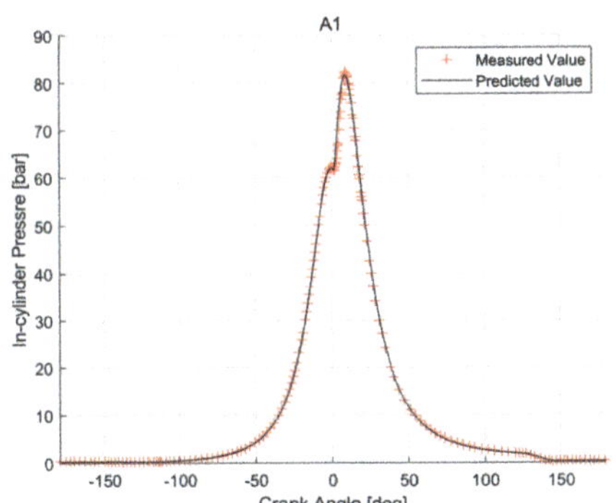

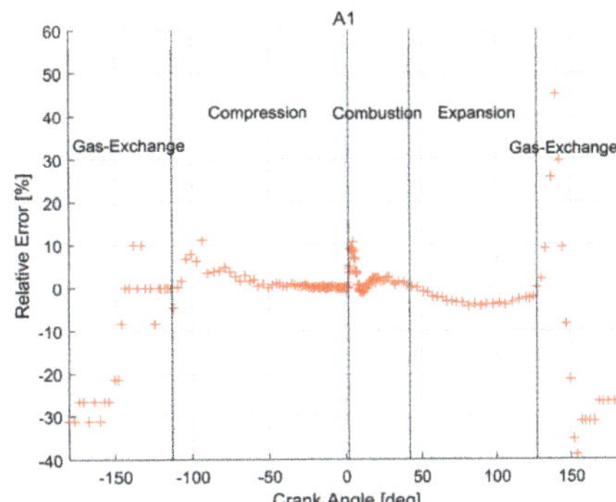

Figure 20. *Cont.*

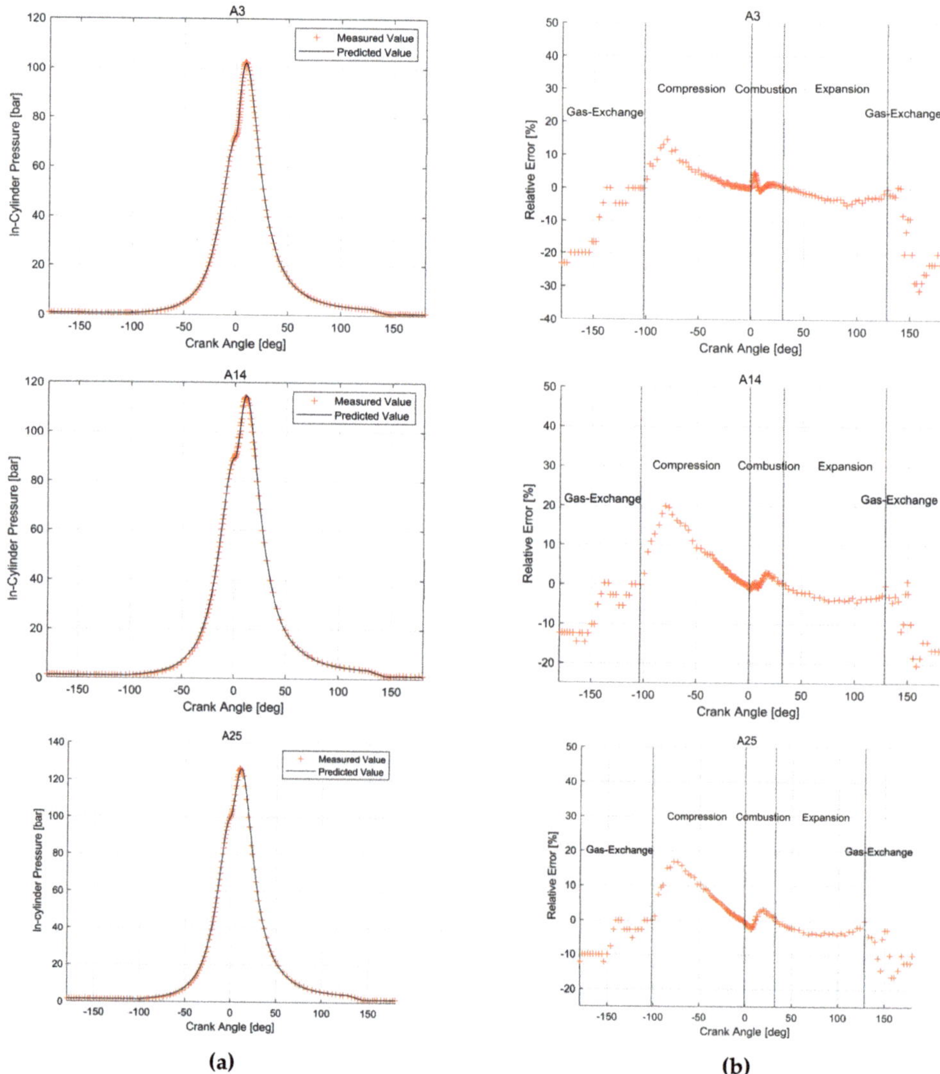

Figure 20. Prediction result and error distribution of the cylinder pressure analytic model: (**a**) predictive results; (**b**) error distribution.

For practical engineering practice, the marine engineers normally judge the engine's operation status by checking the compression pressure p_{comp}, maximum pressure p_{max} and its crank angle position $\theta_{p,max}$. Table 5 compares the predicted and measured results of p_{comp}, p_{max} and $\theta_{p,max}$. As can be observed from this table, the cylinder pressure analytic model is capable of predicting p_{comp} and p_{max} satisfactorily with relative errors less than 1%. Although the relative errors of $\theta_{p,max}$ are relatively higher, the absolute errors are generally less than one crank angle, indicating that the predictive accuracy can be accepted to some extent.

Table 5. Prediction error of the cylinder pressure analytic model.

		p_{comp} (bar)	p_{max} (bar)	$\theta_{p_{max}}$ (deg)
A1	PD^1	62.1	81.6	9.0
	MD^2	62.2	82.3	9.5
	$RD(\%)^3$	−0.16	−0.85	−5.26
A3	PD	73.9	102.2	9.0
	MD	73.8	102.7	8.6
	RD (%)	0.14	−0.49	4.65
A14	PD	89.5	114.5	11.1
	MD	89.7	114.1	10.3
	RD (%)	−0.22	0.35	7.77
A25	PD	100.3	125.8	10.8
	MD	101.3	126.2	9.7
	RD (%)	−0.99	−0.32	11.34

[1] Predicted value; [2] Measured value; [3] Relative error.

The inputs of the analytic model, such as scavenging air pressure and temperature and engine speed, are completely from the MVEM and the calculation process does not influence the MVEM; however, it should be noted that the MVEM and the cylinder pressure analytic model have different simulation speed, which must be handled when they are coupled together and applied in the engine room simulator. Aiming at this problem, two threads are adopted to depart the analytic model from the MVEM as shown in Figure 21. At steady state conditions, the simulation results of the MVEM are the same in every engine cycle, meaning that the engine performance parameters transformed to the analytic model also remain unchanged, therefore, the in-cylinder pressure trace simulated by the analytic model will also be the same in every engine cycle; on the other hand, the in-cylinder pressure trace is not required to be updated in real-time in the engine room simulator, and the "Pressure-Crank Angle" diagram or the "Pressure-Volume" diagram need to be displayed only when the user switches to the corresponding simulation interfaces. Based on the two factors, at steady state conditions, the in-cylinder pressure trace only needs to be calculated for one engine cycle by the analytic model, therefore, the running speed of the whole model can be as fast as the MVEM.

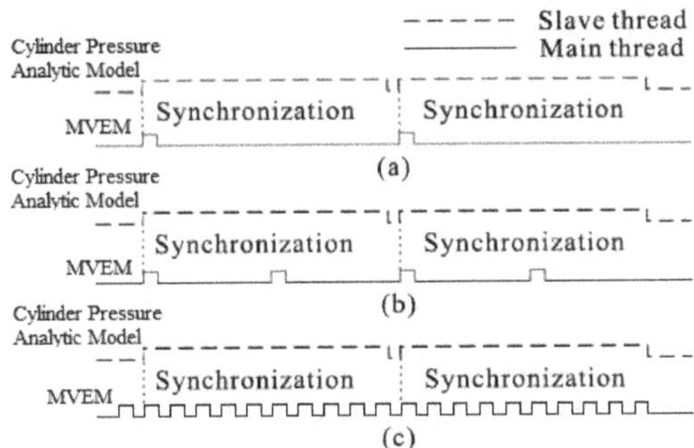

Figure 21. Synchronization approach for the MVEM and the cylinder pressure analytic model: (**a**) $X = 0$; (**b**) $X = 1$; (**c**) the calculation frequency of the MVEM is many times the analytic model.

During the transient process, the engine performance parameters and the in-cylinder pressure trace are different in every engine cycle; on the other hand, the MVEM and the cylinder pressure analytic model have different simulation speed. Therefore, in order to keep the simulation results in sync, the MVEM has to wait for some time before analytic model completes the calculation of an engine cycle as shown in Figure 21a, as a result, the simulation speed of the whole engine model is determined by the cylinder pressure analytic model, which is actually slower than the MVEM. To accelerate the simulation speed of the whole model, the idea of abandoning engine cycles is adopted. Figure 21b presents the case where the calculation frequency of the MVEM is two times the analytic model, meaning that the MVEM has already completed the calculation of two engine cycles when the analytic model only completes the calculation of one engine cycle. Figure 21c presents the case where the calculation frequency of the MVEM is many times the analytic model. By adopting this method, we can keep the MVEM and the cylinder pressure analytic model in sync, furthermore, the simulation speed of the whole model will get faster if more engine cycles are abandoned by the analytic model. The relationship between the number of abandoned cycles X and the reduced time Y can be expressed as $Y=X/(X+1)$. It should be noted that the synchronization mentioned above is a fake synchronization during the transient process and this is because that the current simulated in-cylinder pressure trace simulated by the analytic model does not correspond to the current engine operating conditions outputted by the MVEM but falls behind to some extent.

To assess the improvement in simulation speed of the extended MVEM developed in this paper, it is compared with the merged model developed by Tang et al., where the 0-D model was adopted to calculate the in-cylinder pressure and the MVEM was used to simulate other engine performance parameters with the similar synchronization approach as shown in Figure 21 [5]. The comparison of actual execution time for a 100 s simulation time is presented in Table 6.

Table 6. Comparison of simulation speed.

Model	Abandoned Cycles	Execution Time (s)	Simulation Time (s)
Tang et al. [5]	49	1.002	100
This paper	4	0.946	100

As can be found from Table 6, the engine model developed in this paper is able to achieve similar actual execution time of about one second but only four cycles are abandoned, whereas it is 49 for the merged model developed by Tang et al. [5]. This is due to the fact that the cylinder pressure analytic model runs much faster than the 0-D model and therefore similar actual execution time can be obtained by abandoning less number of engine cycles. As a result, the model developed in this paper can capture the transient response of the in-cylinder pressure trace more accurately.

5. Conclusions

In this article, a marine two stroke diesel engine MVEM with in-cylinder pressure trace predictive capability and a novel compressor model was developed.

A previous study carried out by the first author indicated that none of the compressor empirical mass flow rate models existing in the literature appears to achieve satisfactory predictive accuracy in the whole operating area. To solve this problem, the compressor whole operating area was divided into design, LS, HS and LPR operating areas by defining zonal division standard with appropriate functions, and then the compressor mass flow rate model that achieved the best predictive accuracy was selected for each operating area. In addition, an appropriate blending function was applied when the operating point enters into the LRP area from other areas to avoid the possible discontinuity.

The zonal compressor isentropic efficiency model proposed in this paper was based on the Hadef model. By dividing the whole "Mass flow rate—actual specific enthalpy change" plane into several zones by using the iso-speed curves available in the performance map, the working characteristics of compressor within different speed range can be captured much more accurately. It was revealed from the simulation results that with respect to the original Hadef model, the predictive accuracy of the zonal isentropic efficiency model was effectively improved. In addition, satisfactory extrapolation accuracy was obtained with this zonal compressor isentropic efficiency model.

As can be found from the trajectory of the compressor operating points during the transient process, the compressor model proposed in this paper is able to extrapolate to the LS and HS off-design operating areas reasonably and robustly; in addition, by comparing with the measured data provided in the engine shop trial report, the predictive accuracy of the engine performance parameters relevant to the turbocharger were all satisfactory at each steady state loading condition. Based on these simulation results, the fidelity of the proposed compressor model was validated. Due to the significant influence of the compressor on the engine steady state performance and transient response, the compressor model proposed in this paper is very helpful for improving the MVEM's predictive accuracy in the whole engine operating envelope.

By adapting the cylinder pressure analytic model to the 7S80ME-C9.2 marine two-stroke diesel engine, the MVEM is able to predict the in-cylinder pressure trace without impairing its merit in running speed. For achieving satisfactory predictive accuracy, the model parameters of the analytic model were finely calibrated by using the measured data and a 0-D engine model. At steady state condition, the extended MVEM runs as fast as the MVEM. During the transient process, the extended MVEM is able to achieve the running speed as similar as the MG model (0D + MVEM) proposed by Tang et al. [5] but captures the transient response of the in-cylinder pressure trace more accurately.

Although some sub-models of the MVEM are needed to be re-calibrated if another engine is to be modeled, they can be calibrated readily by only using the engine shop trial report and relevant performance maps. As a result of the lack of sufficient engine measured data, 2-D look-up tables were developed to calculate Wiebe function model parameters as well as the compression and expansion polytropic index. Although rapid calculation speed and satisfactory interpolation accuracy can be achieved with the 2-D look-up table, its extrapolative ability is big issue. In the future study, more measured engine data will be gathered to develop correlations between the model parameters and the engine performance parameters to enhance their extrapolative ability.

Author Contributions: H.S. contributed to the development of mathematical models, analysis of simulation results and the writing of the manuscript. J.Z. provided research conditions, team support and financial support. B.Y. and B.J. gave some useful suggestions. All authors have read and agreed to the published version of the manuscript.

Funding: This work was supported by the National Natural Science Foundation of China, grant number 51479017, and the Fundamental Research Funds for the Central Universities, grant number 3132019315.

Acknowledgments: The authors are especially grateful to Tang, who provided important technical support.

Conflicts of Interest: The authors declare no conflict of interest.

Nomenclature

a	Wiebe function efficiency factor (–)/model parameter (–)	n	polytropic index (–)/number (–)
b	model parameter (–)	N	rotational speed (RPM)
A	area (m^2)	p	pressure (pa)
C	flow coefficient (–)	P	power (W)
c_p	constant pressure specific heat (J/kg K)	$\bar{p}$	mean effective pressure (pa)
c_v	constant volume specific heat (J/kg K)	Q	torque (N m)
h	specific heat (J/kg)	$\dot{Q}_{ht}$	heat transfer rate (W)
H_{LHV}	fuel lower heating value (J/kg)	R	gas constant (J/kg K)
J	moment of inertia (kg m^2)	T	temperature (K)
K_p	propeller torque coefficient (–)	V_d	cylinder displacement volume (m^3)
k	model parameter (–)	$\dot{V}$	volume flow rate (m^3/s)
$\dot{m}$	mass flow rate (kg/s)	V	cylinder instantaneous volume (m^3)
m	Wiebe function form factor (–)/mass (kg)		

Greek symbols

γ	specific heat ratio (–)
θ	crank angle (deg)
ζ	fuel chemical energy proportion in the exhaust gas (–)
η	efficiency (–)
Π	pressure ratio (–)

Γ	expansion ratio (–)
ω	angular velocity (rad/s)
ϕ	fuel air equivalent ratio (–)
χ	mass fraction burned (–)

Subscripts

a	air
ac	air cooler
af	air filter
amb	ambient
act	actual
b	brake
bl	blower
c	compressor
cw	cooling water
comp	compression
comb	combustion
CD	combustion duration
d	diffusive
e	exhaust manifold/engine/exhaust gas
ell	ellipse
exp	expansion
exh	exhaust
f	fuel/fricative

i	indicated
in	inlet
is	isentropic
lb	lower bound
m	motored
out	outlet
p	propeller/premixed
pe	post exhaust
ref	reference
s	scavenging manifold
sur	surging
sh	shaft
scav	scavenging
soc	start of combustion
t	turbine
tc	turbocharger
z	cylinder

References

1. Theotokatos, G.; Guan, C.; Chen, H.; Lazakis, I. Development of an extended mean value engine model for predicting the marine two-stroke engine operation at varying settings. *Energy* **2018**, *143*, 533–545. [CrossRef]
2. Mavrelos, C.; Theotokatos, G. Numerical investigation of a premixed combustion large marine two-stroke dual fuel engine for optimising engine settings via parametric runs. *Energy Convers. Manag.* **2018**, *160*, 48–59. [CrossRef]
3. Stoumpos, S.; Theotokatos, G.; Boulougouris, E.; Vassalos, D.; Lazakis, I.; Livanos, G. Marine dual fuel engine modelling and parametric investigation of engine settings effect on performance-emissions trade-offs. *Ocean Eng.* **2018**, *157*, 376–386. [CrossRef]
4. Eriksson, L.; Nielsen, L. *Modeling and Control of Engines and Drivelines*; John Wiley and Sons Ltd.: Hoboken, NJ, USA, 2014.
5. Tang, Y.Y.; Zhang, J.D.; Gan, H.B.; Jia, B.Z.; Xia, Y. Development of a real-time two-stroke diesel engine model with in-cylinder pressure prediction capability. *Appl. Energy* **2017**, *194*, 55–70. [CrossRef]
6. Altosole, M.; Campora, U.; Figari, M.; Laviola, M.; Martelli, M. A diesel engine modelling approach for ship propulsion real-time simulators. *J. Mar. Sci. Eng.* **2019**, *7*, 138. [CrossRef]
7. Nikzadfar, K.; Shamekhi, A.H. An extended mean value model (EMVM) for control-oriented modeling of diesel engines transient performance and emissions. *Fuel* **2015**, *154*, 275–292. [CrossRef]
8. Theotokatos, G. On the cycle mean value modelling of a large two-stroke marine diesel engine. *Proc. Inst. Mech. Eng. Part M J. Eng. Marit. Environ.* **2010**, *224*, 193–205. [CrossRef]
9. Baldi, F.; Theotokatos, G.; Andersson, K. Development of a combined mean value-zero dimensional model and application for a large marine four-stroke diesel engine simulation. *Appl. Energy* **2015**, *154*, 402–415. [CrossRef]
10. Eriksson, L.; Andersson, I. An analytic model for cylinder pressure in a four stroke SI engine. In Proceedings of the SAE 2002 World Congress, Detroit, MI, USA, 4–7 March 2002.
11. Shen, H.S.; Zhang, J.D.; Yang, B.C.; Jia, B.Z. Development of an educational virtual reality training system for marine engineers. *Comput. Appl. Eng. Educ.* **2019**, *27*, 580–602. [CrossRef]

12. Theotokatos, G.; Tzelepis, V. A computational study on the performance and emission parameters mapping of a ship propulsion system. *J. Eng. Marit. Environ.* **2015**, *229*, 58–76. [CrossRef]

13. Tu, H. Investigation on the Performance Optimization Method for Large Containership Propulsion Engine at Slow Steaming Conditions. Ph.D. Thesis, Wuhan University of Technology, Wuhan, China, 2015.

14. Llamas, X.; Eriksson, L. Parameterizing compact and extensible compressor models using orthogonal distance minimization. *J. Eng. Gas Turbines Power* **2017**, *139*, 012601. [CrossRef]

15. Shen, H.S.; Zhang, C.; Zhang, J.D.; Yang, B.C.; Jia, B.Z. Applicable and comparative research of compressor mass flow rate and isentropic efficiency empirical models to marine large-scale compressor. *Energies* **2020**, *13*, 47. [CrossRef]

16. Hadef, J.E.; Colin, G.; Chamaillard, Y.; Talon, V. Physical-based algorithms for interpolation and extrapolation of turbocharger data maps. *SAE Int. J. Engines* **2012**, *5*, 363–378. [CrossRef]

17. Baratta, M.; Misul, D. Development and assessment of a new methodology for end of combustion detection and its application to cycle resolved heat release analysis in IC engines. *Appl. Energy* **2012**, *98*, 174–189. [CrossRef]

18. Brunt, M.F.; Platts, K.C. Calculation of heat release in direct injection diesel engines. In Proceedings of the SAE 1999 International Congress and Exposition, Detroit, MI, USA, 1–4 March 1999.

19. Kharroubi, K.; Söğüt, O.S. Modelling of the propulsion plant of a large container ship by the partition of the cycle the main diesel engine. *J. Eng. Marit. Environ.* **2019**. [CrossRef]

20. Eriksson, L.; Thomasson, A. Cylinder state estimation from measured cylinder pressure traces—A survey. *IFAC PapersOnLine* **2017**, *50*, 11029–11039. [CrossRef]

21. Oh, S.; Min, K.; Sunwoo, M. Real-time start of a combustion detection algorithm using initial heat release for direct injection diesel engines. *Appl. Therm. Eng.* **2015**, *89*, 332–345. [CrossRef]

22. Choe, D.; Lee, M.; Lee, K.; Sunwoo, M.; Jin, K.K.; Chang, K.; Shin, K. SOC detection of controlled auto-ignition engine. In Proceedings of the 14th Asia Pacific Automotive Engineering Conference, Hollywood, CA, USA, 5–8 August 2007.

23. Zhu, G.; Daniels, C.F.; Winkelman, J. MBT timing detection and its closed-loop control using in-cylinder pressure signal. In Proceedings of the 2003 Powertrain and Fluid Systems Conference and Exhibition, Pittsburgh, PA, USA, 27–30 October 2003.

24. Ghojel, J.I. Review of the development and applications of the Wiebe function: A tribute to the contribution of Ivan Wiebe to engine research. *Int. J. Engine Res.* **2010**, *11*, 297–312. [CrossRef]

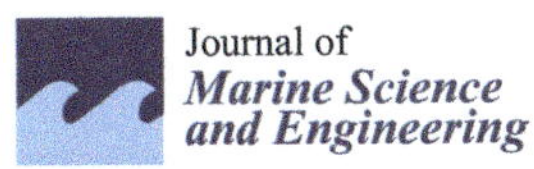

Journal of
Marine Science and Engineering

Article

Effective Mistuning Identification Method of Integrated Bladed Discs of Marine Engine Turbochargers

Václav Píštěk [1],*, Pavel Kučera [1], Oleksij Fomin [2] and Alyona Lovska [3]

[1] Institute of Automotive Engineering, Brno University of Technology, Technická 2896/2,
    616 69 Brno, Czech Republic; kucera@fme.vutbr.cz
[2] State University of Infrastructure and Technologies, Kyrylivska str. 9, 04071 Kyiv, Ukraine;
    fominaleksejviktorovic@gmail.com
[3] Ukrainian State University of Railway Transport, Feierbakh sq. 7, 61050 Kharkiv, Ukraine;
    alyonalovskaya.vagons@gmail.com
* Correspondence: pistek.v@fme.vutbr.cz; Tel.: +420-541-142-271

Received: 3 May 2020; Accepted: 22 May 2020; Published: 25 May 2020

Abstract: Radial turbine and compressor wheels form essential cornerstones of modern internal combustion engines in terms of economy, efficiency and, in particular, environmental compatibility. As a result of the introduction of exhaust gas turbochargers in the extremely important global market for diesel engines, higher engine efficiencies are possible, which in turn reduce fuel consumption. The associated reduced exhaust emissions can answer questions that results from environmentally relevant aspects of the engine development. In shipping, the international Maritime Organisation (IMO) prescribes the step-by-step reduction of nitrogen oxide and other types of emissions. To reduce these emissions, various systems are being developed, in which turbochargers are an important part. The requirements for the reliability and service life of turbochargers are constantly increasing. Turbocharger blade vibration is one of the most important problems to be solved when designing the rotors. In the case of real rotors, so-called mistuning arises, which is a slight deviation of the properties of the individual blades from the design parameters. The article deals with an effective method of mistuning identification for cases of integrated bladed discs of marine engine turbochargers. Unlike approaches that use costly scanning laser Doppler vibrometers, this method is based on using only a simple laser vibrometer in combination with a computational model of the integrated bladed disc. The added value of this method is, in particular, a significant reduction in the cost of laboratory equipment and a reduction in the time required to obtain the results.

Keywords: marine; engine; turbocharger; bladed disc; measurement; laser; simulation

---

1. Introduction

The principle of operation of the exhaust gas turbocharger is to use the unused kinetic and thermal energy of the ejected engine exhaust. The mostly single-stage turbine is set in rotation by the exhaust gas flowing out and thus drives the compressor. This increases the fluid pressure of the outflowing air which is then pressed into the engine. The increased cylinder filling of the driven engine enables better thermal efficiency during combustion. Turbochargers are also an important part of various systems for reducing emissions from internal combustion engines [1–10]. Due to the increased boost pressure of the working medium, either more powerful engines with similar dimensions or similarly powerful engines with significantly smaller dimensions are possible.

In general, particularly high pressure ratios of the compressor blades are required to achieve the required performance increases. The high aerodynamic demands require a very detailed design of blade geometries. Ever thinner blades combined with the low mechanical damping due to the integral construction make the vibration-proof design of turbomachines more difficult. Regardless of the widespread usage of turbomachines, documented damage cases indicate that structural dynamic issues have not been conclusively resolved. Imperfections that already occur during production have a significant influence on the dynamic behaviour of real components. The deviation from the original design is referred to as a "mistuned bladed disc effect" and has been discussed by many researchers in the field of internal combustion engines and aircraft turbine engines [11–14]. Researchers [11,12] have described a method that allows mistuning distributions to be determined indirectly from vibration measurements without asking about the actual cause of the mistuning.

The largely random mistuning reduces the life of the components that are permanently loaded by centrifugal forces, flow deflection, unsteady pressure fluctuations in the flow and temperature gradients. The purely numerical prediction of the resonance strength is almost impossible for the entire operating range due to the numerous operating points. The vibration safety of a new development is therefore detected in regard to the qualification tests. As a rule, these consist of driving through the operating area of the exhaust gas turbocharger on a combustion chamber or an engine test benches. The measured vibrations of the system are included in the fatigue strength analysis. Considerable safety factors are taken into account. If detailed statements on the real structural behaviour of a design were available at an early stage, the safety factors of some resonance points could be reduced. As a result, designs would be feasible that are closer to the permissible mechanical load limit. Only sound assessments guarantee a solid and resonance-proof turbocharger design.

Knowledge of the real deviation between adjacent blades is necessary for the robust consideration of random mistuning. The parameters can be determined for a single, actually existing rotor using a suitable experimental method. A variety of approaches have now been documented. Some authors [15,16] measured the surface of the manufactured blades using white light stripe projection. In doing so, they directly countered the geometric manufacturing tolerance as a presumed cause of the mistuning. Another publication [17] continued this development and demonstrated good agreement between the properties of optically recorded blades and experimental vibration measurements. Some research studies have also appeared in the field of aircraft and automotive turbochargers [18,19], but their results are not very applicable, particularly with regard to the point of identification of the mistuning and the cause of the mistuning of marine engine turbocharger bladed discs.

Further research to identify mistuning has been done mainly in the field of aircraft engines [20]. This reflects the current trend of applying the integrated structures of blade and disc. Contemporary technology enables this type of structure, which has the advantage of effectively reducing both the weight and the number of engine parts. Contrariwise, there are difficulties with mistuning identification. The reason is that with such a structure, separation of each blade from the disc is not possible, as with the dovetail type blade of conventional compressors. Therefore, it is not possible to know the natural frequency of each blade without disc coupling effect. Some authors dealt with the mistuning problems of small rotors of automotive turbochargers [21,22]. One paper [23] presented the experimental and numerical studies of last-stage low-pressure mistuned steam turbine bladed discs during run-down. The tip-timing method was used to find the mistuned bladed disc modes and frequencies.

Based on this situation, a marine engine turbocharger rotor was selected as the subject of the research (see Figure 1). The aim of the solution was to identify mistuning of actually produced turbine rotor of a marine engine turbocharger and to design an effective experimental method for this purpose. Some authors used a laser scanning system to obtain the modal information for different types of mechanical structures [24–27]. In this system, a vibration response against the excitation input is obtained at each reference point using a laser Doppler vibrometer. This process is repeated for the same excitation input, the reference point being changed by scanning the laser Doppler vibrometer. After completion of the scan, the frequency response functions obtained for all reference points are correlated and the modal displacement contour of the entire measured surface is created. This procedure requires the application of a very expensive laser scanning technique; moreover, the processing of the measured results is time consuming. Therefore, a procedure was proposed that uses only a simple laser vibrometer and an FE computational model of the turbine rotor.

Figure 1. Marine engine turbocharger rotor.

2. Modelling of Tuned Blade Disc Systems

Modelling using finite elements makes dynamic analysis of complex components considerably easier. The vibration behaviour of turbocharger blade discs can be described by the differential motion equation:

$$M\ddot{q} + R\dot{q} + Kq = F(t),\tag{1}$$

where M is mass matrix, R—damping matrix, K—stiffness matrix, F—vector of force, q—vector of generalised coordinates and t—time in the continuous structure being discretised by a finite number of degrees of freedom. The characteristic values of the natural vibration behaviour are the eigenvalues, i.e., natural frequencies and eigenvectors, i.e., mode shapes. These can be determined by solving the homogeneous and undamped system according to Equation (1). If the vibration behaviour of the bladed disc is to be described as precisely as possible, a very fine discretization may be required. When using FEM, models with several million degrees of freedom are quickly created. Such model sizes are seamlessly manageable currently, but the question of reducing the degrees of freedom of the model arises without losing accuracy in the system description. In addition to classic reduction methods, the exploitation of symmetry properties is essential. With an ideal bladed disc, all sectors are identical, i.e., they have the same mechanical properties, so that no distinction among the stiffness matrices of the individual sectors is required.

The cyclic symmetry of the blade disc basically allows the analysis to be restricted to one sector without loss of information. Figure 2 shows one sector of the overall blade disc structure. The element size represents a central setting parameter that has a direct influence on the calculation results. The number of elements increases with an increasing level of model details. This is associated with an increasing number of modes to be calculated, which in turn increase the computational demands. In the case of the radial turbine investigated here, a fine and structured mesh of the blade surface is recommended. The disc body and shaft can be modelled using larger elements. The FE model of the entire turbine blade disc is created by the periodic expansion of the sector and subsequent fusion of the sector boundaries. The FE mesh was created at ANSA software and it had to be done thoroughly in connection with the subsequent creation of a rotation model. There, it is important to properly connect the nodes on both sides of the segment to create the whole rotational cyclic model. At the same time, it is necessary to achieve just the same segments that are really connected correctly. The size of the elements used was carefully optimised by a series of sensitivity analysis.

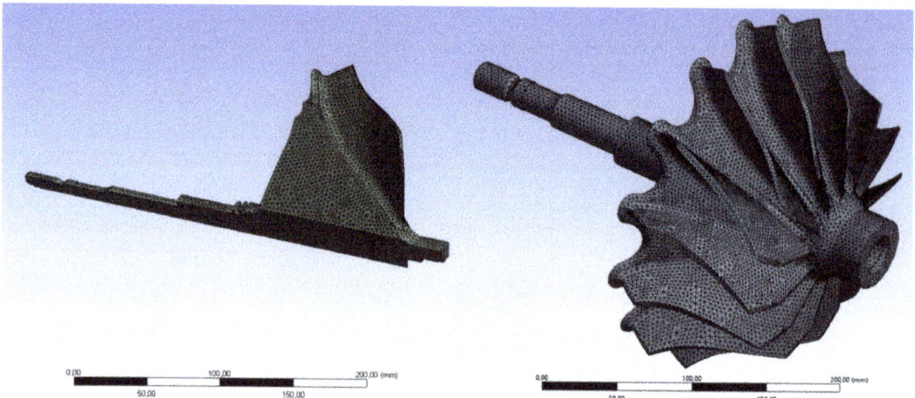

Figure 2. An illustration of one sector and whole mesh of the turbine bladed disc structure.

The turbine rotor is supported in bearings during operation. In modern exhaust gas turbochargers, two plain bearings prevent radial displacements. In addition, a thrust bearing is fitted. The bearing influence plays a negligible role with regard to the blade-dominated vibration forms. In order to achieve the best possible correlation, no boundary conditions are specified for the computational modal analysis.

3. Measurement and Data Processing

3.1. Measurement

The turbine rotor was mounted on a special highly flexible jig, so that the effect of boundary conditions was negligible. In the sense of a classic modal analysis, a small modal hammer excites the individual rotor blades at a suitable position. A laser Doppler vibrometer measures the blade vibration speeds contactless. As part of the experimental investigation regarding the blade vibrations, only the vibration response of one suitable selected point per blade was measured. This is acceptable if the measurements match the expected mode shapes of the computational model prediction.

The measuring chain was assembled for an experiment from POLYTEC Sensor Head OFV-505, Vibrometer Controller OFV-5000 and cDAQ-9179 with modules NI-9229 and NI-9234. The laser head was positioned 438 mm from the top of the turbocharger blade according to the datasheet (Figure 3). The beam was directed perpendicularly to the scanning area.

With the individual blade mistuning measurement, the deviation from the mean value of all blades due to the material property or geometry deviations is converted into a calculation value in an experimental way. To decouple the oscillation movement of an individual blade when measuring, all other blades are additionally detuned by attaching additional masses of appropriate size to the blade tips (see Figure 4 right). The small modal hammer Brüel & Kjær Type 8204 (Figure 4 left) excites the blade to vibrate, while the laser vibrometer measures the vibration speed at the tip of the blade. Through a suitable choice of size and position of the additional masses as well as through variation of the measuring and excitation point, the mistuning behaviour in the frequency range of the blade dominated mode shapes can be identified. The measurement was carried out without and with additional masses on the blades to make clear their positive effect.

Figure 3. Laser equipment for blade vibration measurement.

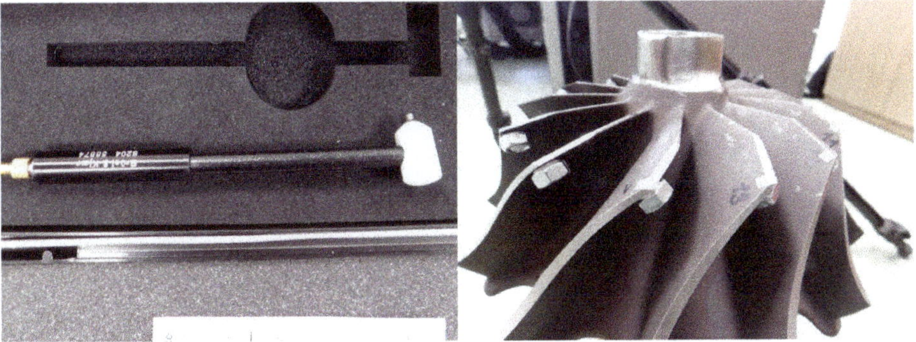

Figure 4. Turbocharger turbine wheel with additional masses on blades and the used modal hammer.

Data monitoring was performed using a programmed script in NI LabVIEW software and 20 second sections were saved. Their data includes at least five responses from the hammer blow to the blade. At the same time, the experiment was checked for a double hammer blow to prevent any adverse effects of the subsequently processed data. Thereafter, five responses lasting 0.2 s and the corresponding time records of the modal hammer force pulses were selected for frequency analysis.

3.2. Data Processing

Data processing from the experiment was done using a script in Matlab software. Several main parts were included. The first was the use of an exponential window. Windowing is a common necessary signal processing technique used in modern data acquisition systems [28,29]. The exponential window is a time domain weighting function that has been elaborated for use with transient-type signals, such as those of impact testing. Used correctly, the exponential window can minimise leakage errors on lightly damped signals and can also improve the signal-to-noise ratio. The effect of the exponential window (Figure 5) is to increase the apparent damping of the measured system. The exponential window [30] is an exponential function as defined in equation (2) where the parameter

f is the last point value of the window, n is the number of samples and i is index of the exponential function:

$$y_i = e^{\frac{\ln f}{n-1}i}, i = 1 \ldots n \tag{2}$$

The time variable for the exponential function starts at zero, regardless if a pretrigger delay is used in the data acquisition. When impact testing lightly damped blade structures, the purpose of the exponential window is to reduce the effects of leakage by forcing the data to meet the requirements of a completely observed transient response signal more closely. By definition, a fully observed transient signal must begin and end within the measured time record. For lightly damped blade systems, the response of the structure will typically continue beyond the time period of data collection as shown in Figure 6a. Since the response does not decay to near zero at the end of the time record, the exponential window is applied to reduce the signal at the end of the time record to approximately 1%. The signal according to Figure 6a after the window has been applied shows Figure 6b. The windowed signal more closely represents an observable transient.

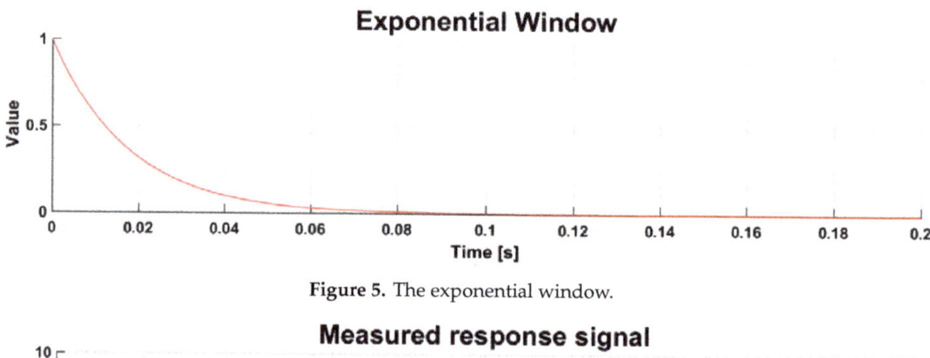

Figure 5. The exponential window.

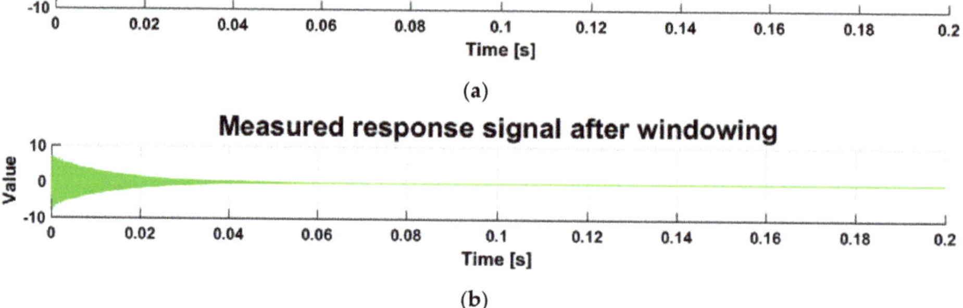

Figure 6. Measured response signal of a lightly damped blade system for impact excitation and its modification by the exponential window.

This data was processed by a fast Fourier transform (FFT) algorithm in a complex form to evaluate the transfer function of the system. That is, the FFT analysis was performed for the impact force course and its corresponding measured response. Subsequently, the ratio of the output to the input of the system was used to obtain the transfer function:

$$Y_i = ABS\left(\frac{FFT_{out}}{FFT_{force}}\right), i = 1 \ldots n, \tag{3}$$

where FFT_{out} is the resulting FFT analysis complex vector from the response data and FFT_{force} is the resulting FFT analysis complex vector from the force pulse data. The calculation works using complex arithmetic. The measurement of each turbocharger blade was performed five times to reduce possible measurement errors, and these five measurements were included in the data processing, at the end of which five calculated FFT spectra were always averaged. Since the time window was 0.2 s, the frequency resolution of the FFT is 5 Hz, and thus, the uncertainty of determination of each spectral component is 2.5 Hz, which is completely satisfactory for the determination of the blade disc mistuning.

4. Results and Discussion

4.1. Finite Elements Method

Modal analysis of one turbocharger turbine segment was performed using FEM, as shown in Figure 7. Boundary conditions with zero displacement and rotation in all directions were applied to the sides of the segment. Thus, the natural frequencies associated with the turbocharger blade were obtained. In this way, the mode shapes were obtained. Figure 7 shows mode 2 for illustration.

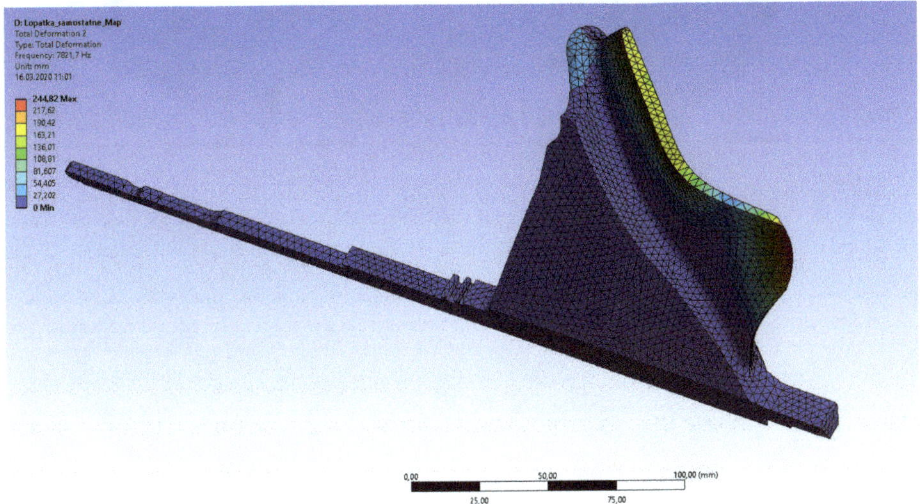

Figure 7. Example of a mode shape of a turbocharger turbine blade segment (mode 2).

The calculated natural frequencies of the blade are presented in Table 1. The interest in the frequency range was up to 20,000 Hz, where the first seven modes are located. The calculated natural frequencies show approximate natural frequency values of the blade-dominant mode shapes of the rotational symmetric turbine disc.

Table 1. Calculated natural frequencies of the blade segment.

Mode [−]	Frequency [Hz]
1	4764
2	7822
3	9229
4	11,484
5	14,287
6	15,190
7	18,730

Subsequent forms of mesh models were used in the next step (Figure 2). The second analysis was performed from one segment as a rotational cyclic modal analysis. It results in a nodal diameter map of the tuned turbine rotor. However, this is not the main subject of this article.

Subsequently, a third FE model was created from a rotated model of one segment. The modal properties of this three-dimensional (3D) turbine FE model are identical to the previous rotational cyclic model. This model will further serve for analyses with different material properties of the blades and for analysing the mistuning effect. An example of the blade-dominated mode shapes for modes 1–6 is shown in Figure 8.

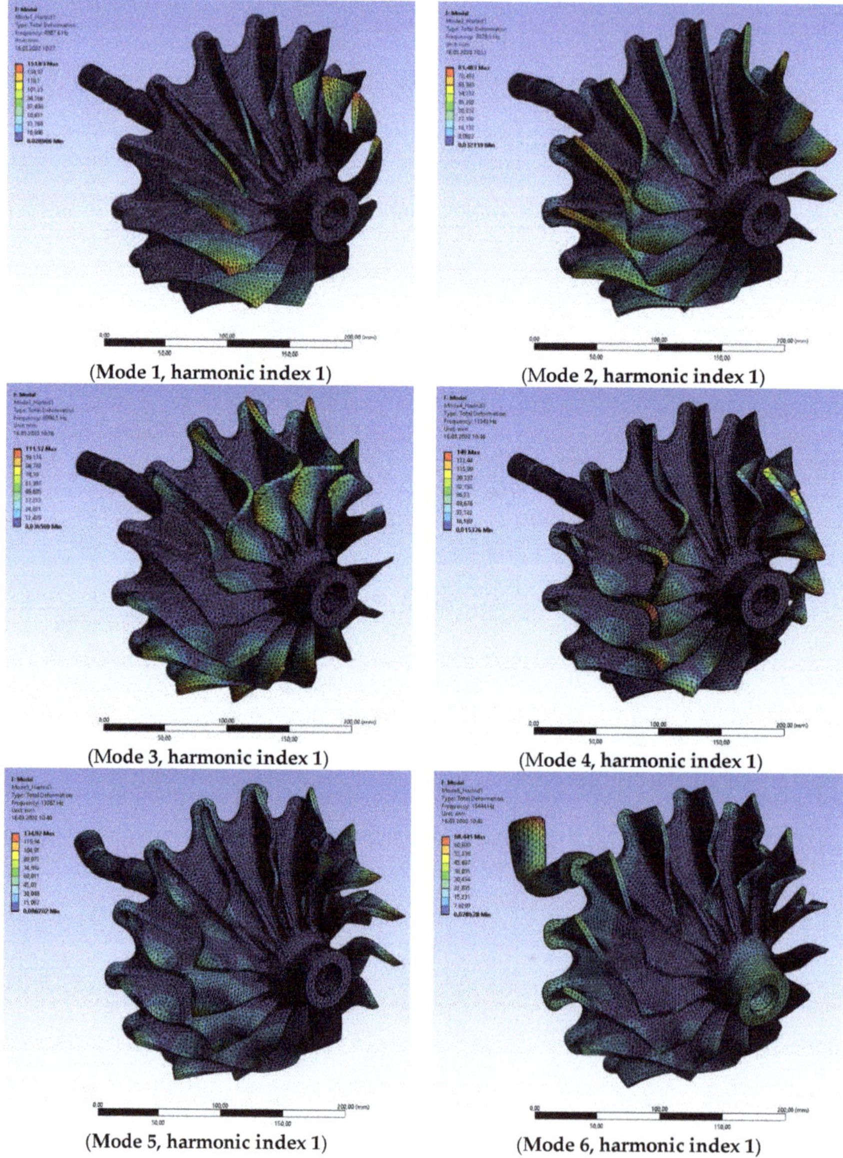

(Mode 1, harmonic index 1)

(Mode 2, harmonic index 1)

(Mode 3, harmonic index 1)

(Mode 4, harmonic index 1)

(Mode 5, harmonic index 1)

(Mode 6, harmonic index 1)

Figure 8. Examples of computed modal shapes of the turbocharger turbine wheel.

The material properties have been the same thus far for all segments to verify the following measurement results and possibly debug the model. The mistuned 3D model will be used for research of dangerous stress concentrators due to different characteristics of the individual blades.

4.2. Measuremet

The transfer functions of the individual blades were obtained by the measurements using subsequent the mathematical procedure according to Equation (3). The resulting transfer function of the first blade is shown in Figure 9. In this case, no additional masses on other blade tips were used. Therefore, this figure is shown here to illustrate the difference from the resulting transfer function in Figure 10 showing the measurement variant when placing additional masses on other blades.

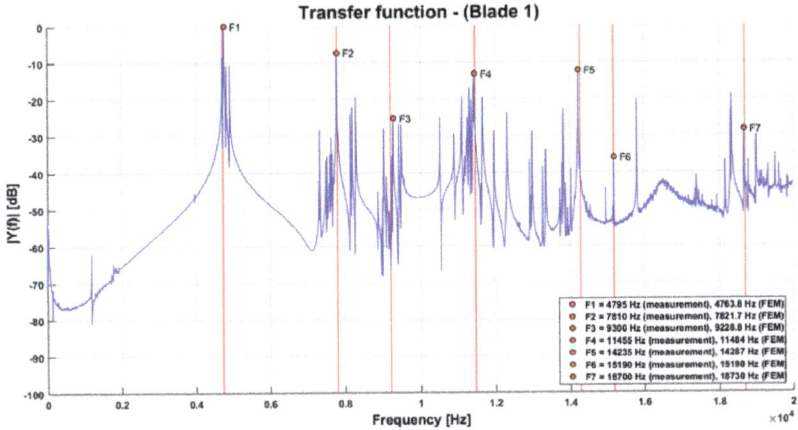

Figure 9. Transfer function and natural frequencies from FE model—without additional masses.

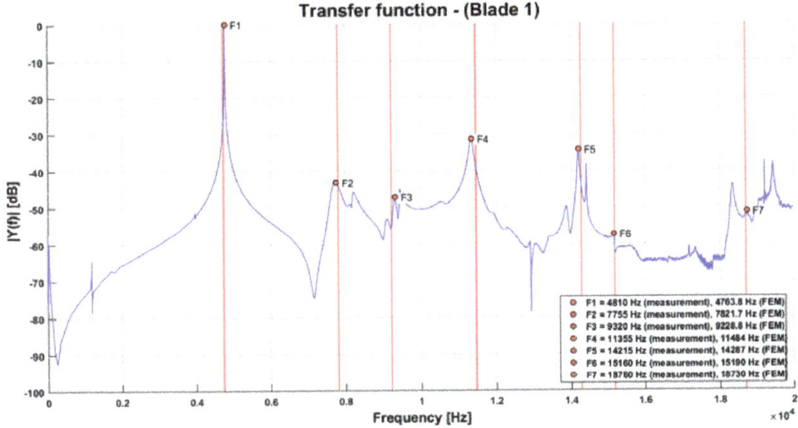

Figure 10. Transfer function and natural frequencies from FE model—with additional masses.

This method is very effective in suppressing the influence of the slightly different natural frequencies of the unmeasured blades and makes it possible to obtain the natural frequencies of individual blades of a real turbine wheel and to create a mistuned 3D model for further detailed dynamic analyses.

In Figure 10, red lines are visible to symbolise the natural frequencies of the turbocharger turbine blade from the FE simulation. Compliance with the measurement is very good and it is obvious that the method using FE model is well prepared for further utilization with much simpler demands than using expensive 3D scanning laser. Furthermore, applying this method, the user can concentrate on the natural blade dominant frequencies only, without having to deal with other spectrum peaks that belong entirely or predominantly to the turbine bladed disc.

5. Conclusions

In the present article, the vibration characteristics of a radial turbine bladed disc of a marine engine turbocharger were investigated. Starting with the analysis of structural dynamic properties of cyclic symmetry components, basic procedures were transferred to special problems of bladed discs of marine turbocharger radial turbines and expanded. In particular, the assignment of individual peaks in the measured transmission functions to the corresponding mode shapes was performed using a carefully designed FE model, without the need to apply an expensive scanning vibrometer. In addition, the distribution of the strain energy between the disc and the blade calculated by the FE model is a next parameter that allows the systematic assessment each of the large number of modal shapes present in this complicated structure. The mode shapes with predominant strain energy in the blade are thus referred to as blade-dominated. Depending on the criterion of distribution of the strain energy among the blades and the disc, some modes can be called mixed. The transfer functions shown above also record these mixed modes such as matching peaks, but their effect on blade stress is not significant. Disc dominant modes do not occur in the examined frequency range of the analysed marine engine turbocharger radial turbine. The reason for this is the massive disc and the rigid connection of the blades to them.

Based on the results of the measurements described above, it was possible to create a 3D mistuned FE model of the turbine bladed disc. Creation of this model is based on individual adaptation of the modulus of elasticity to individual blades using optimization algorithms. The vibration behaviour of the real bladed discs can be achieved with sufficient accuracy by such 3D models, unlike methods that are based on the use of substructures and do not describe the frequency splitting due to mistuning with the necessary accuracy. Since the description of the optimization algorithm and the creation of the 3D model of the mistuned turbine wheel would go far beyond the possible size of this article, this topic will be the subject of a separate publication.

With regard to industrial application, it should be noted that an improvement in manufacturing accuracy is only conditionally recommended, since increasing the manufacturing accuracy of the blades would entail unacceptable financial costs. Rather, it is important to determine the inclination of individual vibration modes in order to tend large increases in stress. It should be noted that large stress increases are always accompanied by strongly localised modes. However, strongly localised mode shapes can only occur if the tuned bladed disc system has a sufficient number of node diameter vibrations in a relative frequency proximity. These cases can be identified already in the early stages of development if a suitable FE blade disc model is used. Together with the described effective mistuning identification method of integrated bladed discs, the development of new marine engine turbochargers can be significantly accelerated. Another added value of the method is the possibility of its relatively easy automation, both in terms of measurement and its subsequent processing and evaluation. The turbocharger manufacturer can thus obtain complete statistical information on the parameters of the blade discs in relation to the mistuning caused by various influences in the production process.

Author Contributions: Conceptualization, V.P.; methodology, V.P. and P.K.; software, P.K. and O.F.; validation, P.K. and A.L.; writing—review and editing, V.P. and P.K. All authors have read and agreed to the published version of the manuscript.

Funding: The authors gratefully acknowledge funding from the Specific research on BUT FSI-S-20-6267.

Acknowledgments: The authors thank the Brno University of Technology for support.

Conflicts of Interest: The authors declare no conflict of interest.

References

1. DieselNet. Emission Standards: EU: Cars and Light Trucks. Available online: https://dieselnet.com/standards/eu/ld.php (accessed on 17 March 2020).
2. DieselNet. Emission Standards: EU: Heavy-Duty Truck and Bus Engines. Available online: https://dieselnet.com/standards/eu/hd.php (accessed on 17 March 2020).
3. European Committee for Standardization. Automotive Fuels—Diesel—Requirements and Test Methods, EN 590:2009. Available online: http://www.envirochem.hu/www.envirochem.hu/documents/EN_590_2009_hhV05.pdf (accessed on 17 March 2020).
4. International Organization for Standardization, ISO 8217:2017. Available online: https://www.iso.org/standard/64247.html (accessed on 17 March 2020).
5. Murakami, M.; Tanaka, H.; Fukuyama, M.; Fujibayashi, T.; Okazaki, S.; Shibata, J.; Tanaka, H.; Hiraiwa, S.; Kobayashi, R.; Ezu, Y. Development of marine SCR system (separate type) for large two-stroke diesel engines. *Hitz Tech. Rev.* **2014**, *75*, 2–9. (In Japanese)
6. Winebrake, J.J.; Corbett, J.J.; Green, E.H.; Lauer, A.; Eyring, V. Mitigating the health impacts of pollution from oceangoing shipping: An assessment of low-sulfur fuel mandates. *Environ. Sci. Technol.* **2009**, *43*, 4776–4782. [CrossRef] [PubMed]
7. Okubo, M.; Kuwahara, T.; Yoshida, K.; Kawai, M.; Hanamoto, K.; Sato, K.; Kuroki, T. Total marine diesel emission control technology using nonthermal plasma hybrid Process. In Proceedings of the 27th CIMAC World Congress on Combustion Engine Technology, Shanghai, China, 13–16 May 2013; Paper No.153. p. 14.
8. Herdzik, J. Emissions from marine engines versus IMO certification and requirements of tier 3. *J. Kones Powertrain Transp.* **2011**, *18*, 161–167.
9. Puškár, M.; Brestovič, T.; Jasminská, N. Numerical simulation and experimental analysis of acoustic wave influences on brake mean effective pressure in thrust-ejector inlet pipe of combustion engine. *Int. J. Veh. Des.* **2015**, *67*, 63–76. [CrossRef]
10. Sinay, J.; Puškár, M.; Kopas, M. Reduction of the NOx emissions in vehicle diesel engine in order to fulfill future rules concerning emissions released into air. *Sci. Total Environ.* **2018**, *624*, 1421–1428. [CrossRef] [PubMed]
11. Feiner, D.M.; Griffin, J.H. Mistuning identification of bladed disks using a fundamental mistuning model—Part I: Theory. *J. Turbomach.* **2004**, *132*, 150–158. [CrossRef]
12. Feiner, D.M.; Griffin, J.H. Mistuning Identification of bladed disks using a fundamental mistuning model—Part II: Application. *J. Turbomach.* **2004**, *126*, 159–165. [CrossRef]
13. Giersch, T.; Beirow, B.; Popig, F.; Kühhorn, A. FSI-based forced response analyses of a mistuned high pressure compressor blisk. In Proceedings of the 10th International Conference on Vibration in Rotating Machinery, London, UK, 11–13 September 2012.
14. Giersch, T.; Hönisch, P.; Beirow, B. Arnold kühhorn forced response analyses of mistuned radial inflow turbines. *J. Turbomach.* **2013**, *135*, 9. [CrossRef]
15. Schoenenborn, H.; Grossmann, D.; Satzger, W.; Zisik, H. Determination of blade-alone frequencies of a blisk for mistuning analysis based on optical measurements. In Proceedings of the ASME Turbo Expo, Orlando, FL, USA, 8 June 2009; pp. 221–229.
16. Schoenenborn, H.; Junge, M.; Retze, U. Contribution to free and forced vibration analysis of an intentionally mistuned blisk. In Proceedings of the ASME Turbo Expo, Munich, Germany, 11 June 2012; pp. 1111–1120.
17. Maywald, T. Modellierung und Simulation Integral Gefertigter Verdichterlaufräder auf der Grundlage einer Dreidimensionale Oberflächenvermessung. Ph.D. Thesis, Brandenburgische Technische Universität Cottbus – Senftenberg, Cottbus, German, 2018.
18. Giersch, T.; Figaschewsky, F.; Hönisch, P.; Kühhorn, A.; Schrape, S. Numerical Analysis and validation of the rotor blade vibration response induced by high pressure compressor deep surge. In Proceedings of the ASME Turbo Expo 2014: Turbine Technical Conference and Exposition, Düsseldorf, Germany, 16–20 June 2014; pp. 1–12.
19. Nguyen-Schäfer, H. *Rotordynamics of Automotive Turbochargers*; Springer: Berlin/Heidelberg, Germany, 2012.

20. Figaschewsky, F.; Kühhorn, A.; Beirow, B.; Nipkau, J.; Giersch, T.; Power, B. Design and analysis of an intentional mistuning experiment reducing flutter susceptibility and minimizing forced response of a jet engine fan. In Proceedings of the ASME Turbo Expo 2017: Turbomachinery Technical Conference and Exposition, Charlotte, CA, USA, 26–30 June 2017; pp. 1–13.

21. Kulkarni, A.; Larue, G. Vibratory response characterization of a radial turbine wheel for automotive turbocharger application. In Proceedings of the ASME Turbo Expo 2008: Power for Land, Sea, and Air, Berlin, Germany, 9–13 June 2008; pp. 583–591.

22. Hemberger, D.; Filsinger, D.; Bauer, H.-J. Mistuning modeling and its validation for small turbine wheels. In Proceedings of the ASME Turbo Expo 2013: Turbine Technical Conference and Exposition, San Antonio, TX, USA, 3–7 June 2013; pp. 1–11.

23. Rzadkowski, R.; Kubitz, L.; Maziarz, M.; Troka, P.; Dominiczak, K.; Szczepanik, R. Tip-Timing measurements and numerical analysis of last-stage steam turbine mistuned bladed disc during run-down. *J. Vib. Eng. Technol.* **2020**, *8*, 409–415. [CrossRef]

24. Hattori, H. Study on mistuning identification of vehicle turbocharger turbine BLISK. In Proceedings of the ASME Turbo Expo 2014: Turbine Technical Conference and Exposition, Düsseldorf, Germany, 16–20 June 2014; pp. 1–10.

25. Drápal, L.; Šopík, L. Influence of crankshaft counterweights upon engine block load. In Proceedings of the 20th International Scientific Conference Transport Means, Juodkranté, Lithuania, 5–7 October 2016; pp. 809–814.

26. Prokop, A.; Řehák, K. Virtual prototype application to heavy-duty vehicle gearbox concept. In Proceedings of the Engineering Mechanics 2017, Svratka, Czech Republic, 15–18 May 2017; pp. 810–813.

27. Zeizinger, L.; Pokorný, P. The consequence of using orthotropic material in a mechanical system. *Vibroengineering Procedia* **2018**, *18*, 128–131. [CrossRef]

28. Tůma, J.; Kočí, P. Calculation of a shock response spectrum. In Proceedings of the 12th International Carpathian Control Conference (ICCC), Velke Karlovice, Czech Republic, 25–28 May 2011; pp. 408–413.

29. Tůma, J.; Kočí, P. Calculation of a shock response spectrum. In Proceedings of the Colloquium Dynamics of Machines 2009, Prague, Czech Republic, 3–4 February 2009; pp. 408–413.

30. National Instruments, Exponential Window VI. Available online: https://zone.ni.com/reference/en-XX/help/371361R-01/lvanls/exponential_window/ (accessed on 17 March 2020).

Journal of
*Marine Science
and Engineering*

Article

Numerical Analysis of NOx Reduction Using Ammonia Injection and Comparison with Water Injection

María Isabel Lamas Galdo [1,*], Laura Castro-Santos [1] and Carlos G. Rodriguez Vidal [2]

[1]   Escola Politécnica Superior, Universidade da Coruña, 15403 Ferrol, Spain; laura.castro.santos@udc.es
[2]   Norplan Engineering S.L., 15570 Naron, Spain; c.rodriguez.vidal@udc.es
*   Correspondence: isabel.lamas.galdo@udc.es; Tel.: +34-881013896

Received: 8 January 2020; Accepted: 7 February 2020; Published: 11 February 2020

Abstract: This work analyzes NO_x reduction in a marine diesel engine using ammonia injection directly into the cylinder and compares this procedure with water injection. A numerical model based on the so-called inert species method was applied. It was verified that ammonia injection can provide almost 80% NO_x reduction for the conditions analyzed. Furthermore, it was found that the effectiveness of the chemical effect using ammonia is extremely dependent on the injection timing. The optimum NO_x reduction was obtained when ammonia is injected during the expansion stroke, while the optimum injection timing using water is near top dead center. Chemical, thermal, and dilution effects of both ammonia and water injection were compared. The chemical effect was dominant in the case of ammonia injection. On the other hand, water injection reduces NO_x through dilution and, more significantly, through a thermal effect.

Keywords: CFD; NO_x; engine; ammonia; water injection

---

1. Introduction

Nowadays, diesel engines rule the transportation sector and power most of the ships in the world. These engines are efficient compared with other thermal machines but emit harmful species such as nitrogen oxides (NO_x), soot, carbon dioxide (CO_2), sulfur oxides (SO_x), carbon monoxide (CO), and non-burnt hydrocarbons (HC). Between these, NO_x is a harmful component that must be reduced since it produces acidification of rain, photochemical smog, greenhouse effects, ozone depletion, and respiratory diseases. Several international, national, and regional policies have been developed to limit NO_x and other pollutants. In the marine field, the European Commission and the Environmental Protection Agency limit emissions in the European Union and the United States, respectively. On an international level, the International Maritime Organization (IMO) maintains a comprehensive regulatory framework for shipping. In 1973, the IMO adopted Marpol 73/78, the International Convention for the Prevention of Pollution from Ships, designed to reduce marine pollution. In particular, Marpol Annex VI limits NO_x emissions for marine ships depending on the manufacturing data, engine speed, and working geographical area.

Due to these increasingly restrictive regulations, several NO_x reduction methods have been developed in recent years. One of them is the utilization of alternative fuels. The main alternative marine fuels may be found in two forms: liquid fuels including ethanol, methanol, bio-liquid fuel, and biodiesel; and gaseous fuels, including propane, hydrogen, and natural gas [1–4].

Operating under diesel, there are two procedures to reduce NO_x, which are primary and secondary measures. The former reduces the amount of NO_x during combustion, while the latter focuses on removing NO_x from the exhaust gases through downstream cleaning techniques. It is well known that the main factors that influence NO_x formation are the temperatures reached in the combustion process and the amount of time in which the combustion gases remain at high temperatures [5–7]. Based on this, primary measures focus on addressing these factors and reducing the concentrations of oxygen and nitrogen [8,9]. Well-known primary measures are exhaust gas recirculation (EGR), Miller timing, common rail, modification of injection and other parameters of the engine, and water addition. Water can be introduced as a fuel-water emulsion injected via the fuel valve, through separate nozzles or by humidifying the scavenge air. Despite the extensive research on primary measures along the recent years, a procedure to reduce NO_x without decreasing emission of other pollutants and/or consumption has not effectively been developed. In this regard, secondary measures reduce NO_x from the flue gas through downstream cleaning techniques. Many applications have been undertaken to reduce NO_x by selective catalytic reduction (SCR) and selective non-catalytic reduction (SNCR). The disadvantages of SCR are its price, poor durability of catalysts, and deposition of particulate on the catalyst. These disadvantages are not present in SNCR, but this procedure is limited to a narrow temperature range with optimal temperatures that are much higher than those characteristic of flue gas from diesel engines [10]. This limitation constitutes a drawback for practical applications in exhaust gases from diesel engines. As SCR reducing agents, ammonia (NH_3), urea, and cyanuric acid have been extensively employed. SNCR using ammonia, urea, and cyanuric acid are known as $DeNO_x$ [11,12], NO_xOUT [13,14], and RAPRENO [15–17], respectively. Between these, this work focuses on NO_x reduction using ammonia. The NO_x reduction capabilities of ammonia were discovered in the seventies by Lyon [18], who found that ammonia selectively reduces NO_x without a catalyst over the temperature range of 1100–1400 K. Typical exhaust gas temperatures from marine engines, around 300–450 °C [19], remain considerably lower than this optimal temperature range for NO_x reduction. Comprehensive investigations have been reported about SNCR analyzing parameters such as temperature, the molar ratio (NH_3/NO) [20], residence time, oxygen level, initial NO_x, combustibles, and so on [21,22], verifying that the most important factor for NO_x reduction is the temperature. Based on this result, Miyamoto et al. [23] proposed to reduce NO_x emissions by injecting ammonia or urea directly into the cylinder. They found an optimum NO_x reduction at injection timing 90° CA ATDC (crankshaft angle after top dead center), i.e., during the expansion stroke, under temperatures between 1100–1600 K. Nam and Gibbs [24] analyzed direct injection of urea and ammonia using a flow reactor which simulates a single cylinder diesel engine, while Nam and Gibbs [25] analyzed the influence of injection temperature, the molar ratio NH_3/NO, residence time, and combustion products, focusing on kinetic parameters. Larbi and Bessrour [26] developed an analytical model to analyze ammonia injection and concluded that the temperature and thus injection timing is critical. In fact, if ammonia is injected near TDC (top dead center), it performs as a fuel instead of as a NO_x reducing agent, since ammonia can also be employed as a fuel [27,28].

These aforementioned studies delivered interesting knowledge about ammonia injection, but an experimental analysis cannot provide complete information about the governing effects. In this regard, Computational Fluid Dynamics (CFD) offers an alternative method to analyze the performance and emissions on engines. In the field of medium and large marine engines, CFD is especially useful because an experimental setup is extremely expensive and a downscale model sometimes is not accurate enough. In particular, the so-called artificial inert species method allows us to investigate several chemical and physical effects separately. This method was initiated by Guo [29], who used an artificial inert component with the same properties as hydrogen to analyze the chemical, dilution and thermal effects of hydrogen addition on a HCCI engine. Voshtani et al. [30] and Neshat et al. [31] analyzed these chemical, dilution, and thermal effects on a blended fuel of isooctane and n-heptane. Subsequently, they studied these effects on reformer gas addition [32] and water addition [33].

This work presents a CFD analysis to study NO_x reduction in a commercial marine engine, the Wärtsila 6 L 46. The NO_x reduction procedure is based on ammonia injection during the expansion stroke. The artificial inert species method was applied to characterize thermal, dilution and chemical effects of ammonia injection. In addition, ammonia injection was compared with water (H_2O) injection.

2. Materials and Methods

This section describes the engine analyzed and the numerical model employed to study the performance and emissions.

2.1. Description of the Engine Analyzed

As mentioned above, the engine analyzed is the Wärtsilä 6 L 46 (Wärtsilä Corporation, Finland) [34,35], diesel, six-cylinder, four-stroke, water-cooled, and turbocharged. Each cylinder of this engine has two intake and two exhaust valves and a fuel injector with 20 holes is situated at the center of the cylinder head. This is a direct injection engine, i.e., the fuel is injected directly into the cylinder. The injection pump provides injection pressures up to 1500 bar. Optionally, this engine includes the possibility to incorporate direct water injection (DWI) to incorporate water at 400 bar from an external pump unit to each injection. The injector is thus equipped with a dual nozzle with separate needles for water and fuel. This system was employed in the numerical model to simulate water or ammonia injection.

In the present work, a comprehensive analysis was performed in a Wärtsilä 6 L 46 installed on a tuna fishing vessel. Many parameters were characterized at different loads, such as in-cylinder pressure, consumption, indicated and effective power, scavenging air pressure and temperature, exhaust gas pressure and temperature, lubricating oil pressure and temperature, cooling water temperature, emissions, etc. Although this engine is designed to operate under heavy fuel oil, marine diesel oil operation is also possible. Since these data were taken on board and near the coast, marine diesel oil was employed. The viscosity and density of this fuel are 12.5 mm^2/s and 885 kg/m^3 at 15 °C and its sulfur content 0.89%. For instance, Figure 1 indicates the results of the in-cylinder pressure along the operating cycle, at 100% load. The engine performance analyzer MALIN 6000 was employed to characterize the in-cylinder pressure. This pressure transducer is connected to the bleed valve, located at the engine head, which acts as an indicator channel. It worth mentioning that the experimental pressure trace can be distorted due to pressure waves in the channel [36], and that no algorithm was applied to correct this drawback.

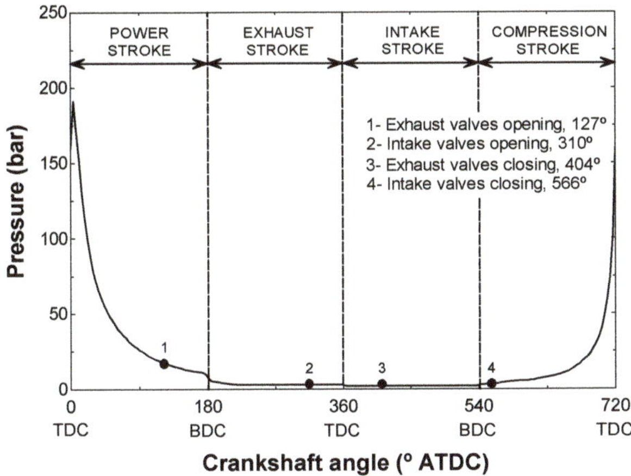

Figure 1. In-cylinder pressure at 100% load, experimentally measured.

Other experimental data at different loads are indicated in Table 1. In particular, the speed, power, mean indicated pressure (MIP), maximum pressure, specific fuel consumption (SFC), and emissions. NO$_x$, CO, HC, and CO$_2$ emissions were analyzed using the Gasboard-3000 series (Wuhan Cubic) gas analyzers, particularly Gasboard-3030 for HC and Gasboard-3000 for NO, CO, and CO$_2$. The load, speed, and SFOC were taken from the engine monitoring system.

Table 1. Experimental data.

Load (%)		25	35	50	75	100
Speed (rpm)		500	500	500	500	500
Power (kW)		2047.6	2367.8	2923.6	4051.1	5430.1
MIP (bar)		8.1	13.9	17.6	20.0	22.5
P_{max} (bar)		102.7	138.0	160.4	175.8	182.3
SFC (g/kWh)		173.2	171.9	169.8	169.5	172.1
Emissions	NO_x (ppm)	1048	1092	1149	1167	1128
	HC (ppm)	510	485	448	466	515
	CO (ppm)	261	255	247	268	292
	CO_2 (%)	3.5	4.6	6.9	7.9	8.3

2.2. Numerical Model

The open software OpenFOAM was employed in the present work. The mesh is indicated in Figure 2. In order to implement the movement of the piston and valves, a deforming mesh was used. In particular, Figure 2a represents the tridimensional mesh, Figure 2b a cross-section at BDC (bottom dead center), i.e., 180° or 540° CA ATDC, and Figure 2c a cross-section at TDC, i.e., 0° or 360° CA ATDC. Several meshes with different elements were tested in order to verify that the results are independent of the mesh size. Table 2 indicates the error obtained between experimental and numerical results of pressure and fractions using a mesh with 501,769 elements at BDC (mesh 1), as well as 802,527 elements (mesh 2) and 1,264,873 (mesh 3). As can be seen, there is no difference between the meshes 2 and 3. For this reason, the mesh 2 was chosen for the present work.

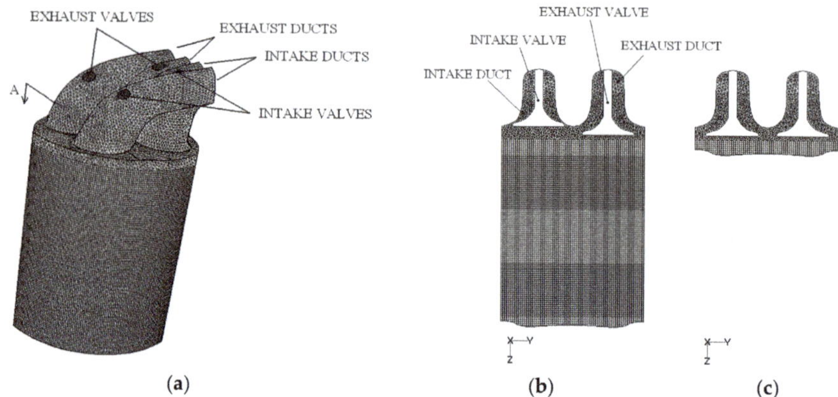

Figure 2. (**a**) Tridimensional mesh at BDC; (**b**) Cross-section mesh at BDC; (**c**) Cross-section mesh at TDC.

Table 2. Error (%) at 100% load obtained using different mesh sizes.

	P	NO_x	CO	HC	CO_2
Mesh 1	4.2	5.1	8.1	6.5	4.7
Mesh 2	4.1	4.9	7.9	6.4	4.6
Mesh 3	4.1	4.9	7.9	6.4	4.6

A new in-house solver was programmed using C++. Briefly, this solver is based on the RANS (Reynolds-averaged Navier Stokes) equations of conservation of mass, momentum, and energy. The k-ε turbulence model was chosen.

The standard Kelvin-Helmholtz and Rayleigh-Taylor breakup (KH-RT) model [37] was employed for fuel droplet breakup, and the Dukowicz model [38] for the heat-up and evaporation. A comprehensible analysis about the adequacy of breakup models can be found in the literature. Compared to other breakup models such as WAVE, TAB (Taylor Analogy Breakup), etc., the KH-RT model is more suitable for the high injection pressures that take place in diesel engines [39]. As indicated previously, ammonia and water injection were modeled through an injector equipped with a dual nozzle with separate needles for water/ammonia and fuel.

In order to solve the chemical kinetics, a reaction mechanism was programmed by adding the three kinetic schemes described in Sections 2.1–2.3 for combustion (131 reactions and 41 species), NO_x formation (43 reactions and 20 species) and NO_x reduction (131 reactions and 41 species), respectively. Several additional equations must be added to model chemical kinetics. Given a set of N species and m reactions, Equation (1), the local mass fraction of each species, f_k, can be expressed by using Equation (2).

$$\sum_{k=1}^{N} v'_{kj} M_k \overset{k_j}{\longleftrightarrow} \sum_{k=1}^{N} v''_{kj} M_k \quad j = 1, 2, \ldots, m \tag{1}$$

$$\frac{\partial}{\partial t}(\rho f_k) + \frac{\partial}{\partial x_i}(\rho u_i f_k) = \frac{\partial}{\partial x_i}\left(\frac{\mu_t}{Sc_t}\frac{\partial f_k}{\partial x_i}\right) + S_k \tag{2}$$

In the equations above, ρ is the density, u the velocity, v'_{kj} the stoichiometric coefficients of the reactant species M_k in the reaction j, v'_{kj} the stoichiometric coefficients of the product species M_k in the reaction j, Sc_t the turbulent Smidth number, and S_k the net rate of production of the species M_k by chemical reaction, given by the molecular weight multiplied by the production rate of the species, Equation (3).

$$S_k = MW_k \frac{d[M_k]}{dt} \tag{3}$$

where MW_k is the molecular weight of the species M_k and $[M_k]$ its concentration. The net progress rate is given by the production of the species M_k minus the destruction of the species M_k along the m reactions:

$$\frac{d[M_k]}{dt} = \sum_{j=1}^{m}\left\{(v'_{kj} - v'_{kj})\left[k_{fj}\prod_{k=1}^{N}[M_k]^{v'_{kj}} - k_{bj}\prod_{k=1}^{N}[M_k]^{v'_{kj}}\right]\right\} \tag{4}$$

where k_{fj} and k_{fb} are the forward and backward reaction rate constants for each reaction j.

The simulation started at 360° CA ATDC and the whole cycle was analyzed. The initial pressure was 1.31 bar, obtained from experimental measurements. Concerning boundary conditions, the intake valve pressure and temperature after the turbocharger were 2.72 bar and 514 K, respectively. The heat transfer from the cylinder to the cooling water was modeled as a combined convection-radiation type, Equation (5). Previous investigations [40] demonstrated the accuracy of this type of boundary condition in comparison with adiabatic or constant temperature:

$$q = h(T_{gas} - T_{water}) \tag{5}$$

where q is the heat transferred, T_{gas} the in-cylinder temperature, T_{water} the cooling water temperature (78 °C), and h the heat transfer coefficient, given by the following expression [41]:

$$h = 10.4 \, kb^{-1/4} \, (u_{piston}/v)^{3/4} \tag{6}$$

where b is the cylinder bore, k the thermal conductivity of the gas, u_{piston} the mean piston speed, and v the kinematic viscosity of the gas. Substituting values into the above equation yields $h = 4151 \ W/m^2K$.

2.2.1. Combustion Kinetic Scheme

The fuel was treated as n-heptane. The kinetic scheme of Ra and Reitz [42], based on 131 reactions and 41 species, was employed for combustion. Another common approach to treat combustion in CFD is to assume that the kinetics is so fast that chemical species remain at equilibrium due to the high temperatures. Nevertheless, previous works [43,44] indicated that a kinetic scheme is more accurate than the equilibrium hypothesis, since the cooling during the expansion process and dilution with the excess air elongates the time needed to achieve equilibrium. Indeed, several studies about diesel engines verified that the measured CO emissions are higher than those provided by the equilibrium concentrations. The reason is that one of the sources of CO in diesel engines are lean regions which are not able to burn properly [45,46]. This happens when the local turbulent and diffusion time scales are much smaller than the time required to achieve equilibrium. In these cases, the chemical equilibrium hypothesis leads us to overestimate the levels of the minor species. For these reasons, another procedure developed in the present work was the implementation of a chemical kinetic model. Figure 3 represents the CO and HC emissions experimentally and numerically obtained using chemical equilibrium and the kinetic model. As can be seen, the kinetic model improves the results. Regarding CO2 emissions, these remain practically inalterable so are not included in the figure. According to the improvement obtained using the kinetic model compared to the equilibrium assumption, the kinetic model developed by Ra and Reitz [42] was employed in the remainder of the present work.

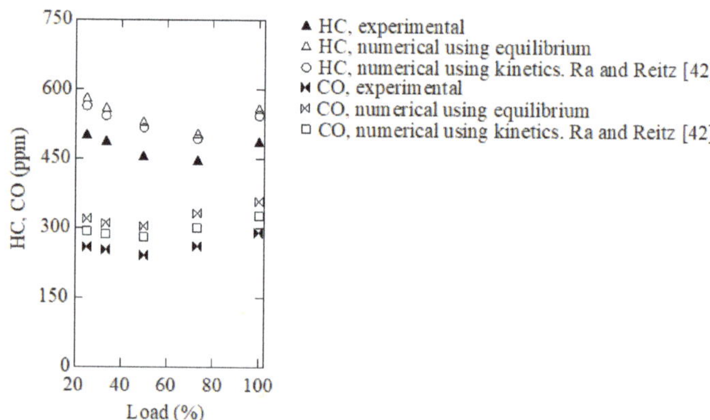

Figure 3. HC and CO emissions experimentally and numerically obtained.

2.2.2. NO$_x$ Formation Kinetic Scheme

The numerical model employs the kinetic scheme of Yang et al. [47], based on 43 reactions and 20 species. In CFD, it is common to employ the so called extended Zeldovich mechanism [48,49], based on 3 reactions and 7 species. Nevertheless, previous works [44,50] compared several kinetic schemes using experimental results and concluded that the model of Yang et al. provides satisfactorily accurate results.

2.2.3. NO$_x$ Reduction Kinetic Scheme

The kinetic scheme chosen for NO$_x$ reduction is the one proposed by Miller and Glarborg [51], based on 134 reactions and 24 species. The accuracy of this and other kinetic schemes was also compared with experimental measurements elsewhere [43,52], concluding that the model of Miller and Glarborg [51] provides satisfactorily accurate results.

2.3. Validation of the Overall Numerical Model

The overall numerical model, summarized in Table 3, was validated using experimental measurements. The emissions and consumption obtained experimentally and numerically at several loads are shown in Figure 4. This figure includes CO, CO_2, NO_x, HC, and SFC. As can be seen, a reasonable correspondence between numerical and experimental results was obtained. The in-cylinder pressure obtained experimentally and numerically at 100% load is shown in Figure 5. This figure also indicates a satisfactory correspondence between experimental and numerical results. Other loads also provided satisfactory concordance between experimental and numerical results, and thus are not represented again.

Table 3. Numerical models.

Turbulence Model	k-ε
Evaporation model	Dukowicz
Breakup model	KH-RT
Combustion model	Ra and Reitz
NO_x formation model	Yang et al.
NO_x reduction model	Miller and Glarborg

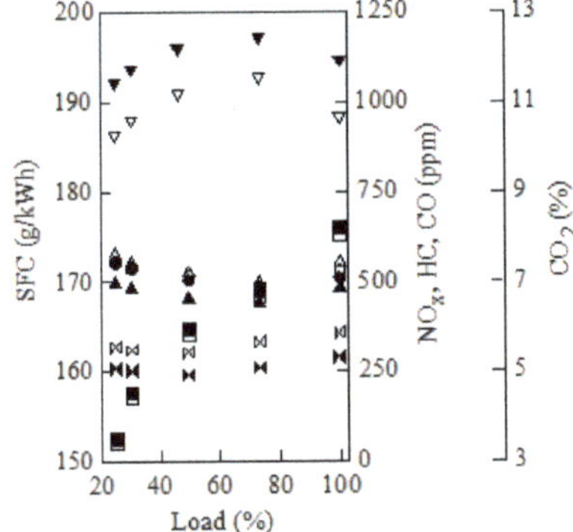

Figure 4. NO_x, HC, CO and CO_2 emissions as well as BSFC experimentally and numerically obtained.

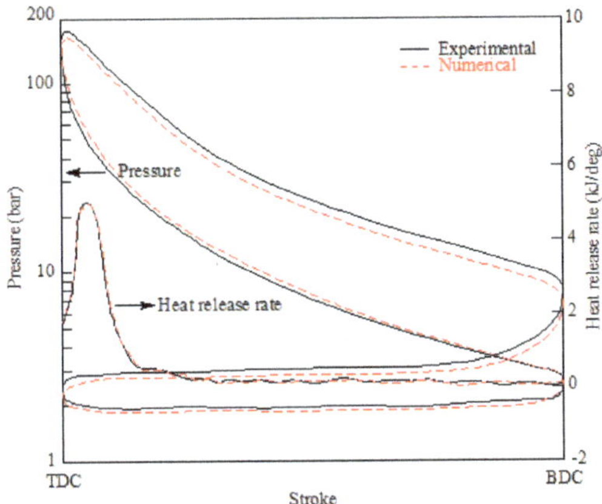

Figure 5. In-cylinder pressure experimentally and numerically obtained at 100% load.

3. Results and Discussion

Once the numerical model was validated, it was employed to analyze NO_x reduction through ammonia and water injection directly into the cylinder. This section presents the results obtained. First of all, results about NO_x reduction are exposed and then results obtained from the artificial inert species method.

3.1. NO_x Reduction

First of all, it is necessary to determine the appropriate quantity of ammonia and water. Regarding ammonia, its main disadvantage is the non-reacted ammonia slip in the exhaust gas. Ammonia is highly toxic, and thus it is important to maintain an un-reacted ammonia slip to the exhaust that is as low as possible. Figure 6 represents the NO_x reduction as well as the ammonia slip in the exhaust gas against the ammonia to fuel ratio, Equation (7), at 100% load. In this figure, the ammonia injection took place 58.4° CA ATDC. This value was chosen because it provides the maximum NO_x reduction, as will be shown below. The effect on CO, HC, and SFC remained practically negligible, so these are not represented in this figure. As can be seen, NO_x reduction improves with the ammonia to fuel ratio, with a tendency to level off around 4%. Ammonia to fuel ratios higher than 3% provide a few additional NO_x reductions with a considerable increment of un-reacted ammonia emitted to the atmosphere, and NO_x reduction drops again for higher ratios since ammonia itself oxidizes to NO. For this reason, an ammonia to fuel ratio of 4% was employed in the remainder of the present work.

$$\text{Ammonia to fuel ratio } (\%) = \frac{\text{mass of ammonia}}{\text{mass of fuel}} \, 100 \tag{7}$$

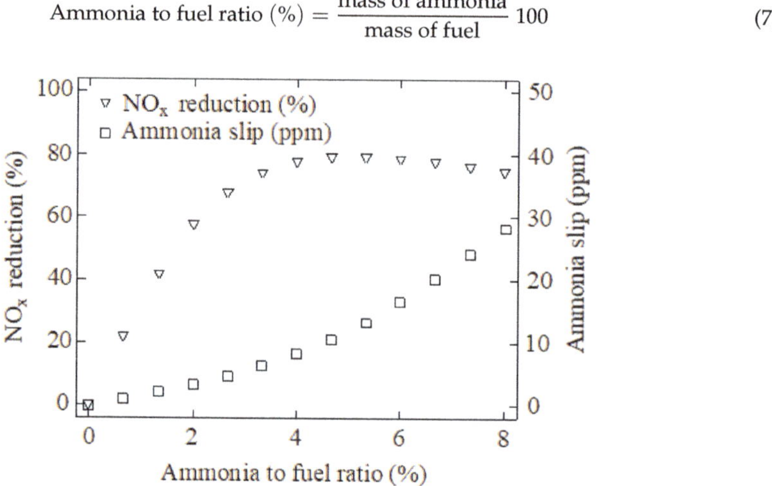

Figure 6. NO_x reduction and ammonia slip against the ammonia to fuel ratio. Ammonia injection timing: 58.4° CA ATDC.

As indicated previously, the present work focuses on water injection through a dual nozzle with separated needles for water and fuel. Using this DWI system, typical water to fuel ratios, Equation (8), in practical applications are within the range 40%–70% [20]. Figure 7 shows the NO_x reduction, as well as the effect on SFC, CO, and HC for water to fuel ratios from 0 to 100% at 100% load. As can be seen in this figure, the water to fuel ratio improves NO_x reduction, but increments both consumption and emissions of CO and HC. For this reason and taking into account usual practical applications, a water to fuel ratio of 70% was employed in the remainder of the present work.

$$\text{Water to fuel ratio } (\%) = \frac{\text{mass of water}}{\text{mass of fuel}} \, 100 \tag{8}$$

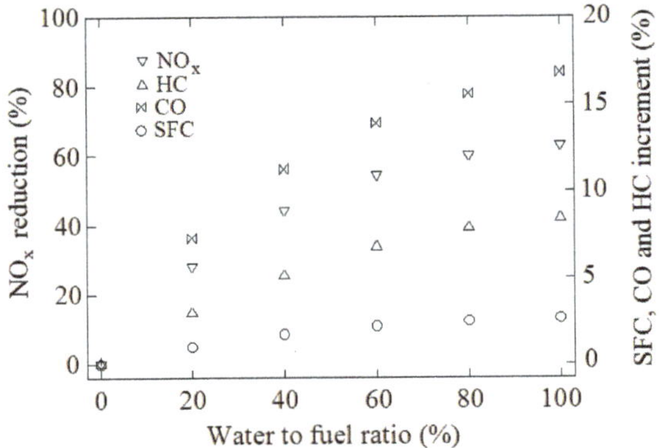

Figure 7. NO_x reduction against the water to fuel ratio. Water injection timing: −2.1° CA ATDC.

Figure 8 shows the effect of the injection timing on NO_x reduction. Ammonia and water were compared using 4% ammonia to fuel ratio and 70% water to fuel ratio. As can be seen in this figure, using water injection a maximum 57.1% NO_x reduction was obtained at −2.1° CA ATDC. On the other hand, if ammonia is injected around TDC, then the NO_x reduction is considerably smaller than when using a water injection. Nevertheless, at 58.4° CA ATDC, NO_x reduction reaches 78.1% using ammonia. As mentioned in the introduction, NO_x reduction using ammonia is very sensitive to the temperature. Injected near TDC, ammonia is not efficient due to the excessive in-cylinder temperatures. Nevertheless, at 58.4° CA ATDC, the in-cylinder temperatures reduce to the optimal values required for NO_x reduction using ammonia. Instead of 100% loads, at lower loads the in-cylinder temperatures are lower too and thus the optimum injection time takes place before 58.4° CA ATDC.

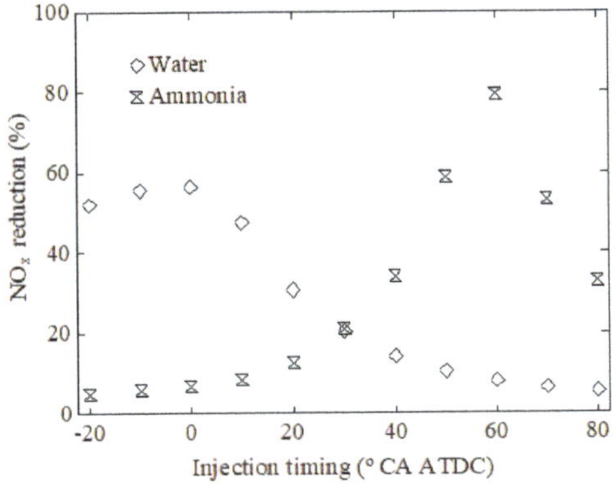

Figure 8. NO_x reduction against the injection timing using ammonia and water. Ammonia to fuel ratio: 4%, water to fuel ratio: 70%.

Figure 9 shows the maximum temperature for the base case without a water or ammonia injection, with a 4% ammonia to fuel ratio and with a 70% water to fuel ratio at 100% load. In these simulations, both ammonia and water were injected at −2.1° CA ATDC. As can be seen, water promotes a reduction in the combustion temperatures. The maximum temperature is lowered 93.2 °C if 70% water is injected at −2.1° CA ATDC. On the other hand, ammonia increases the maximum temperature 8.4 °C if this is injected at −2.1° CA ATDC. This explains the effect on CO, HC, and SFC. As indicated above, ammonia has a negligible effect on these parameters and water increases them. Water reduces the combustion temperature due to the increment in the specific heat capacity of the cylinder gases (water has higher specific heat capacity than air) and lowers the concentration of oxygen, which reduces the availability of oxygen for the NO_x forming reactions. The main effect is a reduction in NO_x emissions due to the lower temperatures, but water injection also promotes incomplete combustion and thus increases both CO and HC emissions as well as SFC. SFC is increased due to the lower pressures, which promotes lower power. On the other hand, when injected near TDC, ammonia acts as a fuel and slightly increases the combustion temperature with a negligible effect on CO, HC and SFC. In the next section, the chemical effect of water and ammonia will be analyzed too using the artificial inert species method.

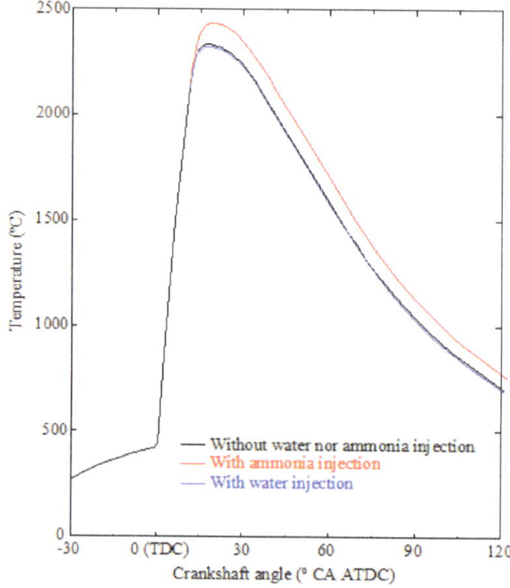

Figure 9. Maximum temperature without water nor ammonia injection; with a 4% ammonia to fuel ratio and an injection timing of −2.1° CA ATDC; water to fuel ratio: 70% and injection timing: −2.1° CA ATDC.

3.2. Artificial Inert Species Method

In this section the so-called artificial inert species method [29–33] is applied to analyze the chemical, thermal, and dilution effects of both ammonia and water injection. The chemical effect is promoted by the chemical reactions. The thermal effect is promoted by the properties, specially the high specific heat capacity of ammonia and water which increases the heat absorption. The dilution effect is promoted by the presence of the additive, which reduces the possibility of reaction between fuel and air.

In the case of water injection, two artificial species were added: inertH$_2$O and inertAIR. The species inertH$_2$O has the same properties as water but does not participate in the chemical reactions. On the other hand, the species inertAIR has the same properties as air but does not participate in the chemical reactions. According to this, the difference between the results using water and inertH$_2$O represents the chemical effect. The dilution effect of water injection is represented by the difference between the results using inertAIR and the base case without water. Finally, the difference between the results using inertH$_2$O and inertAIR leads to the thermal effect of water injection. Figure 10 illustrates the contribution of thermal, dilution, and chemical effects of water injection for a water injection timing of −2.1° CA ATDC at a 100% load. As can be seen, the chemical effect is negligible. Water injection reduces NO_x by the dilution and, more significantly, thermal effects. The thermal effect is important since water absorbs heat due to its high specific heat capacity. Since water has a higher specific heat capacity than air, the specific heat capacity of the cylinder gases is increased, leading to a reduction in the in-cylinder temperature and thus to the NO_x emissions. Figure 10 also illustrates the importance of the dilution effect. The presence of water reduces the interaction between fuel and air and thus deteriorate the combustion process.

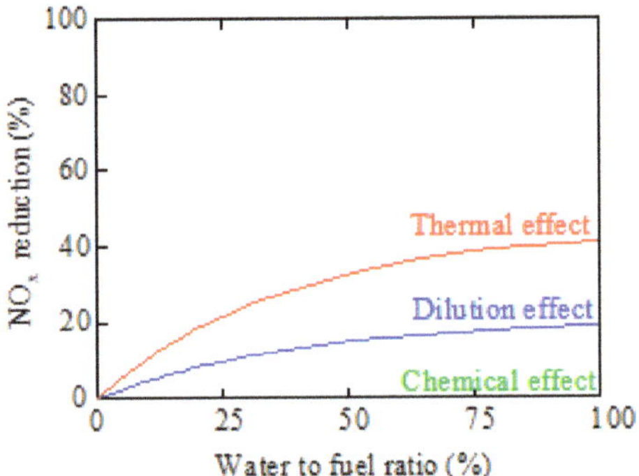

Figure 10. Chemical, thermal, and dilution effects of water injection on NO_x emissions. Water injection timing: −2.1° CA ATDC.

In the case of ammonia injection, two artificial species were added: inertNH$_3$ and inertAIR. The species inertNH$_3$ has the same properties as ammonia but does not participate in the chemical reactions. On the other hand, the species inertAIR has the same properties as air but does not participate in the chemical reactions. According to this, the difference of the results using ammonia and inertNH$_3$ represents the chemical effect of ammonia injection. The dilution effect of ammonia injection is represented by the difference between the results using inertAIR and the base case without ammonia. Finally, the difference between the results using inertNH$_3$ and inertAIR leads to the thermal effect of ammonia. Figure 11 illustrates the contribution of thermal, dilution, and chemical effects of ammonia injection for an ammonia injection timing of 58.4° CA ATDC at 100% load. As can be seen, the chemical effect is the only one responsible for NO_x reduction, while the thermal and dilution effects are negligible.

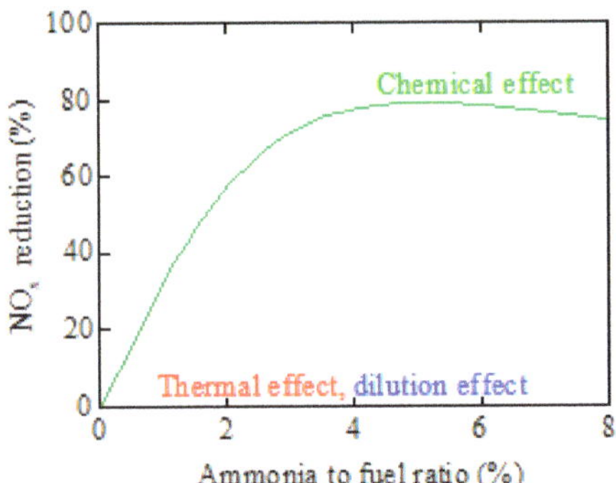

Figure 11. Chemical, thermal, and dilution effects of ammonia injection on NO_x emissions. Ammonia injection timing: 58.4° CA ATDC.

If ammonia is injected near TDC, particularly at −2.1° CA ATDC, the NO_x reduction is noticeably lower, Figure 12. The thermal and dilution effects become more important and the chemical effect is reduced.

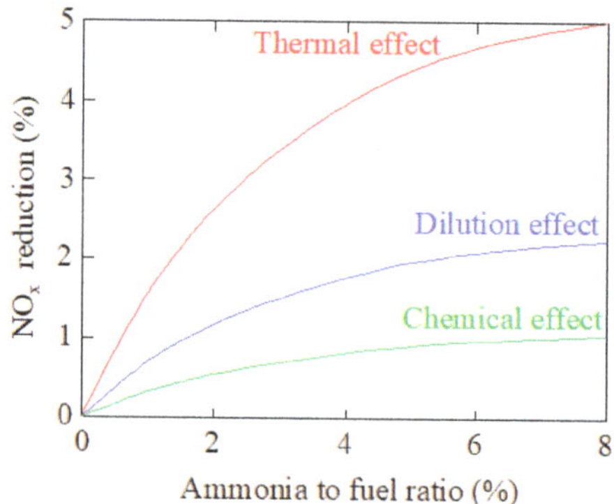

Figure 12. Chemical, thermal, and dilution effects of ammonia injection on NO$_x$ emissions. Ammonia injection timing: −2.1° CA ATDC.

4. Conclusions

This work presents a CFD analysis to study NO$_x$ reduction in a commercial marine engine, the Wärtsila 6 L 46. The NO$_x$ reduction is based on ammonia injection directly into the cylinder, and this measure was compared with water injection. The so-called artificial inert species method was employed. Inert species with the same properties of water, ammonia, and air were used to characterize the chemical, thermal, and dilution effects of water and ammonia injection.

It was found that the chemical effect using ammonia injection is extremely dependent on the injection timing. The optimum NO$_x$ reduction using ammonia is obtained when this is injected during the expansion stroke, leading to a significant chemical effect, and negligible thermal and dilution effects. The optimum NO$_x$ reduction using water is obtained when this is injected near TDC. Injected near TDC, water promotes NO$_x$ reduction by the dilution and, more significantly, the thermal effect. If ammonia is injected near TDC, the thermal and dilution effects become more significant but the global NO$_x$ reduction is noticeably lower than the values obtained when using water.

Author Contributions: Conceptualization, M.I.L.G. and C.G.R.V.; methodology, M.I.L.G. and C.G.R.V.; software, M.I.L.G. and C.G.R.V.; validation, M.I.L.G. and C.G.R.V.; formal analysis, M.I.L.G. and L.C.-S.; investigation, M.I.L.G. and C.G.R.V.; resources, M.I.L.G. and C.G.R.V.; writing—original draft preparation, M.I.L.G. and L.C.-S.; writing—review and editing, M.I.L.G. and L.C.-S. All authors have read and agreed to the published version of the manuscript.

Funding: This research received no external funding.

Acknowledgments: The authors would like to express their gratitude to Talleres Pineiro S.L., sale and repair of marine engines, as well as to Norplan Engineering S.L.

Conflicts of Interest: The authors declare no conflict of interest.

References

1. Ammar, N.R.; Seddiek, I.S. Eco-environmental analysis of ship emission control methods: Case study RO-RO cargo vessel. *Ocean Eng.* **2017**, *137*, 166–173. [CrossRef]
2. Seddiek, I.S.; ElGohary, M.M.; Ammar, N.R. The hydrogen-fuelled internal combustion engine for marine application with a case study. *Brodogradnja* **2015**, *66*, 23–38.
3. ElGohary, M.M.; Seddiek, I.S. Utilization of alternative marine fuels for gas turbine power plant onboard ships. *Int. J. Naval Archit. Ocean Eng.* **2013**, *5*, 21–32. [CrossRef]
4. Lamas, M.I.; Rodríguez, C.G.; Telmo, J.; Rodríguez, J.D. Numerical analysis of emissions from marine engines using alternative fuels. *Polish Marit. Res.* **2015**, *22*, 48–52. [CrossRef]
5. Kuiken, K. *Diesel Engines for Ship Propulsion and Power Plants*, 3rd ed.; Target Global Energy Training: The Netherlands, 2017.

6. Di Sarli, V. Stability and emissions of a lean pre-mixed combustor with rich catalytic/lean-burn pilot. *Int. J. Chem. React. Eng.* **2014**, *12*, 77–89. [CrossRef]

7. Sencic, T.; Mrzljak, V.; Blecich, P.; Bonefacic, I. 2D CFD simulation of water injection strategies in a large marine engine. *J. Mar. Sci. Eng.* **2019**, *7*, 296. [CrossRef]

8. Lamas, M.I.; Rodríguez, C.G.; Rodríguez, J.D.; Telmo, J. Internal modifications to reduce pollutant emissions from marine engines. A numerical approach. *Int. J. Naval Archit. Mar. Eng.* **2013**, *5*, 493–501. [CrossRef]

9. Lamas, M.I.; Rodríguez, C.G.; Aas, H.P. Computational fluid dynamics analysis of NOx and other pollutants in the MAN B&W 7S50MC marine engine and effect of EGR and water addition. *Int. J. Marit. Eng.* **2013**, *155*, A81–A88.

10. Puskar, M.; Bigos, P.; Balazikova, M.; Petkova, V. The measurement method solving the problems of engine output characteristics caused by change in atmospheric conditions on the principle of the theory of optimal temperature range of exhaust system. *Meas. J. Int. Meas. Confed.* **2013**, *46*, 467–475. [CrossRef]

11. Rota, R.; Zanoelo, E.F.; Antos, D.; Morbidelli, M.; Carra, S. Analysis of the thermal DeNOx process at high partial pressure of reactants. *Chem. Eng. Sci.* **2000**, *55*, 1041–1051. [CrossRef]

12. Glarborg, P.; Dam-Johansen, K.; Miller, J.A.; Kee, R.J.; Coltrin, M.E. Modeling the thermal DeNOx process in flow reactors. Surface effects and nitrous oxide formation. *Int. J. Chem. Kinet.* **1994**, *26*, 421–436. [CrossRef]

13. Rota, R.; Antos, D.; Zanoelo, E.F.; Morbidelli, M. Experimental and modeling analysis of the NOxOUT process. *Chem. Eng. Sci.* **2002**, *57*, 27–38. [CrossRef]

14. Birkhold, F.; Meinghast, U.; Wassermann, P. *Analysis of the Injection of Urea-Water-Solution for Automotive SCR DeNOx-Systems: Modeling of Two-Phase Flow and Spray/Wall Interaction*; SAE Technical Paper 2006-01-0643; SAE International: Warrendale, PA, USA, 2006. [CrossRef]

15. Javed, M.T.; Irfan, N.; Gibbs, B.M. Control of combustion-generated nitrogen oxides by selective non-catalytic reduction. *J. Environ. Manag.* **2007**, *83*, 251–289. [CrossRef] [PubMed]

16. Caton, J.A.; Xia, Z. The selective non-catalytic removal (SNCR) of nitric oxides from engine exhaust streams: Comparison of three processes. *Trans. ASME* **2004**, *126*, 234–240. [CrossRef]

17. Miller, J.A.; Bowman, C. Mechanism and modeling of nitrogen chemistry in combustion. *Prog. Energy Combust. Sci.* **1989**, *15*, 287–338. [CrossRef]

18. Lyon, R.K. Method for the Reduction of the Concentration of NO in Combustion Effluents Using Ammonia. U.S. Patent 3,900,554, 19 August 1975.

19. Woodyard, D. *Pounder's Marine Engines and Gas Turbines*, 9th ed.; Elsevier: Oxford, UK, 2009.

20. Kimball-Linne, M.A.; Hanson, R.K. Combustion-driven flow reactor studies of thermal DeNOx reaction kinetics. *Combust. Flame* **1986**, *64*, 337–351. [CrossRef]

21. Lyon, R.K.; Benn, D. Kinetics of the NO-NH3-O2 reaction. *Symp. Combust.* **1979**, *17*, 601–610. [CrossRef]

22. Kasuya, F.; Glarborg, P.; Johnsson, J.E.; Dam-Johansen, K. The thermal DeNOx process: Influence of partial pressures and temperature. *Chem. Eng. Sci.* **1995**, *50*, 1455–1466. [CrossRef]

23. Miyamoto, N.; Ogawa, H.; Wang, J.; Shudo, T.; Yamazaki, K. Diesel NOx reduction with ammonium deoxidizing agents directly injected into the cylinder. *Int. J. Veh. Des.* **1995**, *16*, 71–79. [CrossRef]

24. Nam, C.M.; Gibbs, B.M. Selective non-catalytic reduction of NOx under diesel engine conditions. *Proc. Combust. Inst.* **2000**, *28*, 1203–1209. [CrossRef]

25. Nam, C.M.; Gibbs, B.M. Application of the thermal DeNOx process to diesel engine DeNOx: An experimental and kinetic modelling study. *Fuel* **2002**, *81*, 1359–1367. [CrossRef]

26. Larbi, N.; Bessrour, J. Measurement and simulation of pollutant emissions from marine diesel combustion engine and their reduction by ammonia injection. *Adv. Mech. Eng.* **2010**, *41*, 898–906. [CrossRef]

27. Heywood, J. *Internal Combustion Engine Fundamentals*, 2nd ed.; McGraw-Hill Education: New York, NY, USA, 2018.

28. Lesmana, H.; Zhang, Z.; Li, X.; Zhu, M.; Xu, X.; Zhang, D. NH3 as a transport fuel in internal combustion engines: A technical review. *J. Energy Resour. Technol.* **2019**, *141*, 070703–070714. [CrossRef]

29. Guo, H.; Neill, W.S. The effect of hydrogen addition on combustion and emission characteristics of an n-heptane fueled HCCI engine. *Int. J. Hydrog. Energy* **2013**, *38*, 11429–11437. [CrossRef]

30. Voshtani, S.; Reyhanian, M.; Ehteram, M.; Hosseini, V. Investigating various effects of reformer gas enrichment on a natural gas-fueled HCCI combustion engine. *Int. J. Hydrog. Energy* **2014**, *39*, 19799–19809. [CrossRef]

31. Neshat, E.; Saray, R.K.; Hosseini, V. Effect of reformer gas blending on homogeneous charge compression ignition combustion of primary reference fuels using multi zone model and semi detailed chemical-kinetic mechanism. *Appl. Energy* **2016**, *179*, 463–478. [CrossRef]

32. Neshat, E.; Saray, R.K.; Parsa, S. Numerical analysis of the effects of reformer gas on supercharged n-heptane HCCI combustion. *Fuel* **2017**, *200*, 488–498. [CrossRef]

33. Neshat, E.; Bajestani, A.V.; Honnery, D. Advanced numerical analyses on thermal, chemical and dilution effects of water addition on diesel engine performance and emissions utilizing artificial inert species. *Fuel* **2019**, *242*, 596–606. [CrossRef]

34. Lamas, M.I.; Rodríguez, C.G.; Rebollido, J.M. Numerical model to study the valve overlap period in the Wärtsilä 6L46 four-stroke marine engine. *Polish Marit. Res.* **2012**, *19*, 31–37. [CrossRef]

35. Lamas, M.I.; Rodríguez, C.G. Numerical model to study the combustion process and emissions in the Wärtsilä 6L 46 four-stroke marine engine. *Polish Marit. Res.* **2013**, *20*, 61–66. [CrossRef]

36. Polanowski, S.; Pawletko, R.; Witkowski, K. Influence of the indicator channel and indicator valve on the heat release characteristics of medium speed marine diesel engines. *Key Eng. Mater.* **2014**, *588*, 149–156. [CrossRef]

37. Ricart, L.M.; Xin, J.; Bower, G.R.; Reitz, R.D. *In-Cylinder Measurement and Modeling of Liquid Fuel Spray Penetration in a Heavy-Duty Diesel Engine*; SAE Technical Paper 971591; SAE International: Warrendale, PA, USA, 1997. [CrossRef]

38. Dukowicz, J.K. A particle-fluid numerical model for liquid sprays. *J. Comput. Phys.* **1980**, *35*, 229–253. [CrossRef]

39. Weber, J. *Optimization Methods for the Mixture Formation and Combustion Process in Diesel Engines*, 1st ed.; Cuvillier Velag: Göttingen, Germany, 2008.

40. Sigurdsson, E.; Ingvorsen, K.M.; Jensen, M.V.; Mayer, S.; Matlok, S.; Walther, J.H. Numerical analysis of the scavenge flow and convective heat transfer in large two-stroke marine diesel engines. *Appl. Energy* **2014**, *123*, 37–46. [CrossRef]

41. Taylor, C.F. *The Internal Combustion Engine in Theory and Practice*, 2nd ed; MIT Press: Cambridge, MA, USA, 1985.

42. Ra, Y.; Reitz, R. A reduced chemical kinetic model for IC engine combustion simulations with primary reference fuels. *Combust. Flame* **2008**, *155*, 713–738. [CrossRef]

43. Lamas, M.; Rodriguez, C.G. Numerical model to analyze NOx reduction by ammonia injection in diesel-hydrogen engines. *Int. J. Hydrog. Energy* **2017**, *42*, 26132–26141. [CrossRef]

44. Lamas, M.I.; de Dios Rodriguez, J.; Castro-Santos, L.; Carral, L.M. Effect of multiple injection strategies on emissions and performance in the Wärtsilä 6L 46 marine engine. A numerical approach. *J. Clean. Prod.* **2019**, *206*, 1–10. [CrossRef]

45. Sher, E. *Handbook of Air Pollution from Internal Combustion Engines*; Academic Press: Boston, MA, USA, 1998.

46. Lapuerta, M.; Hernández, J.; Armas, O. *Kinetic Modelling of Gaseous Emissions in a Diesel Engine*; SAE Technical Paper 2000-01-2939; SAE International: Warrendale, PA, USA, 2000. [CrossRef]

47. Yang, H.; Krishnan, S.R.; Srinivasan, K.K.; Midkiff, K.C. Modeling of NOx emissions using a super-extended Zeldovich mechanism. In Proceedings of the ICEF03 2003 Fall Technical Conference of the ASME Internal Combustion Engine Division, Erie, PA, USA, 7–10 September 2003.

48. Zeldovich, Y.B.; Sadovnikov, D.A.; Kamenetskii, F. *Oxidation of Nitrogen in Combustion*; Institute of Chemical Physics: Moscow-Leningrad, Russia, 1947.

49. Lavoie, G.A.; Heywood, J.B.; Keck, J.C. Experimental and theoretical investigation of nitric oxide formation in internal combustion engines. *Combust. Sci. Technol.* **1970**, *1*, 313–326. [CrossRef]

50. Lamas, M.I.; Rodriguez, C.G. NOx reduction in diesel-hydrogen engines using different strategies of ammonia injection. *Energies* **2019**, *12*, 1255. [CrossRef]

51. Miller, J.A.; Glarborg, P. Modeling the formation of N2O and NO2 in the thermal DeNOx process. *Springer Ser. Chem. Phys.* **1996**, *61*, 318–333.

52. Lamas, M.I.; Rodriguez, C.G.; Rodriguez, J.D.; Telmo, J. Computational fluid dynamics of NOx reduction by ammonia injection in the MAN B&W 7S50MC marine engine. *Int. J. Marit. Eng.* **2014**, *156*, A213–A220. [CrossRef]

Journal of
Marine Science and Engineering

Article

Research of the Effectiveness of Selected Methods of Reducing Toxic Exhaust Emissions of Marine Diesel Engines

Kazimierz Witkowski

Mechanical Faculty, Faculty of Marine Engineering, Gdynia Maritime University, Morska Street 83, 81-225 Gdynia, Poland; k.witkowski@wm.umg.edu.pl

Received: 19 May 2020; Accepted: 17 June 2020; Published: 20 June 2020

Abstract: The article's applications are very important, as it is only a dozen or so years since the current issues of protection of the atmosphere against emissions of toxic compounds from ships. The issue was discussed against the background of binding legal norms, including rules introduced by the IMO (International Maritime Organization) in the context of the MARPOL Convention (International Convention for the Prevention of Pollution from Ships), Annex VI, with the main goal to significantly strengthen the emission limits in light of technological improvements. Taking these standards into account, effective methods should be implemented to reduce toxic compounds' emissions to the atmosphere, including nitrogen oxides NO_x and carbon dioxide CO_2. The purpose of the article was, based on the results of our own research, to indicate the impact of the effectiveness of selected methods on reducing the level of nitrogen oxides and carbon dioxide emitted by the marine engine. The laboratory tests were carried out with the use of the one-cylinder two stroke, crosshead supercharged diesel engine. Methods of reducing their emissions in the study were adopted, including supplying the engine with fuel mixtures of marine diesel oil (MDO) and rapeseed oil ester (RME)-(MDO/RME mixtures) and changing the fuel injection parameters and the advance angles of fuel injection. The supply of the engine during the tests and the mixtures of marine diesel oil (MDO) and rape oil esters (RMEs) caused a clear drop in emissions of nitrogen oxides and carbon dioxide, particularly for a higher engine load, as has been shown. The decrease of the injection advance angle unambiguously makes the NO_x content in exhaust gas lower.

Keywords: ships diesel engines; exhaust gas emission; fuel mixtures; rapeseed oil methyl ester; marine diesel oil; fuel injection parameters

---

1. Introduction

Global warming, i.e., in the last approximately 50 years, the observed gradual increase in the average temperature at the Earth's surface, is a phenomenon caused by the influence of humans on the intensification of the greenhouse effect.

It is estimated that over the last century, the average temperature at the Earth's surface has increased by around 0.74 °C (±0.18) [1], but 2018 IPCC (Intergovernmental Panel on Climate Change) data indicate a 1.5 degree increase in this temperature [2].

The industrial revolution is most often associated with global warming. The official position of the IPCC (Intergovernmental Panel on Climate Change—the Intergovernmental Panel on Climate Change) says that "most of the observed increase in global average temperatures since the mid-twentieth century is probably due to the increase in anthropogenic greenhouse gas concentrations" [3].

Greenhouse gases are commonly deemed inter alia as water vapor, carbon dioxide, methane, chlorofluorocarbons, nitrous oxide, and halons. Of these, marine engine exhaust gas includes more than 5% water vapor and approximately 5% carbon dioxide, which is an integral product of the combustion of fossil fuels. Since the amount of carbon dioxide emitted is proportional to the quantities of unburnt fuel, it can be concluded that the piston engine commonly used as a source of ship propulsion is the friendliest for the atmosphere from the known conventional solutions. This is influenced by the efficiency of the reciprocating engines, which is the highest among all heat engines [4,5].

It is estimated that the total amount of carbon dioxide that is emitted into the atmosphere by the burning of fuels is about 26,583 million tons per year. Only approximately 2% of this figure represents CO_2 from ship engines used for the propulsion of ships at −521 million tons per year. These data are shown in Figure 1. Despite these favorable data for maritime transport, the aim is now to drastically reduce the negative impact of maritime transport on atmospheric pollution.

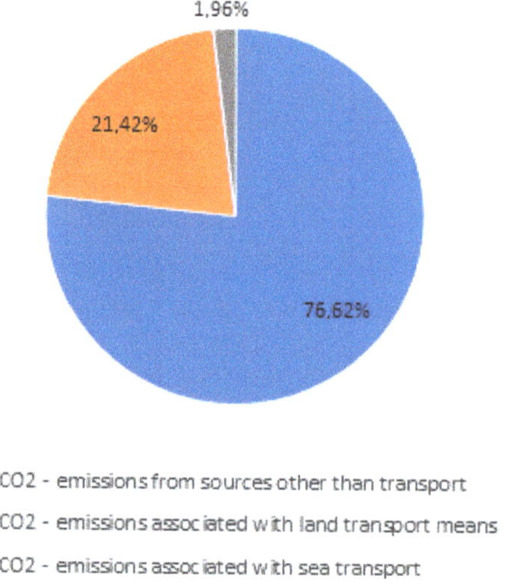

Figure 1. Percentage share of global emissions of CO_2 by source of origin (own compilation according to [1]).

Atmosphere protection against pollution from sea vessels is one of the most important areas of human ecological activity, which has its own history as well as some achievements. The most crucial ones include 73/78 MARPOL Convention (International Convention for the Prevention of Pollution from Ships) referring to prevention against marine environmental pollution, and later amendments to the Convention with Annex VI dealing with reducing the emission of nitric oxides and sulphur oxides into the atmosphere by sea vessel engines. It also prohibits deliberate emissions of ozone-depleting substances, regulates ship-board incineration, and the emissions of volatile organic compounds from tankers (MARPOL Annex VI). Annex VI entry came into force on May 2005. Already, in July 2005, the Marine Environmental Protection Committee (MEPC) agreed to revise MARPOL Annex VI, with the main goal of significantly strengthening the emission limits in light of technological improvements. In October 2008, MEPC adopted the revised Annex VI and the associated NO_x Technical Code 2008, which entered into force on 1 July 2010. They concerned a progressive reduction globally in emissions of oxides of nitrogen (NO_x), sulphur oxides (SO_x), and particulate matter (PM), and a reduction of the emission in emission control areas (ECAs). ECAs are the Baltic Sea, North Sea, and North American and United States Caribbean Sea.

2. Exhaust Emission by Marine Diesel Engine

Exhaust emitted by marine diesel engines contains a number of combustion products that are noxious to the environment. The composition of these gases depends on the content of working liquids delivered to the engine, that is on the air, fuel, and lubricating oil (see Figure 2), and the combustion process. Exhaust gases emitted from marine diesel engines comprise nitrogen (N_2), oxygen (O_2), carbon dioxide (CO_2) and water vapor (H_2O), and pollutants, including nitrogen oxides (NO_x), sulphur oxides (SO_x), carbon monoxide (CO), hydrocarbons (HC), and particulate matter (PM).

Specific emission of the toxic components (kg/kWh) of exhaust gases is a basic coefficient of atmospheric pollution. Typical analysis of the exhaust gases from a modern low-speed two-stroke marine diesel engine is shown in Figure 2.

2.1. Control of NO_x Emissions

Nitrogen oxides (NO_x) are formed from nitrogen and oxygen at high temperatures of combustion in the cylinder. NO_x emissions are considered carcinogenic compounds and contribute to the formation of photochemical smog and acid rain.

A global approach to the control of NO_x emissions has been undertaken by the IMO through Annex VI to MARPOL 73/78.

Annex VI applies to engines with power over 130 kW installed on new ships built after 1 January 2000 (the date the keel was laid) and pre-built engines that are subject to significant technical changes.

The starting NO_x emission level recommended by the IMO (dependent on the rotational speed of the engine crankshaft—n) is as follows:

–   17 g/kWh when the diesel engine n is less than 130 rpm;
– $45 \times n^{-0.2}$ g/kWh when $2000 > n > 130$; and
–   9.84 g/kWh when $n > 2000$ rpm.

Amendments agreed by IMO in 2008 will set progressively tighter NO_x emission standards for new engines, depending on the date of their installation (see also Table 1).

Table 1. The maximum content of NO_x by MARPOL Annex VI.

Year	The Maximum Content of NO_x in the Exhaust Gas		
	$n < 130$	$130 \leq n < 2000$	$n \geq 2000$
2000	17.0	$45 \times n^{-0.2}$	9.8
2011	14.4	$44 \times n^{-0.23}$	7.9
2016 *	3.4	$9 \times n^{-0.2}$	1.96

* The maximum content of NO_x in areas of special control. In areas of common border with the values of 2011.

Tier I applies to diesel engines installed on ships constructed on or after 1 January 2000 and prior to 1 January 2011, and represents the 17 g/kWh NO_x emissions standard stipulated in the original Annex VI.

Tier II covers engines installed in a ship constructed on or after 1 January 2011, and reduces the NO_x emission limit to 14.4 g/kWh.

Tier III, covering engines installed in a ship constructed on or after 1 January 2016, reduces the NO_x emissions limit to 3.4 g/kWh when the ship is operating in a designated ECA. Outside such an area, Tier II limits will apply.

Much tougher curbs on NO_x and other emissions are set by regional authorities, such as California's Air Resources Board, and Sweden has introduced a system of differentiated ports and fairway dues, making ships with higher NO_x emissions pay higher fees than more environment-friendly tonnage of a similar size.

To show compliance, an engine has to be certified according to the NO_x technical code and delivered with an Engine International Air Pollution Prevention (EIAPP) certificate of compliance. The certification process includes NO_x measurement for the engine type concerned, stamping of components that affect NO_x formation, and a technical file, which is delivered with the engine.

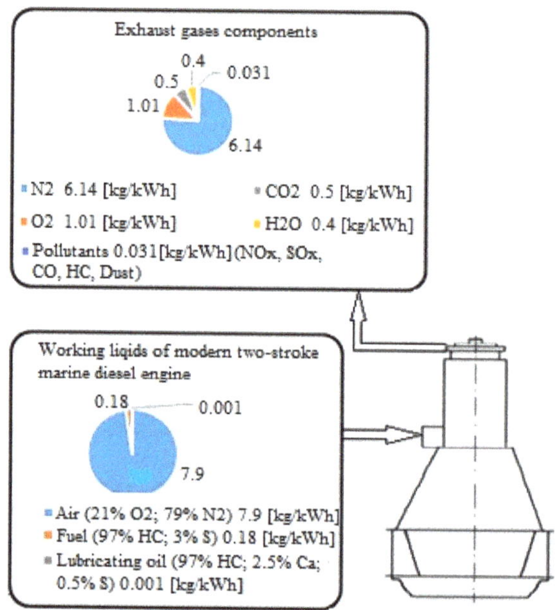

Figure 2. Exhaust gas components of a modern two-stroke marine diesel engine (based on [4]).

NO_x technical code-certified engines have a technical file, which includes the applicable survey regime, termed the onboard NO_x verification procedure. The associated parameter check method effectively stipulates the engine components and range of settings to be adopted to ensure that NO_x emissions from the given engine, under reference conditions, will be maintained within the certified value.

2.2. The Overall Amount of Global Sulphur Oxide Emissions at the Sea and in the Port Areas

Studies on sulphur pollution showed that in 1990, SO_x emissions from ships contributed around 4% to the total in Europe. In 2001, such emissions represented around 12% of the total and could rise to as high as 18%.

The simplest approach to reducing SO_x emissions is to burn bunkers with a low sulphur content. A global heavy fuel oil sulphur content cap of 4.5% and a fuel sulphur limit of 1.5% in certain designated sulphur emission control areas (SECAs), such as the Baltic Sea, North Sea, and English Channel, are currently mandated by the International Maritime Organization (IMO) to reduce SO_x pollution at the sea and in the port areas. In 2008, the IMO approved further amendments to curb SO_x emissions (see also Figure 3):

- The fuel sulphur limit applicable in emission control areas (ECA)s from 1 March 2010 would be 1% (10,000 ppm), reduced from the existing 1.5% content (15,000 ppm).
- The global fuel sulphur cap would be reduced to 3.5% (35,000 ppm), reduced from the existing 4.5% (45,000 ppm), effective from 1 January 2012.
- The fuel sulphur limit applicable in ECAs from 1 January 2015 would be 0.1% (1000 ppm).
- The global fuel sulphur cap would be reduced to 0.5% (5000 ppm) effective from 1 January 2020. Currently, all limits apply.

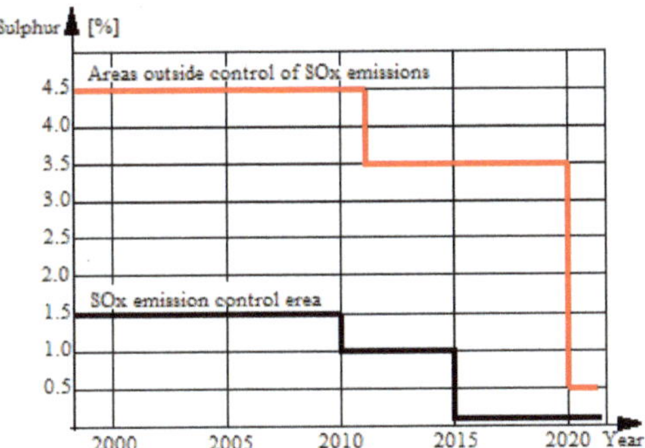

Figure 3. The maximum sulphur content of marine fuel by the MARPOL Convention (author's drawing based on the data with [6]).

2.3. Control of CO_2 Emissions

In total, 6% (0.5 kg/kWh) of the exhaust gas emission from marine diesel engines is carbon dioxide. Although it is nontoxic itself, carbon dioxide contributes to the greenhouse effect (global warming and climate change) and hence to changes in the Earth's atmosphere. This gas is an inevitable product of combustion of all fossil fuels, but emissions from diesel engines, thanks to their thermal efficiency, are the lowest of the all heat engines. A lower fuel consumption translates to reduced carbon dioxide emissions since the amount produced is directly proportional to the volume of fuel used, and therefore to the engine or plant efficiency.

International concern over the atmospheric effect of carbon dioxide has stimulated measures and plans to curb the growth of such emissions, and the marine industry must be prepared for future legislation. There are currently no mandatory regulations on carbon dioxide emissions from shipping, but they are expected. Under international agreements, such as the Kyoto Protocol and the European Union's accord on greenhouse gases, many governments are committed to substantial reductions in the total emissions of carbon dioxide.

The Conference of Parties to the International Convention for the Prevention of Pollution from Ships, 1973, as modified by the Protocol of 1978 relating thereto, held from 15 to 26 September 1997 in conjunction with the Marine Environment Protection Committee's 40th session, adopted Conference resolution 8 on CO_2 emissions from ships. The Marine Environment Protection Committee, at its 59th session (13 to 17 July 2009), agreed to circulate the guidelines for voluntary use of the ship energy efficiency operational indicator (EEOI) as set out in the annex. This document constitutes the guidelines for the use of an energy efficiency operational indicator (EEOI) for ships. It sets out:

– What the objectives of the IMO CO_2 emissions indicator are;
– How a ship's CO_2 performance should be measured; and
– How the index could be used to promote low-emission shipping, in order to help limit the impact of shipping on global climate change.

2.4. Reduction of NO_x Emissions

Fuel brought to marine engine cylinders contains potential and chemical energy, which changes by a combustion process into the thermal energy. Analysis of the combustion process in the cylinder and the reactions that are involved in nitric oxide (NO) has identified three main sources of NO of which some are converted to nitrogen dioxide (NO_2) to give the NO_x mixture: Thermal NO, fuel source, and prompt NO. Thermal nitric oxides are produced in exhaust in temperatures higher than 1500 K. Prompt nitric oxides are produced in the flame front, in which there is deficiency of oxygen.

The range of thermal nitric oxide emissions depends on the flame temperature, partial oxygen pressure in exhaust (air excess coefficient), time of nitrogen, and oxygen particles' presence in high temperatures. The higher the temperature in which the oxidation and concentration of oxygen take place in the reaction area, the more intensive the reaction of oxidation and the oxygen concentration that occurs. During fuel combustion in the engine cylinder, the temperature exceeds 1600 K so the conditions for nitric oxide creation are perfect.

The second group comprises nitric oxides crested from nitrogen compounds included in fuel, which form nitric oxide by oxidation of combustible components, and then molecule nitrogen. The stage of nitrogen conversion from fuel into NO_x depends on the fuel type and it ranges from 20% to 80%. Over 90% of oxides of NO_x produced in internal combustion engines constitute NO. The oxidation of NO into NO_2 takes place in the engine outlet channels, with the presence of oxygen contained in the channels and in the atmosphere. The term nitric oxides comprises both compounds.

Heavy fuel oil burnt in marine engines (HFO) contains greater amounts of nitric compounds than marine diesel oil (MDO). The nitric oxide emissions are greater in the case of heavy fuel than in MDO.

The reduction of the exhaust emission of NO_x of modern marine diesel engines can be achieved through:

- Primary measures;
- Secondary measures; and
- Fuel modifications.

Primary reduction of exhaust emission takes place by influencing the fuel combustion process in the engine cylinder. The purpose of this action is to attack the problem at its source during the process of exhaust formation. In practice, in order to reduce nitric oxide emission, the following steps are taken:

- Change of air parameters;
- Change of fuel injection parameters;
- Supplying water to the cylinder; and
- Exhaust gas recirculation.

Secondary measures are necessary to apply external treatment of exhaust gases after they have left the engine cylinders. In practice, the devices available for this purpose include selective catalytic reduction (SCR)—SCR converters.

The reduction of NO_x emission through fuel modification is achieved, inter alia burning in the engine, by fuel/water emulsion. The combustion of alternative fuels, including diesel oil mixture with vegetable oils or their esters, may also be taken into account.

3. Vegetable Oils

Contemporary main diesel engines of sea-going ships are commonly supplied with heavy fuels oil (HFO). This very often also concerns auxiliary engines, especially electric generating sets. However, on many ships, the electric generating sets are still fed with marine diesel oil (MDO). Additionally, most diesel engines installed on small ships run on MDO.

Increasingly more attention is paid to alternative fuels, also called substitute, renewable, or unconventional fuels, because of the permanently increasing demand of marine diesel oil, their prices, and ecological requirements.

Unconventional fuels for supplying diesel engines are inter alia alcohol, ethanol, vegetable and mineral oils, fatty acid methyl esters, and diethyl ether [7]. In the research project planned by the author, supplying a ship with diesel engine only with a mixture of marine diesel oil (MDO) and rapeseed oil methyl ester (RME) was accounted for. It can be observed that some parameters of rape oil, their ester, and diesel oils show similar values. However, their density, kinematic viscosity, and flow temperature values differ from those of diesel oils. These data are listed in Table 2 [7].

Table 2. Comparison of some parameters of diesel oil, rape oil, and methyl ester of higher acid of rape oil (RME) [7].

Parameters	Unit	Diesel Oil	Rape Oil	Esters (RME)
Density at 15 °C	kg/m^3	820 ÷ 860	920	860 ÷ 900
Kinematic viscosity at 40 °C	mm^2/s	1.5 ÷ 4.5	30.0÷43.0	4.3 ÷ 6.3
Cetane number	–	45 ÷ 55	~51	49 ÷ 56
Gross caloric value	MJ/kg	42 ÷ 45	37.1 ÷ 37.5	37 ÷ 39
Flow temperature	°C	<−15	−6	−5 ÷ −8

As far as rape oils are concerned (interesting in the case of Poland), their density and viscosity is distinctly higher, which can make supplying diesel with them difficult; however, their positive features are practically no sulphur content and their bio-degradation ability. The results of research on the application of only vegetable oil for supplying diesel engines (mainly in the automotive industry, and not on ships) show worse cylinder filling, worse supplying, and greater lengths of injected oil jets, associated with their large viscosity and density [7]. The expected phenomena associated with supplying diesel engines with rape oil are disturbing, namely, the often occurrence of clogging sprayer nozzles in the injector, and troubles with starting engines at low ambient temperature. There is also the seizing of precise pairs of injection pumps, and great susceptibility to the formation of carbon deposits on piston heads, ring grooves, valve, and valve seats.

Due to substantial difficulties in applying rape oil only, as well as due to the limitation in using esters for running diesel engines, an alternative is to use mixtures of diesel oils and vegetable oil esters. In this way, it is expected to decrease the density and viscosity of the mixture to relative to those of a given ester. Tests on mechanical vehicles running on a mixture (20% rape oil/80% diesel oil) did not reveal any detrimental consequences [8].

Therefore, the author decided to carry out research on a marine diesel engine, to which the MDO/RME mixture containing up to 20% RME was fed.

Laboratory Tests

Object of tests. The tests were carried out with the use of the one-cylinder two stroke crosshead supercharged diesel engine, which is an element of the test stand adapted to investigations on emission exhaust gas components (Figure 4). The Wimmer MRU/2D analyzer with measuring accuracy of ≤±5% and a resolution of 1 ppm was used to measure the composition of the exhaust gases.

To supply the engine during the tests in question, the marine diesel oil (MDO) and its mixtures with rape oil esters (RME) of the following proportions were prepared:

- 15% of RME in MDO; and
- 20% of RME in MDO.

The MDO had a density of 831 kg/m^3 and the RME of 883 kg/m^3. As a result of the mixing, the biofuel had a density of 839 kg/m^3 in the first case and 840 kg/m^3 in the second case.

Test program. The tests were carried out within the broad range of the engine's loading, namely: 40%, 50%, 60%, 70%, and 80% M/M_n (set torque of engine M/nominal torque of engine M_n) and for a constant rotational speed of the engine, set at 220 rpm. At a given rotational speed and successively set loads, measurements of the engine's exhaust gas content during combusting by the engine were realized: The MDO alone, and the two above specified mixtures (i.e., 15% of RME in MDO, and 20% of RME in MDO). The results obtained from the tests during the supply of the engine with the MDO alone was assumed as the reference point for determination of the influence of combustion of the MDO/RME mixtures on the engine's exhaust gas content.

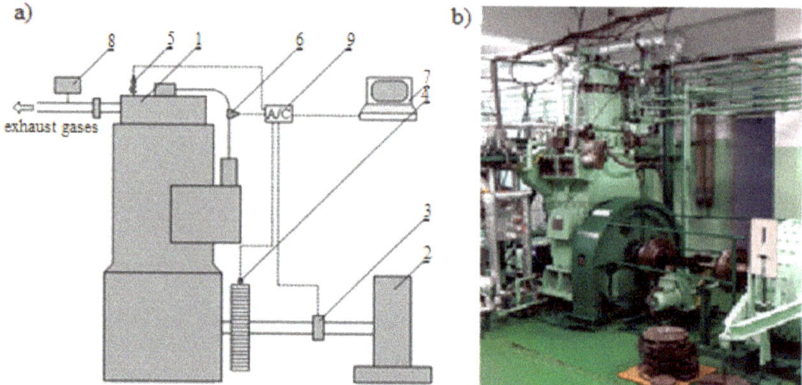

Figure 4. Test stand: (**a**) block diagram: 1—L-22 diesel engine, 2—water brake, 3—torsiometer, 4—gauge for crankshaft position marking and rotational speed measuring, 5—combustion pressure transducer, 6—injection pressure transducer, 7—computer, 8—exhaust gas analyzer, 9—analog/digital converter; (**b**) engine L22-view.

Test results and their analysis. The selected tests results are presented in Figures 5 and 6. On the basis of the exhaust gas analysis, it can be stated that combustion of MDO with 15% addition of RME caused, on average, a drop of the NO_x content by 35% (for higher engine loads—70% and 80% M/M_n) and that during combusting of the MDO with the 20% addition of RME, the drop, on average, exceeded 57% (for higher engine loads—70% and 80% M/M_n), as shown in Figure 5.

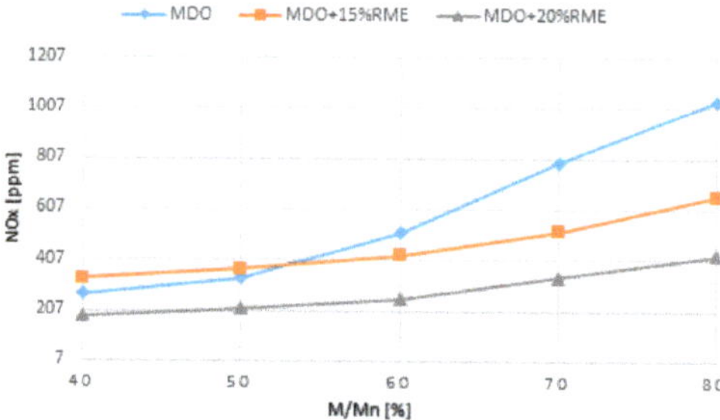

Figure 5. NO_x content in exhaust gas as a function of the engine load for different kinds of fuel at a constant rotational speed n = 220 rpm.

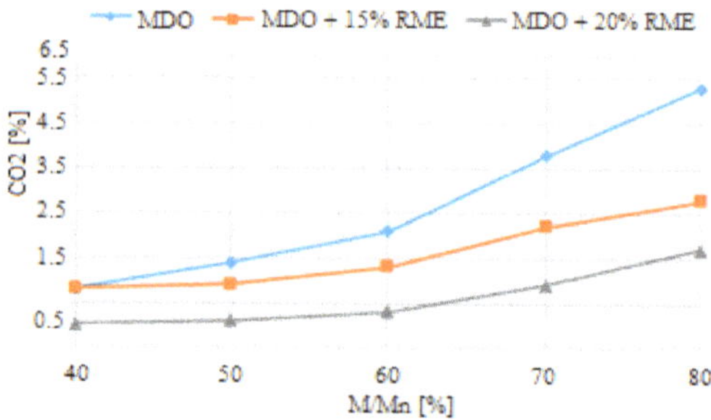

Figure 6. CO_2 content in exhaust gas as a function of the engine load for different kinds of fuel at a constant rotational speed n = 220 rpm.

It can also be noted that the combustion of MDO with the 15% and 20% addition of RME, especially for higher engine loads (70% and 80% M/M$_n$), causes a clear drop in CO$_2$. This drop is exceeded by 33% in the combustion of MDO with the 15% addition of RME and exceeded by 57% in the combustion of MDO with the 20% addition of RME (Figure 6).

4. Change of Fuel Injection Parameters-Change Advance Angles of Fuel Injection

The possibility of reducing NO$_x$ emission by changing the fuel injection parameters has been noted [9]. The fundamental parameter for an injection system is the advance angle of fuel injection. You can expect that change of the advance angle of fuel injection affects the composition of exhaust gases, including toxic component emissions. This thesis is confirmed by the research carried out by the author, described in follow section.

Laboratory Tests

Object of tests. The tests were carried out with the use of the one-cylinder two stroke crosshead supercharged diesel engine (Figure 4), which is an element of the test stand adapted to investigations on emission exhaust gas components.

To supply the engine during the tests in question, the marine diesel oil (MDO) was used and the fuel injection advance angle (injection timing) changed for three selected values:

- −13° (rated value) before the piston top dead center (TDC);
- −10° before the piston top dead center (TDC); and
- −7° before the piston top dead center (TDC).

Test program. The tests were carried out within the broad range of the engine's loading, namely: 40%, 50%, 60%, 70%, and 80% M/M$_n$ (set torque of engine M/nominal torque of engine M$_n$) and for a constant rotational speed of the engine, set at 220 rpm. At the given rotational speed and successively set loads, measurements of the engine's exhaust gas content were realized for three values of the angle: −13° (rated), −10°, and −7° before the piston top dead center (TDC).

Test results and their analysis. The selected tests results are presented in Figures 7 and 8. Changes to the injection advance angle can be made by adjusting the factory settings. A decrease of the injection advance angle unambiguously makes the NO$_x$ content in exhaust gas lower. A decrease of the injection advance angle of −13 degrees to −7 degrees unambiguously makes the NO$_x$ content in exhaust gas lower (even more than 18%—see Figure 7). However, it should be remembered that both an advance and delay of fuel injection starting, in relation to the values recommended by the producer and set during static adjustment of the engine, influences not only the exhaust gas content but also other important operational parameters of the engine by changing the combustion process quality. Inter alia, an increase in the specific fuel consumption can be expected.

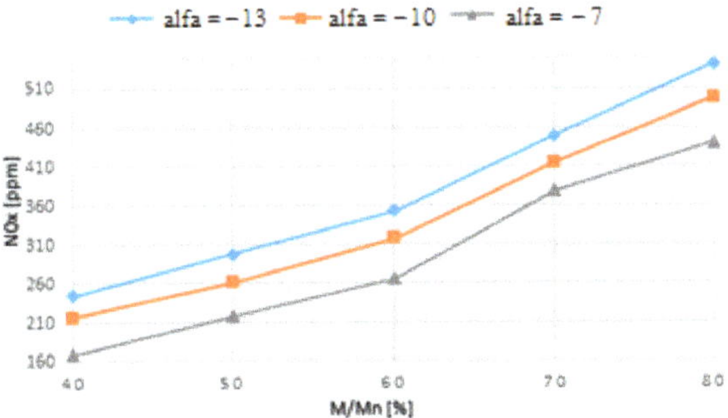

Figure 7. NO$_x$ content in exhaust as a function of the engine load and three different advance angles of fuel injection, for a constant rotational speed n = 220 rpm.

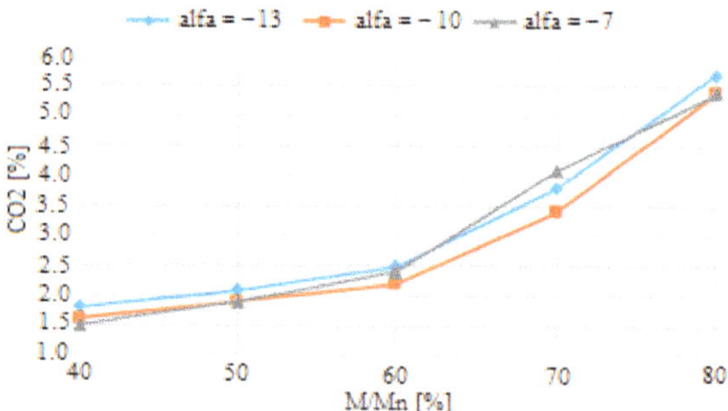

Figure 8. CO_2 content in exhaust as a function of the engine load and three different advance angles of fuel injection, for a constant rotational speed $n = 220$ rpm.

The investigations also revealed (see Figure 8) that the advance angles of fuel injection have no significant impact on the change content of carbon dioxide (CO_2) in exhaust gas.

5. Conclusions

Atmosphere protection against pollution from seagoing vessels is currently one of the most important areas of ecological activity in maritime transport. Various solutions are implemented on ships that allow for a significant reduction of toxic compounds in the exhaust gas.

Among others, fuel modifications made possible a reduction of the emission of toxic components of exhaust gases from modern marine diesel engines. The test results show that for environmental reasons, mixtures of marine diesel oil and rape oil esters (MDO)/(RME) may be supplied to the marine engines.

The supply of the engine during the tests, the mixtures of marine diesel oil (MDO) and rape oil esters (RME), caused a clear drop in emissions of nitrogen oxides and carbon dioxide, particularly for a higher engine load.

There is a possibility of reducing the emission of NO_x through changes of the fuel injection parameters. The effect of changes to the fuel injection advance angle of emission oxides of nitrogen (NO_x) and carbon dioxide (CO_2) was investigated in the test.

The decrease of the injection advance angle unambiguously makes the NO_x content in exhaust gas lower. The investigations also revealed that the advance angles of fuel injection change and they have no significant impact on the change content of carbon dioxide (CO_2) in exhaust gas.

The use of biofuels (a mixtures of marine diesel oil and rapeseed oil esters), or a reduction in the advance injection angle are the possible solutions to be used immediately, without significant investment costs.

Funding: This research received no external funding.

Conflicts of Interest: The authors declare no conflict of interest.

References

1. *Marine Emissions—The Facts, Man Diesel Facts*; MAN Diesel & Turbine: Copenhagen, Denmark, 2007.
2. Summary for Policymakers. In *Global Warming of 1.5 °C*; An IPCC Special Report on the Impacts of Global Warming of 1.5°C Above Pre-Industrial Levels and Related Global Greenhouse Gas Emission Pathways, in the Context of Strengthening the Global Response to the Threat of Climate Change, Sustainable Development, and Efforts to Eradicate Poverty; Masson-Delmotte, V., Zhai, P., Pörtner, H.-O., Roberts, D., Skea, J., Shukla, P.R., Pirani, A., Moufouma-Okia, W., Péan, C., Pidcock, R., et al., Eds.; IPCC: Geneva, Switzerland, 2018.

J. Mar. Sci. Eng. **2020**, *8*, 452

3. Summary for Policymakers. In *Climate Change 2007: The Physical Science Basis*; Contribution of Working Group I to the Fourth Assessment Report of the Intergovernmental Panel on Climate Change; Cambridge University Press: Cambridge, UK, 2007.

4. Piotrowski, I.; Witkowski, K. *Operation of Marine Internal Combustion Engines*; Baltic Surveyors Group Ltd.: Gdynia, Poland, 2012. (In Polish)

5. Woodyard, D. *Pounder's Marine Diesel Engines and Gas Turbines*, 8th ed.; Heinemann: Butterworth, Malaysia; Oxford, UK, 2004.

6. International Maritime Organization. *Articles, Protocols, Annexes, Unified Interpretations of the International Convention for the Prevention of Pollution from Ships, 1973*; MARPOL 73/78, Consolidated Edition, as Modified by the Protocol of 1978 Relating Thereto; IMO: Londyn, UK, 2002.

7. Baczewski, K.; Kałdoński, T. *Fuels for Self-Ignition Engines*; Wydawnictwo Komunikacji i Łączności: Warszawa, Poland, 2004. (In Polish)

8. Szlachta, Z. *Supplying Diesel Engines with Rape Oils*; Wydawnictwo Komunikacji i Łacności: Warszawa, Poland, 2002. (In Polish)

9. Witkowski, K. Research on Influence of Selected Control Parameters of the Injection System Marine Diesel Engine on its Exhaust Gas Toxicity. *J. KONES* **2012**, *19*, 551–556. [CrossRef]

Journal of
Marine Science and Engineering

Article

Performance and Regeneration of Methane Oxidation Catalyst for LNG Ships

Kati Lehtoranta [1,*], Päivi Koponen [1], Hannu Vesala [1], Kauko Kallinen [2] and Teuvo Maunula [2]

[1] VTT Technical Research Centre of Finland, FI-02044 Espoo, Finland; paivi.koponen@vtt.fi (P.K.); hannu.vesala@vtt.fi (H.V.)
[2] Dinex, FI-41330 Vihtavuori, Finland; kki@dinex.fi (K.K.); tma@dinex.fi (T.M.)
* Correspondence: kati.lehtoranta@vtt.fi; Tel.: +358-20-722-5615

Abstract: Liquefied natural gas (LNG) use as marine fuel is increasing. Switching diesel to LNG in ships significantly reduces air pollutants but the methane slip from gas engines can in the worst case outweigh the CO_2 decrease with an unintended effect on climate. In this study, a methane oxidation catalyst (MOC) is investigated with engine experiments in lean-burn conditions. Since the highly efficient catalyst needed to oxidize methane is very sensitive to sulfur poisoning a regeneration using stoichiometric conditions was studied to reactivate the catalyst. In addition, the effect of a special sulfur trap to protect the MOC and ensure long-term performance for methane oxidation was studied. MOC was found to decrease the methane emission up to 70–80% at the exhaust temperature of 550 degrees. This efficiency decreased within time, but the regeneration done once a day was found to recover the efficiency. Moreover, the sulfur trap studied with MOC was shown to protect the MOC against sulfur poisoning to some extent. These results give indication of the possible use of MOC in LNG ships to control methane slip emissions.

Keywords: methane slip; methane oxidation catalyst; LNG; natural gas

Citation: Lehtoranta, K.; Koponen, P.; Vesala, H.; Kallinen, K.; Maunula, T. Performance and Regeneration of Methane Oxidation Catalyst for LNG Ships. *J. Mar. Sci. Eng.* **2021**, *9*, 111. https://doi.org/10.3390/jmse9020111

Received: 21 December 2020
Accepted: 20 January 2021
Published: 22 January 2021

Publisher's Note: MDPI stays neutral with regard to jurisdictional claims in published maps and institutional affiliations.

1. Introduction

The emissions from ships can be a significant source of air pollution in coastal areas and port cities and can have negative impact on human health and climate [1–3]. Therefore, the International Maritime Organization (IMO) has implemented regulations to reduce emissions from ships. So far, these regulations concentrate mainly on emissions of nitrogen oxides (NO_x) and sulfur oxides (SO_x). To answer these requirements emission reduction technologies are needed, namely fuel technologies, combustion technologies and/or exhaust gas after-treatment technologies (see e.g., [4]). One solution is to use natural gas (NG) as a fuel.

Liquefied natural gas (LNG) use as marine fuel is increasing and more and more gas engines, mainly dual fuel, are being installed in ships. With LNG both SO_x and NO_x regulations of IMO can be achieved without any need for after-treatment, since NG is nearly sulfur free resulting in very minor/no SO_x emissions while lower NO_x levels (compared to diesel) can be achieved due to low combustion temperature of natural gas (in lean-burn conditions). In addition, particle emissions from natural gas combustion are low and only minor black carbon is formed from NG combustion [5–7]. Moreover, CO_2 emission is lower with NG use compared to diesel fuels, which is because NG is mainly composed of methane with a higher H/C ratio compared to diesel. The hydrocarbon emissions, on the other hand, are higher with NG compared to diesel fuels [8–11]. Because natural gas is mainly methane, most of the hydrocarbon emissions is also methane. Since methane is a strong greenhouse gas, its emissions should be minimized.

Three different gas engine groups are used for marine applications, namely lean-burn spark-ignited engines, low pressure dual fuel engines and high pressure dual fuel engines [12]. For dual fuel engines, the natural gas and air mix is ignited with a small diesel

187

pilot injection. In addition, the diesel can be used as the main fuel (back-up fuel) if LNG is not available, making this dual fuel concept the most popular one in marine applications.

According to Sharafian et al. [13] LNG use in high pressure dual fuel (HPDF) engines, which are used only for large low-speed oceangoing vessels, can reduce greenhouse gas (GHG) emissions by 10% compared to their heavy fuel oil (HFO)-fueled counterparts. However, the current deployment of medium speed low pressure dual fuel (LPDF) cannot reliably reduce GHG emissions. This is primarily due to the high levels of methane slip from these engines. The methane slip from HPDF engines is reported to be significantly lower compared to LPDF engines [13] but the LPDF is the most popular LNG engine technology with at least 350 ships (e.g., LNG carriers, car/passenger ferries, cruise ships) while HPDF is used in less than 100 ships (e.g., LNG carriers, container ships) [14].

Peng et al. [6] studied the impacts of switching a marine vessel from diesel fuel to natural gas. They showed that the GHG impact of NG compared to diesel is higher especially on lower engine load cases while at higher loads (>75%) the GHG impact is comparable to diesel. The test vessel in their study operated with medium speed dual fuel engines.

ICCT's (The International Council on Clean Transportation) working paper on "The climate implications of using LNG as a marine fuel" also concludes that there is no climate benefit from using LNG, when using a 20-year global warming potential, including upstream emissions, combustion emissions and unburned methane [14]. However, over the 100-year time frame, a life cycle GHG benefit of LNG is reported to be 15% compared to diesel, but this is only for ships with HPDF engines. A life cycle GHG emission study on the use of LNG prepared by Sphera reports GHG emission reductions with LNG operation (compared to HFO fueled ships) 14–21% for 2-stroke slow speed and also 7–15% for 4-stroke medium speed [15].

One key issue in the current and future LNG ships is the control of methane slip. Methane emissions from engines are being reduced, e.g., by better fuel mixing conditions, improvements in combustion chamber design, and by reducing crevices [16,17]. One option is the use of oxidation catalyst. To oxidize methane, a highly efficient catalyst is needed. Although catalysts based on platinum are commonly used for non-methane hydrocarbon and CO oxidation, palladium catalysts have shown good activity for methane oxidation (e.g., [18,19]). Challenge in the development of MOC is the catalyst deactivation since palladium-based catalysts are very sensitive to sulfur poisoning and as little as 1 ppm SO_2 present in the exhaust has already been found to inhibit the oxidation of methane [20,21].

Simplified, when palladium-based MOC is to be used the SO_2 in the exhaust should be minimized and/or a regeneration procedure to recover the catalyst activity is needed.

There are few scientific studies published about the regeneration of sulfur-poisoned methane oxidation catalysts. Arosio et al. [22] studied the regeneration by short CH_4 pulses. Honkanen et al. [23] also used CH_4 conditions to regenerate sulfur-poisoned Pd-based catalyst. According to Kinnunen et al. [24] a sulfur-poisoned catalyst can be regenerated under low-oxygen conditions and Lott et al. [25] also favor rich conditions to achieve efficient regeneration. In addition, H_2 usage has been studied to recover the catalyst activity [26,27]. Most of these regeneration studies are conducted in laboratories with synthetic gases.

To minimize SO_x in the exhaust, the sulfur in the fuel and lubricating oil is to be minimized. However, in the case of natural gas, the sulfur level is already very low (a few ppm). In dual fuel engines, obviously, also the pilot diesel fuel sulfur level is to be minimized, although the pilot fuel use is only one to a few percent of the natural gas use. In this study, we also add a sulfur trap as one choice. A sulfur trap, also called the SO_x trap, is an adsorber catalyst, specifically designed to store sulfur. They have earlier been studied and developed to protect NO_x adsorbers that can adsorb and store NO_x under lean conditions and release it under rich operation but are however poisoned by SO_x present in the exhaust gas [28]. In the present study, the SO_x traps are connected to MOC for the first time (to authors' knowledge).

In this study, the performance and regeneration of one methane oxidation catalyst (MOC) is studied in engine exhaust. Catalyst performance studies with an engine under lean conditions are done by emission measurements upstream and downstream of the catalyst while the regeneration is done by switching on engine-driving mode to stoichiometric for couple of minutes time. In addition to MOC only, we study the effect of a SO_x trap, using it upstream the MOC, to see how it protects the MOC against sulfur poisoning.

2. Methods

The research facility included a passenger car engine that was modified to run with natural gas. The engine with the test facility was presented in detail in Murtonen et al. [29]. The engine was operated with a lean air-to-fuel mixture. For the present study, only one engine-driving mode was used. The exhaust gas flow and temperature (measured upstream of the catalysts) were adjusted independently, and therefore, it was possible e.g., to keep the exhaust gas composition and flow constant while changing only the temperature. In the present study, most of the tests were conducted with exhaust temperature adjusted to 550 °C and exhaust flow to 60 kg/h while initial performance studies were also conducted at lower exhaust temperatures and with different exhaust flows.

The NG was from Nord Stream and was high in CH_4 content (>95%). The sulfur content of the gas was below 1.5 ppm. The lubricating oil used for medium speed LNG engines was selected and used in this smaller test engine also. The lubricating oil sulfur content was 2100 mg/kg, density was 0.85 kg/dm^3 and viscosity at 100 °C was 13.4 mm^2/s.

MOC used platinum-palladium (1:4) as active metals on a tailored coating developed for lean NG applications and supported on a metallic substrate. The volume of MOC was 0.75 dm^3. This means a space velocity (1/h) of 61,500 with exhaust flow of 60 kg/h. In addition to MOC only, similar studies were done with a SO_x trap installed in front of the MOC. The SO_x trap has a property to absorb and release SO_x in controlled conditions and had a volume of 0.75 dm^3. The catalyst set-up is presented in Figure 1.

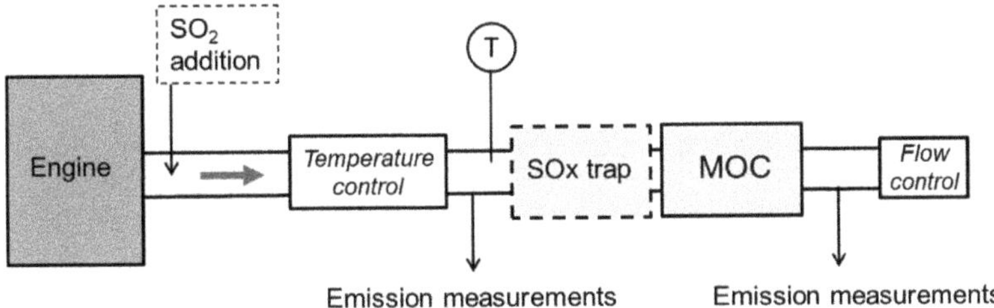

Figure 1. Set-up.

Emission measurements were done upstream and downstream of the catalyst system (see Figure 1). This included a Horiba PG-250 analyzer used to measure NO_x, CO, CO_2, and O_2. CO and CO_2 were measured by non-dispersive infrared, NO_x by chemiluminescence, and O_2 by a paramagnetic measurement cell. Exhaust gas was dried with gas-cooler before it was measured by Horiba.

Online SO_2 emissions were determined by a Rowaco 2030 1 Hz Fourier transfer infrared (FTIR) Spectrometer equipped with an Automated MEGA-1 (miniMEGA) sampling system. The detection limit for SO_2 was 2.5 ppm. This device was used mainly only during the regeneration periods to observe the SO_2 release.

An agilent 490 MicroGC (Gas Chromatography) was used to measure the hydrocarbons and hydrogen (H_2). The detection limits for ethane (C_2H_6), ethylene (C_2H_4), and propane were approximately 2 ppm, for methane (CH_4) 10 ppm, and 100 ppm for H_2.

Multiple gaseous components were measured continuously by two Gasmet DX-4000 Fourier transformation infrared (FTIR) spectrometers simultaneously upstream and downstream of the catalyst system.

Test Program

The experiments were conducted over a 190 h' ageing at the selected driving mode. The driving mode selection was based on the exhaust gas composition with an aim to produce as similar exhaust gas as possible that would be emitted from LNG engine. The engine out exhaust gas composition on the selected driving mode is presented in Table 1 and shows that the NO_x level was approx. 225 ppm and CO 550 ppm, while hydrocarbons constituted of methane with near 1000 ppm level, ethane 15 ppm, and ethylene 30 ppm. In addition, formaldehyde (HCHO) was found in the exhaust with a level of approx. 55 ppm. For comparison and as a reference we have added the exhaust gas composition measured form one medium speed dual fuel marine engine run with NG to Table 2. These were earlier presented in g/kWh [5] and now in concentrations at two engine loads of 85% and 40%. Comparing these values to the values of the present study's selected driving mode (Table 1 "Ageing") we see that these are in the same order of magnitude.

Table 1. Engine out exhaust gas composition (as measured in dry gas). In "Ageing"-mode hydrocarbons are measured by GC and others by Horiba, while in "Regeneration"-mode O_2 and H_2 were measured by GC and others by FTIR. HCHO was measured with FTIR.

Engine Out	CO	CO_2	NO_x	O_2	CH_4	C_2H_6	C_2H_4	H_2	HCHO
	ppm	vol%	ppm	vol%	ppm	ppm	ppm	vol%	ppm
Ageing (normal operation)	549	8.0	226	6.5	970	14.6	30.3	<0.01	55.7
	14.8	0.1	7.1	0.1	125.0	1.4	0.8		4.4
Regeneration	9377	10.2	2437	1.0	862	<2	<2	0.6	37.0
	1825.7	0.2	187.3	0.2	58.6			0.1	3.6

Table 2. Engine out exhaust gas composition from one medium speed dual fuel marine engine run with natural gas as the main fuel [5].

Engine Out	CO	CO_2	NO_x	O_2	CH_4	C_2H_6	C_2H_4	H_2	HCHO
	ppm	vol%	ppm	vol%	ppm	ppm	ppm	vol%	ppm
Marine engine 85% load	342	5.5	337	10.9	1823	41.8	7.2	-	44.7
Marine engine 40% load	650	5.3	380	11.1	3750	89.0	23.1	-	76.0

To ensure that the starting point for the experiments was as similar to possible for all catalyst reactors, they were preconditioned by ageing for 48 h in the selected driving mode with an exhaust gas temperature of 550 °C and exhaust flow of 60 kg/h. After the preconditioning, the initial performance of the MOC was studied by emission measurements. The same driving mode was used. The methane level in this driving mode was near 1000 ppm (Table 1) and in addition, one higher methane level was included in the initial performance studies to see if the methane level influences the performance. This approx. 1500 ppm methane level was achieved in the same driving mode by adding methane to the exhaust prior to the catalyst.

After the initial performance studies, the ageing tests were conducted. Additional SO_2 was fed into the exhaust (see Figure 1) in part of the ageing tests, while one test was also conducted without any additional SO_2. The added SO_2 contributed to a 1 ppm increase in the exhaust gas while any sulfur from the natural gas and lubricating oil, led to a SO_2 level of approximately 0.5 ppm in the exhaust gas. This means SO_2 level of 0.5 ppm without any additional SO_2 and SO_2 level of 1.5 ppm with the added SO_2 in the exhaust. The SO_2 level of 1.5 ppm was confirmed by sampling exhaust gas to a sample bag and analyzing by gas chromatography prior to the ageing tests. In the following ageing tests, the SO_2 feed was then kept constant.

Altogether, three similar experimental campaigns were conducted. Two ageing tests were done with MOC only, meaning one ageing test with additional SO_2 and the other one without SO_2 addition. The third ageing test was conducted with SO_x trap installed in front of the MOC (including the additional SO_2 in the exhaust gas).

During the ageing test, the engine was running without stops and once a day regeneration was done by turning the engine to stoichiometric condition for 5 min time. In Table 1, also the engine out emissions at the regeneration mode are presented. This shows that in addition to the targeted O_2 decrease, other emission components' levels changed as well. The NO_x increased to 2400 ppm and CO to above 9000 ppm in the stoichiometric mode while methane level decreased to 860 ppm. In addition, H_2 with a level of 0.6% was found in the exhaust when running on the nearly stoichiometric mode.

3. Results

3.1. MOC Initial Performance

To oxidize methane, in addition to the effective catalyst, high temperature is required. This was found to be true in the present study also. After preconditioning the catalyst for 48 h in the engine exhaust, the initial performance was defined by measurements upstream and downstream of the catalyst. Figure 2 presents the initial methane conversion as a function of exhaust temperature and at three different exhaust flows. At exhaust temperature of 400 °C, the methane conversion was found to be negligible (only 2%). However, when increasing the temperature, the methane conversion sharply increases, being above 20% at exhaust temperature of 460 °C, above 60% at exhaust temperature of 500 °C and at the highest temperature of 550 °C the methane conversion is approx. 70%. Furthermore, the lower exhaust flow (40 kg/h) studied at 550 °C increased the methane conversion to near 80% which is reasonable since the lower exhaust flows means more time for the catalytic reactions to occur. Roughly, changing the exhaust flow from 80 kg/h to 40 kg/h means similar effect on the performance that could be expected with doubling the catalyst size. In this case, the methane conversion increased from 60% (with 80 kg/h) to 80% (with 40 kg/h).

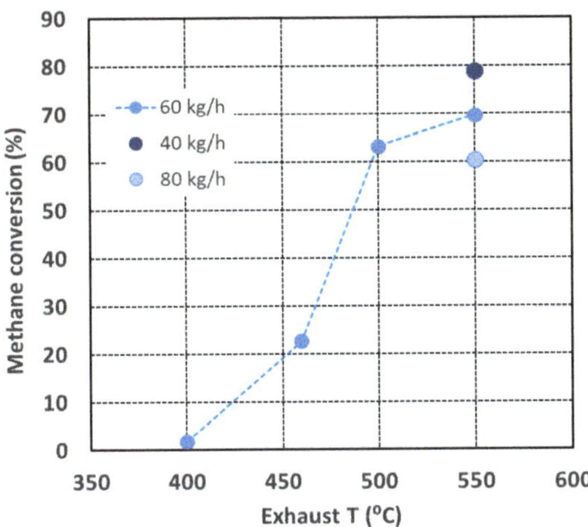

Figure 2. Initial methane conversion (the line is to guide eye only).

In practice, the engine out methane level of 1500 ppm compared to 1000 ppm had no effect on the MOC performance and the difference in methane conversions within these two cases was found to be less than 4%.

The ethane conversion was measured to be approx. 30% at exhaust T of 400 °C while at exhaust T of 460 °C and higher no ethane was found in the exhaust downstream of the MOC meaning near 100% conversion. Ethylene found in the engine out exhaust (see Table 1), was not found in the exhaust downstream of MOC in any of the test conditions meaning near 100% conversion for ethylene already at 400 °C. In addition, CO was in the level of only a few ppm downstream MOC, confirming nearly total CO conversion. From the FTIR measurements also formaldehyde was analyzed and was found to be below 2 ppm downstream of MOC in all test conditions (while the engine out level was approx. 55 ppm) confirming nearly total HCHO conversion as well. The NO$_x$ level was not affected by the MOC, as expected.

3.2. MOC Performance in Ageing

The methane conversion of the MOC was found to decrease rather quickly over time. The conversions calculated form the GC measurement results are presented as a function of time in Figure 3. The GC measurement was done just before the regeneration and then again 3 h after the regeneration.

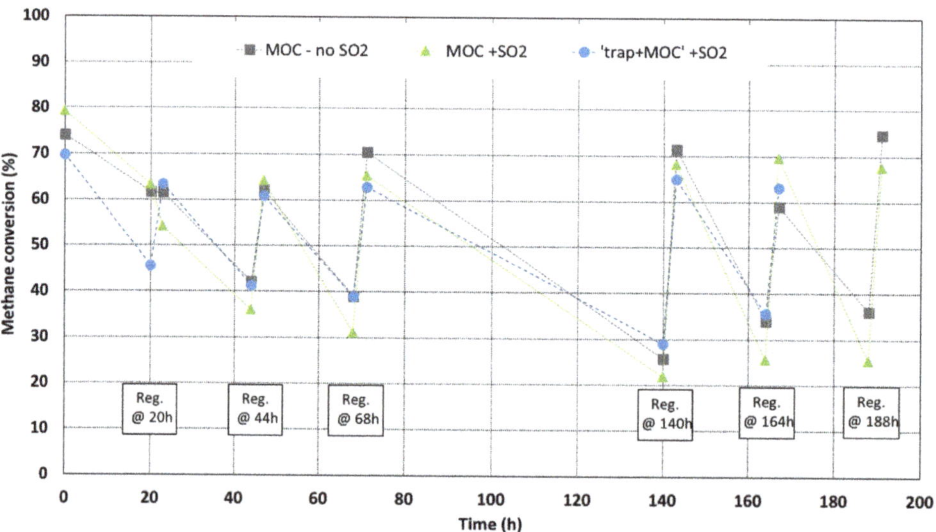

Figure 3. Methane conversion as a function of time, over the ageing experiment, calculated from GC results. GC measurements were done just before regeneration and 3 h after the regeneration. (The lines are to guide eye only).

The conversion was found to decrease by approx. 20 percentage units during the first 20 h of driving (see Figure 3). However, the regeneration, done once a day, was found to recover the methane efficiency of the catalyst. For example, in the case of "SO$_x$ trap + MOC" the methane conversion in the beginning was approx. 70%, after 20 h it decreased below 50%, but after the regeneration the conversion was found to be near 65%. During the following day (20–44 h) the conversion decreased again, now approx. to 40%; however, after the regeneration the conversion increased back to above 60% (see Figure 3, 44 h). Next, the third day was very similar to the second day's performance. After this, the next 2 days were conducted without regenerations (68–140 h), and the regeneration was done on the third day. The conversion decreased, without the daily regenerations, to 30%, but again, after the 5 min regeneration period, the conversion was back to above 60% (see Figure 3, 140 h).

The exhaust SO$_2$ level had a clear influence on the methane catalyst efficiency. When MOC was aged with 1 ppm extra SO$_2$ added to the exhaust, the methane conversion after each 24 h of driving was lower compared to the case where MOC was aged without any extra SO$_2$ in the exhaust (see Figure 3 cases "MOC-no SO$_2$" and "MOC + SO$_2$"). The SO$_x$ trap was found to protect the MOC since when the same ageing was conducted with trap installed upstream of the MOC the methane efficiency was similar to the case with no SO_2 added to the exhaust (Figure 3, cases "trap + MOC + SO_2" and "MOC-no SO_2").

Note. All three MOCs of the tests were similar and were preconditioned for 48 h prior to the actual ageing tests start. However, the methane conversion in the start of the ageing test varied between 70–80% with the MOCs. In addition, for some reason, the first regeneration in the cases of "MOC-no SO_2" and "MOC + SO_2" was not successful, as it seemed not to recover the catalyst efficiency. For this, we do not have any clear explanation. Turning engine to stoichiometric mode was done manually and due to human factors, this might not have been done exactly similarly in all cases. In the case of "trap + MOC" the engine, unintended, faced changes in operation between 68–140 h, which we observed as an increased methane slip, while other emission components were not significantly changed. The methane slip in this case was near 2000 ppm at 140 h and forward. This was also the reason, why this test was ended at 164 h.

To have a closer look at the MOC performance from hour to hour, the methane conversion calculated from the simultaneously measuring FTIR devices' methane results are presented in the Figure 4 roughly for two days period. We started the period to examine only after the first day (due to the unsuccessful first regenerations discussed above). After the second regeneration, i.e., at the 44 h point, all the catalysts resulted in similar methane conversions. The conversion after the regeneration decreases more quickly in the case of "MOC + SO_2" than for the two other cases ("MOC-no SO_2" and "trap + MOC") that seem to behave very similarly. Again, after the regeneration the conversion is recovered, in all cases, and similarly to earlier day, the conversion after the regeneration (68 h) decreases more quickly in the case of "MOC + SO_2" than for the two other cases. The methane conversion is approx. 10–15% units higher when the trap is placed upstream the MOC, meaning there are more active sites for methane to convert in the MOC downstream the trap. This indicates the SO_2 level inside the MOC is lower when trap is involved compared to the MOC only case.

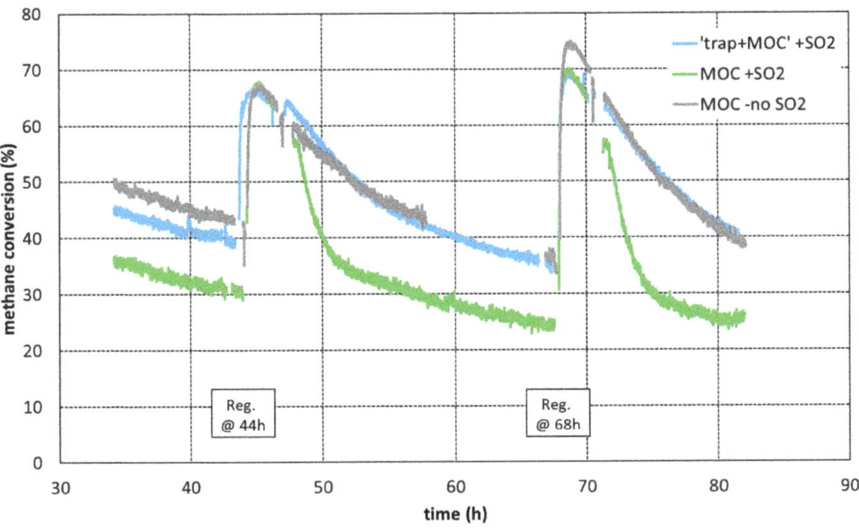

Figure 4. Methane conversion based on two simultaneously measuring FTIRs.

Although the methane conversion was significantly decreased within time, we saw, in practice, no change in the performance regarding the other emission components. The ethane and ethylene levels downstream of MOC stayed close to zero (below the detection limit) throughout the ageing test. CO level was only a few ppm downstream MOC and formaldehyde level stayed also in the level of 2 ppm (or below) throughout the ageing test.

3.3. MOC Performance in Regeneration

The regeneration, recovering the catalyst efficiency for methane oxidation, already indicates SO_2 is most probably released from the catalyst during the regeneration and the SO_2 measurement confirms this. In each case, during the 5 min regeneration a release of SO_2 was detected with the FTIR downstream of the catalyst. The total SO_2 amounts released over the 5 min regeneration periods are collected to Table 3. The lowest SO_2 amounts were released in regeneration of MOC, which was tested without any additional SO_2 in the exhaust. In the second case ("MOC + SO_2") where more SO_2 was present in the exhaust, also more SO_2 was found to be released from the MOC during the regenerations indicating that more SO_2 is collected to MOC also. If we look at for example the third regenerations (Table 3) we see that in the case of "MOC-no SO_2", 0.33 g SO_2 was released while in the case of "MOC + SO_2" 0.54 g SO_2 was released. In the case of "trap + MOC" the SO_2 released during the regeneration was 1.52 g (see Table 2), indicating roughly that the "trap" is contributing to the release with SO_2 amount of 0.98 g. This means that the trap is collecting at least the same amount of SO_2. In addition, this SO_2 release indicates that the trap itself is regenerating at the same time as the MOC. Also, since the methane oxidation efficiency is recovered (Figures 3 and 4), no significant amount of the SO_2 released form trap during the regeneration period is expected to be collected in MOC but only flowing through the MOC.

Table 3. The SO_2 amounts (g) released during regenerations. * Note. 4.reg was done after weekend. In case of Italic labelled values, engine-driving mode was not exactly as intended, see text for more details.

Hours		MOC Only	MOC Only	Trap + MOC
		w/o SO_2 Add	with SO_2 Add	with SO_2 Add
20	1.reg	0.35	0.51	1.29
44	2.reg	0.18	0.54	1.51
68	3.reg	0.33	0.54	1.52
140	4.reg *	0.35	0.58	*2.26*
164	5.reg	*0.09*	0.52	*2.00*
188	6.reg	*0.21*	0.51	

When the extra SO_2 was added to exhaust, the total SO_2 available during 24 h of driving was 4.10 g. Therefore, in the case of "MOC + SO_2" the released SO_2 amount was 13% of the amount available, while in the case of "trap + MOC" the released amount was 37% of the amount available. Since the "trap + MOC" released significantly more SO_2 compared to MOC only, the "trap + MOC" must also be collecting more SO_2. This is what we saw in the catalyst performance and ageing tests, since the methane conversion of "trap + MOC" was higher than in the case of MOC only indicating the poisoning effect of SO_2 was less in the case of "trap + MOC".

The results in Table 3 also show that in the case of "trap + MOC" when regeneration is done after a longer time period (4th regeneration) some more SO_2 is released during the regeneration. One should note that in this case (of "trap + MOC") the engine-driving mode was changed between third and fourth regenerations, meaning that the methane slip increased to near 2000 ppm level in ageing mode (discussed above). The actual regenerations were repeated as similarly as manually possible and now significant changes were observed in the regeneration mode engine out emission levels.

Regarding the regenerations conducted during the first tests (the case of "MOC-no SO_2") the adjustment to stoichiometric mode was not always realized similarly. In the cases of the 5th and the 6th regenerations, changing the driving mode to stoichiometric did not happen as smoothly as for the other cases (therefore we marked those in italic to the table).

During the regeneration, the exhaust gas composition changed significantly (Table 1). However, the MOC also worked during the regeneration. Although the engine out CO level increased to above 9000 ppm in the short regeneration phase (Table 1), the level measured downstream of MOC increased only after approximately 2 min from the start of the regeneration (i.e., turning the engine to stoichiometric driving mode) and did not reach to higher than 3000 ppm (Figure 5A). This also had strong influence on the exhaust temperature since oxidizing high amount of CO contributes to the temperature increase. This was also observed by temperature measurement downstream the MOC showing an increase of 50–60 degrees during the regeneration (Figure 5A). Furthermore, no H_2 was observed downstream the MOC, confirming that also the H_2 (engine out concentration 0.6%, Table 1) was oxidized during the regeneration, influencing exhaust temperature (measured downstream the MOC) as well.

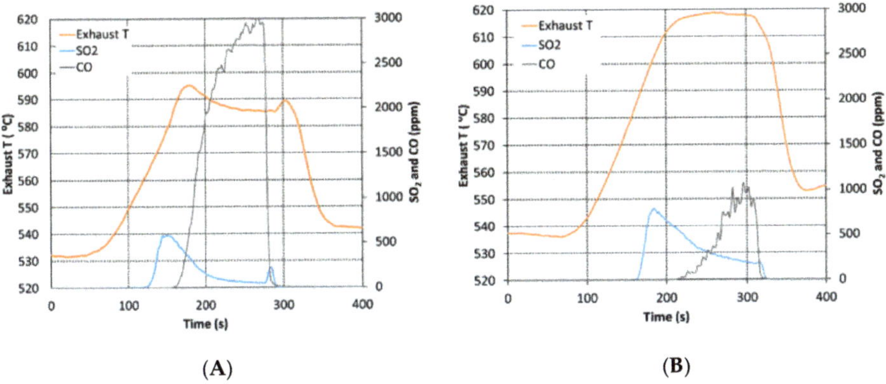

Figure 5. Exhaust temperature downstream the MOC. SO_2 and CO downstream the MOC during the regeneration phase. (**A**) the case of "trap + MOC" for 1st and 3rd regenerations (**B**) the case of "trap + MOC" for 4th regeneration.

Interestingly, the regeneration of "trap + MOC" done after longer time period (68–140 h) resulted in different behavior than all the other regenerations. As discussed above, in this case the methane slip from the engine increased during 68–140 h to a level of 2000 ppm. The temperature increase during regeneration was 20–30 degrees higher in this case (Figure 5B) compared to other regenerations (Figure 5A). The exhaust temperature downstream the catalyst was a few degrees higher in the normal driving mode ("ageing") as well (see Figure 5A,B prior to regeneration start). Since the methane conversion over the MOC was similar for both methane slip levels (normal 1000 ppm versus this higher 2000 ppm), the MOC was oxidizing more methane in this higher slip case and this methane oxidation increase can also result in temperature increase.

The CO level downstream the MOC in this regeneration of Figure 5B (i.e., max. 1000 ppm) was significantly lower than in other regenerations although the engine out CO level was similar in all cases. This more effective CO oxidation can be due to temperature increase while this CO oxidation also influences to the temperature increase itself. In addition, a higher SO_2 release was observed (Figure 5B). The temperature increase might be one reason for this while also the fact that after a longer period without regenerations also more SO_2 is most probably collected to the trap and therefore more SO_2 can be released during the regeneration as well. Also, the SO_2 level at the end of regeneration is on higher level in the regeneration of Figure 5B being approx. 190 ppm while in the case of Figure 5A the SO_2 level at the end of regeneration is approx. 50 ppm. The higher SO_2 level at the end of regeneration might indicate that if the regeneration mode is to be continued for a longer time period also more SO_2 could be released. However, on the other hand, especially in the case of Figure 5A i.e., daily regenerations, most of the SO_2 is released in the first half of the regeneration period. This indicates that shorter regeneration times could be relevant.

4. Discussion

The reports of ICCT and Sphera present average methane slip values of 5.5 g/kWh and 3.8 g/kWh for LPDF engines, respectively [14,15]. However, the methane emissions from actual LNG vessels have not yet been widely studied. Anderson et al. [8] conducted on board studies on an LNG ship resulting in total hydrocarbon emissions of 1.1–6.7 g/kWh (depending on engine load). Ushakov et al. [12] report average methane emission of 5.26 g/kWh based on on-board measurements. Peng et al. [6] also did on-board studies and reported methane levels of 3.7–25.5 g/kWh, the lowest level corresponding to highest engine load and the highest methane level corresponding to engine load of 25%. They also publish the actual methane concentration of 1000 ppm at 90% engine load and higher concentrations at lower engine loads. Lehtoranta et al. [5] reported methane levels from one medium speed engine (dual fuel, natural gas) at two load modes of 85% and 40% resulting in 5.6 g/kWh and 13.8 g/kWh, respectively. These levels correspond to methane concentrations of 1800 ppm (at 85% load) and 3750 ppm (at 40% load) (Table 2). Wei and Geng [30] review also reports total hydrocarbon concentrations above 2000 ppm from natural gas/diesel dual fuel combustion. Comparing these published results, there is a large variation in the methane levels reported, depending, most probably, on different engines (e.g., engine model, size, age) used as well as on engine loadings. Most of the values reported are in g/kWh while there are few publications reporting in concentrations (ppm) also. By comparing these values, it can be concluded that all the methane slips reported are in the order of 1000 ppm or higher. Furthermore, comparing this to the present study, the methane level (1000 ppm and 1500 ppm) in the present study was in the same order of magnitude. However, we used a smaller engine to investigate the performance of MOC but with a target to have similar exhaust gas composition to medium speed DF engines operating with NG as the main fuel. Therefore, we consider the results of the present study to be interpreted as indications of the possibilities to use MOC in LNG vessels.

In the present study, MOC was found to decrease the methane emission by 70–80% at the exhaust temperature of 550 °C. This efficiency, however, decreased significantly within time, even with only 0.5 ppm SO_2 in the exhaust. Regeneration was done once a day and was found to recover the efficiency. This regeneration was done in nearly stoichiometric conditions, the very low O_2 enabling SO_2 release from the MOC while also the temperature increase observed during the regeneration phase might influence the SO_2 release. In addition, the H_2 might play a role here too, since H_2 was found in the exhaust in the regeneration phase and has also been used for methane catalyst regeneration purposes in earlier studies [26,27].

Overall, it seems that the temperature is in significant role, both in methane oxidation efficiency as well as in regeneration. One regeneration performed in the present study resulted in higher exhaust temperature downstream of the catalyst and released higher amount of SO_2 compared to other regenerations.

In addition to MOC, a SO_x trap was studied in connection to MOC, for the first time (to the authors' knowledge). The SO_x trap was shown to protect the MOC against sulfur poisoning to some extent, since tripling the exhaust SO_2 level (from 0.5 ppm to 1.5 ppm) was found to have, in practice, no effect on MOC performance when trap was used upstream the MOC.

Other hydrocarbon emissions (ethane, ethylene) and carbon monoxide are easier to oxidize and were found to be nearly totally diminished by the MOC over the whole ageing test. The NG in the present study was high in methane content, while the fuel gas composition from different suppliers can have lower methane content (see e.g., [12]) and higher share of ethane and heavier hydrocarbons favoring better performance of MOC.

Formaldehyde, HCHO, is defined as carcinogenic substance and contribute also to other severe health effects, such as asthma, see, e.g., [31,32]. Therefore it is important to also consider the HCHO increase by LNG use (e.g., [6]), especially in coastal areas. In the present study, formaldehyde measurements were employed and showed that the

MOC very effectively (by 95–99%) reduced the formaldehyde emissions and practically no decrease was found in the formaldehyde oxidation efficiency during the catalyst ageing studies. This indicates that formaldehyde emissions from NG engines can be diminished with the MOC employment.

The results of the present study give indication of the possible use of MOC in LNG ships to control methane slip emissions. However, the regeneration process in real sized lean-burn marine engine is an issue that needs to be solved. Optimization of the regeneration interval and duration depending on the actual target of application is needed as well. Further catalyst development, regarding the efficiency and sizing, is to be done to have the best possible catalyst suitable to be installed at high-temperature conditions in the exhaust line.

Currently, LNG is a viable marine fuel deployed to substantially reduce pollutant emissions (NO_x, SO_x and particles) from ships. The use of LNG has therefore a notable effect on air quality and human health. In addition, LNG can provide reduction in GHG emissions if methane slip is controlled. This methane slip challenge needs to be solved to maximize LNG's potential to contribute to climate neutrality. The present study shows that MOC can play a role here. As it is an after-treatment system, it has potential both in new vessels as well as to retrofit to existing vessels. However, further studies are needed to solve the optimized solution for MOC and performance on different engine loadings as well in transient loading relevant in vessel operation.

In the long term, the transition from fossil NG to renewable NG is needed. This means also that further studies on, e.g., biogas (LBG) is needed to be able to develop suitable systems for biogas emissions since, e.g., possible biogas impurities may play a major role in the catalyst performance.

Author Contributions: Conceptualization, K.L., H.V., K.K. and T.M.; methodology, K.L., P.K., H.V.; investigation, K.L., P.K., H.V., K.K. and T.M.; original draft preparation, K.L.; writing—review and editing, K.L., P.K., H.V., K.K. and T.M. All authors have read and agreed to the published version of the manuscript.

Funding: This research was part of INTENS-research project, funded by Business Finland and several Finnish companies.

Institutional Review Board Statement: Not applicable.

Informed Consent Statement: Not applicable.

Conflicts of Interest: The authors declare no conflict of interest.

References

1. Eyring, V.; Isaksen, I.S.A.; Berntsen, T.; Collins, W.J.; Corbett, J.J.; Endresen, O.; Grainger, R.G.; Moldanova, J.; Schlager, H.; Stevenson, D.S. Transport impacts on atmosphere and climate: Shipping. *Atmos. Environ.* **2010**, *44*, 4735–4771. [CrossRef]
2. Corbett, J.J.; Winebrake, J.J.; Green, E.H.; Kasibhatla, P.; Eyring, V.; Lauer, A. Mortality from ship emissions: A global assessment. *Environ. Sci. Technol.* **2007**, *41*, 8512–8518. [CrossRef] [PubMed]
3. Viana, M.; Hammingh, P.; Colette, A.; Querol, X.; Degraeuwe, B.; de Vlieger, I.; van Aardenne, J. Impact of maritime transport emissions on coastal air quality in Europe. *Atmos. Environ.* **2014**, *90*, 96–105. [CrossRef]
4. Ni, P.; Wang, X.; Li, H. A review on regulations, current status, effects and reduction strategies of emissions for marine diesel engines. *Fuel* **2020**, *279*, 118477. [CrossRef]
5. Lehtoranta, K.; Aakko-Saksa, P.; Murtonen, T.; Vesala, H.; Ntziachristos, L.; Rönkkö, T.; Karjalainen, P.; Kuittinen, N.; Timonen, H. Particulate Mass and Nonvolatile Particle Number Emissions from Marine Engines Using Low-Sulfur Fuels, Natural Gas, or Scrubbers. *Environ. Sci. Technol.* **2019**, *53*, 3315–3322. [CrossRef]
6. Peng, W.; Yang, J.; Corbin, J.; Trivanovic, U.; Lobo, P.; Kirchen, P.; Rogak, S.; Gagné, S.; Miller, J.W.; Cocker, D. Comprehensive analysis of the air quality impacts of switching a marine vessel from diesel fuel to natural gas. *Environ. Pollut.* **2020**, *266*. [CrossRef]
7. Lehtoranta, K.; Aakko-Saksa, P.; Murtonen, T.; Vesala, H.; Kuittinen, N.; Rönkkö, T.; Ntziachristos, L.; Karjalainen, P.; Timonen, H.; Teinilä, K. Particle and Gaseous Emissions from Marine Engines Utilizing Various Fuels and Aftertreatment Systems. In Proceedings of the 29th CIMAC World Congress on Combustion Engine, Vancouver, BC, Canada, 10–14 June 2019.
8. Anderson, M.; Salo, K.; Fridell, E. Particle- and Gaseous Emissions from an LNG Powered Ship. *Environ. Sci. Technol.* **2015**, *49*, 12568–12575. [CrossRef]

9. Hesterberg, T.W.; Lapin, C.A.; Bunn, W.B. A comparison of emissions from vehicles fueled with diesel or compressed natural gas. *Environ. Sci. Technol.* **2008**, *42*, 6437–6445. [CrossRef]

10. Liu, J.; Yang, F.; Wang, H.; Ouyang, M.; Hao, S. Effects of pilot fuel quantity on the emissions characteristics of a CNG/diesel dual fuel engine with optimized pilot injection timing. *Appl. Energy* **2013**, *110*, 201–206. [CrossRef]

11. Lehtoranta, K.; Murtonen, T.; Vesala, H.; Koponen, P.; Alanen, J.; Simonen, P.; Rönkkö, T.; Timonen, H.; Saarikoski, S.; Maunula, T.; et al. Natural Gas Engine Emission Reduction by Catalysts. *Emiss. Control Sci. Technol.* **2017**, *3*. [CrossRef]

12. Ushakov, S.; Stenersen, D.; Einang, P.M. Methane slip from gas fuelled ships: A comprehensive summary based on measurement data. *J. Mar. Sci. Technol.* **2019**, *24*, 1308–1325. [CrossRef]

13. Sharafian, A.; Blomerus, P.; Mérida, W. Natural gas as a ship fuel: Assessment of greenhouse gas and air pollutant reduction potential. *Energy Policy* **2019**. [CrossRef]

14. Pavlenko, N.; Comer, B.; Zhou, Y.; Clark, N.; Rutherford, D. The climate Implications of Using LNG as a Marine Fuel. In *WORKING PAPER 2020-02*; ICCT: Washington, DC, USA, 2020; Available online: https://www.stand.earth/sites/stand/files/20 200128-ICCT-StandEarth-Climate-Implications-LNG-As-Marine-Fuel.pdf (accessed on 30 January 2020).

15. Sphera Life Cycle GHG Emission Study on the Use of LNG as Marine Fuel Final Report. Report 2019. Available online: https:// sea-lng.org/wp-content/uploads/2020/06/19-04-10_ts-SEA-LNG-and-SGMF-GHG-Analysis-of-LNG_Full_Report_v1.0.pdf (accessed on 7 December 2020).

16. Järvi, A. Methane slip reduction in Wärtsilä lean burn gas engines. In Proceedings of the 26th CIMAC World Congress on Combustion Engines, Bergen, Norway, 14–17 June 2010.

17. Hiltner, J.; Loetz, A.; Fiveland, S. Unburned Hydrocarbon Emissions from Lean Burn Natural Gas Engines—Sources and Solutions. In Proceedings of the 28th CIMAC World congress, Helsinki, Finland, 6–10 June 2016.

18. Burch, R.; Urbano, F.J.; Loader, P.K. Methane combustion over palladium catalysts: The effect of carbon dioxide and water on activity. *Appl. Catal. A Gen.* **1995**, *123*, 173–184. [CrossRef]

19. Gélin, P.; Urfels, L.; Primet, M.; Tena, E. Complete oxidation of methane at low temperature over Pt and Pd catalysts for the abatement of lean-burn natural gas fuelled vehicles emissions: Influence of water and sulphur containing compounds. In *Proceedings of the Catalysis Today*; Elsevier: Amsterdam, The Netherlands, 2003; Volume 83, pp. 45–57.

20. Ottinger, N.; Veele, R.; Xi, Y.; Liu, Z.G. Desulfation of Pd-based Oxidation Catalysts for Lean-burn Natural Gas and Dual-fuel Applications. *SAE Int. J. Engines* **2015**, *8*, 1472–1477. [CrossRef]

21. Lampert, J.K.; Kazi, M.S.; Farrauto, R.J. Palladium catalyst performance for methane emissions abatement from lean burn natural gas vehicles. *Appl. Catal. B Environ.* **1997**, *14*, 211–223. [CrossRef]

22. Arosio, F.; Colussi, S.; Groppi, G.; Trovarelli, A. Regeneration of S-poisoned Pd/Al2O3 catalysts for the combustion of methane. *Catal. Today* **2006**, *117*, 569–576. [CrossRef]

23. Honkanen, M.; Wang, J.; Kärkkäinen, M.; Huuhtanen, M.; Jiang, H.; Kallinen, K.; Keiski, R.L.; Akola, J.; Vippola, M. Regeneration of sulfur-poisoned Pd-based catalyst for natural gas oxidation. *J. Catal.* **2018**, *358*, 253–265. [CrossRef]

24. Kinnunen, N.M.; Hirvi, J.T.; Kallinen, K.; Maunula, T.; Keenan, M.; Suvanto, M. Case study of a modern lean-burn methane combustion catalyst for automotive applications: What are the deactivation and regeneration mechanisms? *Appl. Catal. B Environ.* **2017**, *207*, 114–119. [CrossRef]

25. Lott, P.; Eck, M.; Doronkin, D.E.; Zimina, A.; Tischer, S.; Popescu, R.; Belin, S.; Briois, V.; Casapu, M.; Grunwaldt, J.D.; et al. Understanding sulfur poisoning of bimetallic Pd-Pt methane oxidation catalysts and their regeneration. *Appl. Catal. B Environ.* **2020**, *278*, 119244. [CrossRef]

26. Heikkilä, S.; Sirviö, K.; Niemi, S.; Roslund, P.; Lehtoranta, K.; Pettinen, R.; Vesala, H.; Koponen, P.; Kallinen, K.; Maunula, T.; et al. Methane Catalyst Regeneration with Hydrogen Addition. In Proceedings of the 29th CIMAC World Congress on Combustion Engine, Vancouver, BC, Canada, 10–14 June 2019.

27. Jones, J.M.; Dupont, V.A.; Brydson, R.; Fullerton, D.J.; Nasri, N.S.; Ross, A.B.; Westwood, A.V.K. Sulphur poisoning and regeneration of precious metal catalysed methane combustion. In *Catalysis Today*; Elsevier: Amsterdam, The Netherlands, 2003; Volume 81, pp. 589–601.

28. Bailey, O.H.; Dou, D.; Molinier, M. Sulfur traps for NOx adsorbers: Materials development and maintenance strategies for their application. *SAE Tech. Pap.* **2000**. [CrossRef]

29. Murtonen, T.; Lehtoranta, K.; Korhonen, S.; Vesala, H.; Koponen, P. Imitating emission matrix of large natural catalyst studies in engine laboratory. In Proceedings of the 28th CIMAC World congress, Helsinki, Finland, 6–10 June 2016; p. 107.

30. Wei, L.; Geng, P. A review on natural gas/diesel dual fuel combustion, emissions and performance. *Fuel Process. Technol.* **2016**, *142*, 264–278. [CrossRef]

31. Zhang, X.; Zhao, Y.; Song, J.; Yang, X.; Zhang, J.; Zhang, Y.; Li, R. Differential Health Effects of Constant versus Intermittent Exposure to Formaldehyde in Mice: Implications for Building Ventilation Strategies. *Environ. Sci. Technol.* **2018**, *52*, 1551–1560. [CrossRef] [PubMed]

32. McGwin, G.; Lienert, J.; Kennedy, J.I. Formaldehyde exposure and asthma in children: A systematic review. *Environ. Health Perspect.* **2010**, *118*, 313–317. [CrossRef] [PubMed]

Journal of
Marine Science and Engineering

Article

Optimization of a Multiple Injection System in a Marine Diesel Engine through a Multiple-Criteria Decision-Making Approach

Maria Isabel Lamas [1,*], Laura Castro-Santos [1] and Carlos G. Rodriguez [2]

[1] Higher Polytechnic University College, University of Coruña, 15403 Ferrol, Spain; laura.castro.santos@udc.es
[2] Norplan Engineering S.L., 15570 Naron, Spain; c.rodriguez.vidal@udc.es
* Correspondence: isabellamas@udc.es; Tel.: +34-881-013-896

Received: 25 October 2020; Accepted: 18 November 2020; Published: 20 November 2020

Abstract: In this work, a numerical model was developed to analyze the performance and emissions of a marine diesel engine, the Wärtsilä 6L 46. This model was validated using experimental measurements and was employed to analyze several pre-injection parameters such as pre-injection rate, duration, and starting instant. The modification of these parameters may lead to opposite effects on consumption and/or emissions of nitrogen oxides (NO_x), carbon monoxide (CO), and hydrocarbons (HC). According to this, the main goal of the present work is to employ a multiple-criteria decision-making (MCDM) approach to characterize the most appropriate injection pattern. Since determining the criteria weights significantly influences the overall result of a MCDM problem, a subjective weighting method was compared with four objective weighting methods: entropy, CRITIC (CRiteria Importance Through Intercriteria Correlation), variance, and standard deviation. The results showed the importance of subjectivism over objectivism in MCDM analyses. The CRITIC, variance, and standard deviation methods assigned more importance to NO_x emissions and provided similar results. Nevertheless, the entropy method assigned more importance to consumption and provided a different injection pattern.

Keywords: MCDM; marine engine; injection; emissions; consumption

---

1. Introduction

Pollution levels in recent years have been reaching dangerous limits. Important contributors to global pollution are diesel engines, which are efficient machines but emit important levels of particulate matter (PM), NO_x, CO_2, CO, HC, SO_x, etc. [1–5]. Between these, NO_x and SO_x are characteristic of marine diesel engines [6–10]. According to the International Maritime Organization (IMO), NO_x and SO_x from ships represent 5% and 13% of global NO_x and SO_x emissions, respectively [11]. IMO regulates NO_x and SO_x in the shipping sector. Regarding SO_x, since the sulfur content of the fuel is the reason for SO_x emission, IMO limits the sulfur content of fuels or requires the use of exhaust gas cleaning systems to reduce sulfur emissions [12]. Regarding NO_x, IMO imposes even increasing limitations. According to this, several NO_x reduction procedures have been developed in recent years. Some of them, called primary measurements, operate on the engine performance, such as EGR, water injection, modification of the injection parameters, etc. On the other hand, other NO_x reduction procedures, called secondary measurements, remove this pollutant from exhaust gases by downstream cleaning techniques, such as selective catalytic reduction (SCR). The present work focuses on pre-injection systems. It is well known that pilot injections reduce NO_x noticeably [13–17], but sometimes pilot injections can increase consumption and other pollutants such as smoke or hydrocarbons (HC) [18–22], mainly depending on parameters such as injection time, duration, number of pre-injections, dwelling time, etc. Since these parameters provide conflicting results, a formal tool to establish the most appropriate injection pattern is necessary. According to this, multiple-criteria decision-making (MCDM) approaches constitute a formal tool for handling complex decision-making problems. MCDM methods are complex decision-making tools for choosing the optimal option in cases where there are conflicting criteria. Since the start of the MCDM methods in the 1960s, they were employed in many fields such as sustainability, supply chain management, materials, quality management, GIS, construction and project management, safety and risk management, manufacturing systems, technology, information management, soft computing, tourism management, etc. [23]. One of the handicaps of MCDM methods is the determination of the criteria weights, i.e., the degree of importance for each criterion. It is important to focus on the criteria weights due to their influence on the overall result. According to this, several approaches to define the criteria weights can be found in the literature. Briefly, these approaches can be divided into subjective, i.e., based on the estimations of experts, and objective, i.e., calculated through mathematical expressions. In practice, subjective weights are most commonly used [24]. Contrary to subjective methods, the objective weights are based on mathematical methods and decision-makers have no role in determining the relative importance of criteria. Common objective methods are entropy [25], CRITIC [26], standard deviation [27], variance, mean weight, etc.

The present paper proposes a MCDM approach to select the most appropriate injection pattern using a pilot injection in the marine diesel engine Wärtsilä 6L 46. The pre-injection rate, duration, and starting instants were analyzed and the criteria were specific fuel consumption (SFC) and NO_x, CO, and HC emissions. These emissions and consumption were characterized through CFD (Computational Fluid Dynamics) analyses. Due to the importance of the criteria weights on the overall result, a comparison of several weighting methods was realized. A subjective criteria weighting method was compared to four objective criteria weighting methods: entropy, CRITIC, variance, and standard deviation.

2. Materials and Methods

The Wärtsilä 6L 46 is a four-stroke marine diesel engine, turbocharged and intercooled with direct fuel injection. It has 6 in-line cylinders and each cylinder has 2 inlet and 2 exhaust valves. The standard engine employed in the present work does not implement any NO_x reduction system such as water injection, EGR, SCR, etc. It incorporates a fast and efficient turbocharging system called single pipe exhaust (SPEX). The cooling system is split into high-temperature and low-temperature stages. Other specifications are provided in Table 1.

Table 1. Characteristics of the engine at 100% load.

Parameter	Value
Output	5430 kW
Speed	500 rpm
Piston displacement	96.4 L/cyl
Bore	460 mm
Stroke	580 mm
Speed	500 rpm
Mean effective pressure	22.5 bar
Mean piston speed	9.7 m/s

A CFD analysis previously validated with experimental results [28–34] was carried out using the open software OpenFOAM. The simulation was based on the equations of conservation of mass, momentum, and energy and the numerical details are listed in Table 2.

Table 2. Numerical details.

Parameter	Model
Turbulence model	k-ε
Combustion kinetic scheme	Ra and Reitz [35] (131 reactions and 41 species)
NO_x formation kinetic scheme	Yang et al. [36] (43 reactions and 20 species)
NO_x reduction kinetic scheme	Miller and Glarborg [37] (131 reactions and 24 species)
Fuel heat-up and evaporation model	Dukowicz [38]
Fuel droplet breakup model	Kelvin-Helmholtz and Rayleigh-Taylor [39]

A comparison between the numerical and experimental results is illustrated in Figure 1 and Figure 2. Figure 1 shows the emissions and SFC obtained numerically and experimentally at several loads, and Figure 2 shows the in-cylinder pressure and heat release rate obtained numerically and experimentally at 100% load. As can be seen, both figures show a reasonable correspondence between numerical and experimental results.

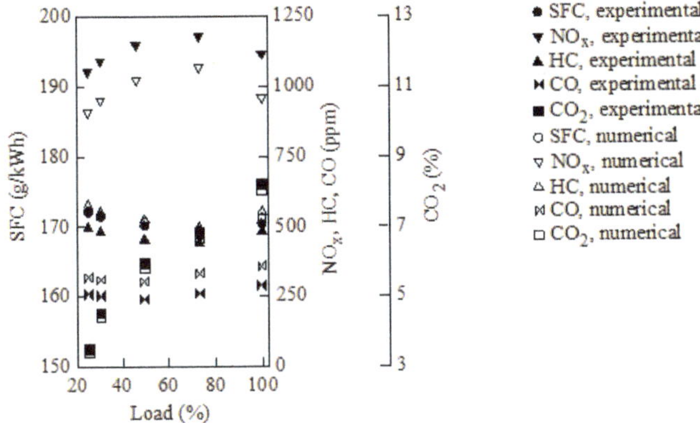

Figure 1. Specific fuel consumption (SFC) and emissions numerically and experimentally obtained at different loads.

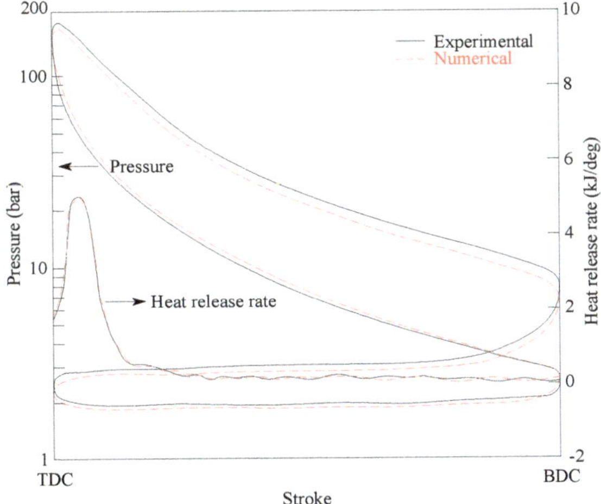

Figure 2. In-cylinder pressure numerically and experimentally obtained at 100% load.

3. Results and Discussion

Once validated, this CFD model was used to provide the data for the MCDM approach. The simulation calculation was carried out simply as a process and as sample results for applying the optimal selection method of multiple injection conditions. The 125 cases illustrated in Figure 3 were analyzed. As can be seen, five pre-injection rates (R): 5%, 10%, 15%, 20%, and 25%; five pre-injection durations (D): 1°, 2°, 3°, 4°, and 5° crank angle (CA); and five pre-injection starting instants (S): −22°, −21°, −20°, −19°, and −18° crank angle after top dead center (CA ATDC), were employed.

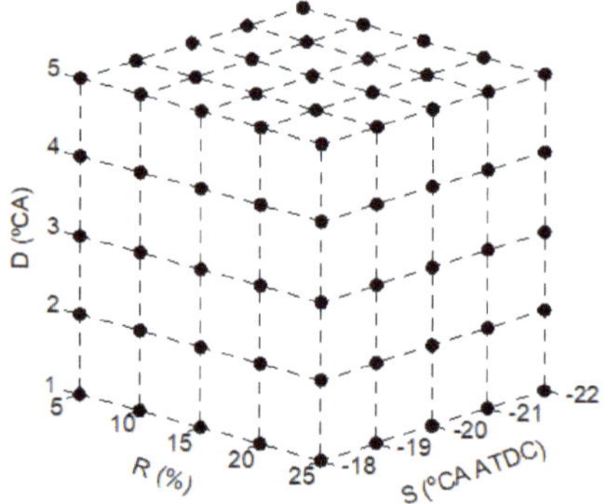

Figure 3. Illustration of the 125 cases analyzed.

Four criteria were analyzed: SFC, NO_x, CO, and HC. It is worth mentioning that PM emissions should be included in this model. They were not included because current numerical methods do not provide enough accuracy regarding PM [40]. Nevertheless, continuous efforts are being made to develop more models which will provide proper accuracy in the near future.

Table 3 outlines the pre-injection rate, duration, and starting instant, as well as the values of SFC, NO_x, CO, and HC provided by the CFD model for the 125 cases. According to this, the data matrix is composed of 125 rows and 4 columns. In the remaining of the present work, each datum of the decision matrix will be represented as X_{ij}, where i is the case and j is the criteria considered.

Table 3. Data for the multiple-criteria decision-making (MCDM) problem.

Case (i)	S (°CA ATDC)	R (%)	D (°CA)	$j = 1$ SFC (g/kWh)	$j = 2$ NO_x (g/kWh)	$j = 3$ CO (g/kWh)	$j = 4$ HC (g/kWh)
1	−22	5	1	190.9	7.38	4.65	5.72
2	−22	5	2	189.0	7.83	4.67	5.73
3	−22	5	3	187.5	8.19	4.70	5.76
4	−22	5	4	186.6	8.43	4.74	5.80
5	−22	5	5	186.1	8.58	4.78	5.84
6	−22	10	1	196.4	6.01	4.70	5.78
7	−22	10	2	193.9	6.57	4.73	5.81
8	−22	10	3	192.1	7.00	4.77	5.84
9	−22	10	4	190.8	7.31	4.81	5.89
10	−22	10	5	190.2	7.49	4.86	5.94
11	−22	15	1	200.4	5.06	4.74	5.83
12	−22	15	2	197.5	5.70	4.77	5.86
13	−22	15	3	195.3	6.19	4.81	5.90
14	−22	15	4	193.9	6.53	4.86	5.95
15	−22	15	5	193.1	6.73	4.92	6.02
16	−22	20	1	203.6	4.32	4.77	5.87
17	−22	20	2	200.3	5.01	4.80	5.90

Table 3. *Cont.*

Case (i)	S (°CA ATDC)	R (%)	D (°CA)	$j = 1$ SFC (g/kWh)	$j = 2$ NO$_x$ (g/kWh)	$j = 3$ CO (g/kWh)	$j = 4$ HC (g/kWh)
				Criterion (j)			
18	−22	20	3	197.9	5.55	4.85	5.94
19	−22	20	4	196.3	5.92	4.91	6.00
20	−22	20	5	195.5	6.14	4.97	6.07
21	−22	25	1	206.3	3.70	4.79	5.90
22	−22	25	2	202.8	4.44	4.83	5.93
23	−22	25	3	200.2	5.01	4.88	5.98
24	−22	25	4	198.4	5.41	4.94	6.05
25	−22	25	5	197.5	5.65	5.02	6.13
26	−21	5	1	185.2	8.65	4.62	5.68
27	−21	5	2	183.3	9.11	4.65	5.70
28	−21	5	3	181.9	9.46	4.67	5.72
29	−21	5	4	180.9	9.71	4.71	5.76
30	−21	5	5	180.4	9.85	4.75	5.80
31	−21	10	1	189.1	7.57	4.67	5.73
32	−21	10	2	186.6	8.14	4.69	5.76
33	−21	10	3	184.7	8.57	4.73	5.79
34	−21	10	4	183.5	8.88	4.77	5.84
35	−21	10	5	182.9	9.05	4.83	5.89
36	−21	15	1	191.9	6.83	4.70	5.77
37	−21	15	2	189.0	7.47	4.73	5.80
38	−21	15	3	186.8	7.96	4.77	5.84
39	−21	15	4	185.4	8.31	4.82	5.89
40	−21	15	5	184.6	8.51	4.88	5.96
41	−21	20	1	194.1	6.25	4.72	5.81
42	−21	20	2	190.9	6.94	4.76	5.84
43	−21	20	3	188.5	7.48	4.80	5.88
44	−21	20	4	186.9	7.86	4.86	5.94
45	−21	20	5	186.1	8.07	4.93	6.01
46	−21	25	1	196.0	5.76	4.74	5.83
47	−21	25	2	192.5	6.50	4.78	5.87
48	−21	25	3	189.9	7.08	4.83	5.91
49	−21	25	4	188.1	7.48	4.89	5.98
50	−21	25	5	187.3	7.71	4.97	6.06
51	−20	5	1	181.2	9.61	4.60	5.65
52	−20	5	2	179.3	10.07	4.62	5.67
53	−20	5	3	177.9	10.42	4.65	5.69
54	−20	5	4	176.9	10.67	4.68	5.73
55	−20	5	5	176.4	10.81	4.73	5.77
56	−20	10	1	183.9	8.76	4.64	5.69
57	−20	10	2	181.4	9.32	4.66	5.72
58	−20	10	3	179.6	9.75	4.70	5.75
59	−20	10	4	178.3	10.06	4.74	5.80
60	−20	10	5	177.7	10.24	4.80	5.85
61	−20	15	1	185.9	8.17	4.66	5.73
62	−20	15	2	183.0	8.81	4.69	5.75
63	−20	15	3	180.8	9.30	4.74	5.79
64	−20	15	4	179.3	9.64	4.79	5.85
65	−20	15	5	178.6	9.84	4.85	5.91
66	−20	20	1	187.4	7.71	4.68	5.75
67	−20	20	2	184.2	8.40	4.72	5.78
68	−20	20	3	181.8	8.93	4.76	5.83
69	−20	20	4	180.2	9.31	4.82	5.89
70	−20	20	5	179.4	9.53	4.89	5.96
71	−20	25	1	188.8	7.32	4.70	5.78
72	−20	25	2	185.3	8.06	4.74	5.81
73	−20	25	3	182.6	8.63	4.79	5.86
74	−20	25	4	180.9	9.04	4.85	5.92

Table 3. *Cont.*

Case (*i*)	S (°CA ATDC)	R (%)	D (°CA)	Criterion (*j*)			
				j = 1 SFC (g/kWh)	*j* = 2 NO$_x$ (g/kWh)	*j* = 3 CO (g/kWh)	*j* = 4 HC (g/kWh)
75	−20	25	5	180.0	9.27	4.93	6.00
76	−19	5	1	178.9	10.26	4.59	5.62
77	−19	5	2	177.0	10.71	4.61	5.64
78	−19	5	3	175.5	11.07	4.63	5.67
79	−19	5	4	174.6	11.32	4.67	5.70
80	−19	5	5	174.1	11.46	4.71	5.75
81	−19	10	1	180.9	9.55	4.62	5.66
82	−19	10	2	178.4	10.12	4.64	5.68
83	−19	10	3	176.5	10.55	4.68	5.72
84	−19	10	4	175.3	10.86	4.72	5.76
85	−19	10	5	174.7	11.03	4.78	5.82
86	−19	15	1	182.4	9.07	4.64	5.69
87	−19	15	2	179.5	9.71	4.67	5.72
88	−19	15	3	177.3	10.20	4.71	5.75
89	−19	15	4	175.8	10.54	4.76	5.81
90	−19	15	5	175.1	10.74	4.82	5.87
91	−19	20	1	183.5	8.69	4.66	5.71
92	−19	20	2	180.3	9.38	4.69	5.74
93	−19	20	3	177.9	9.92	4.74	5.78
94	−19	20	4	176.3	10.29	4.79	5.84
95	−19	20	5	175.5	10.51	4.86	5.92
96	−19	25	1	184.5	8.37	4.67	5.73
97	−19	25	2	181.0	9.11	4.71	5.76
98	−19	25	3	178.4	9.68	4.76	5.81
99	−19	25	4	176.6	10.08	4.82	5.87
100	−19	25	5	175.8	10.32	4.90	5.95
101	−18	5	1	178.2	10.59	4.57	5.60
102	−18	5	2	176.3	11.05	4.60	5.62
103	−18	5	3	174.8	11.40	4.62	5.65
104	−18	5	4	173.9	11.65	4.66	5.68
105	−18	5	5	173.4	11.79	4.70	5.73
106	−18	10	1	180.0	9.97	4.60	5.63
107	−18	10	2	177.5	10.53	4.63	5.66
108	−18	10	3	175.7	10.96	4.67	5.69
109	−18	10	4	174.4	11.27	4.71	5.74
110	−18	10	5	173.8	11.45	4.76	5.79
111	−18	15	1	181.3	9.53	4.62	5.66
112	−18	15	2	178.4	10.17	4.65	5.68
113	−18	15	3	176.2	10.66	4.70	5.72
114	−18	15	4	174.8	11.01	4.75	5.78
115	−18	15	5	174.1	11.21	4.81	5.84
116	−18	20	1	182.4	9.20	4.64	5.68
117	−18	20	2	179.1	9.89	4.67	5.71
118	−18	20	3	176.7	10.42	4.72	5.75
119	−18	20	4	175.1	10.80	4.78	5.81
120	−18	20	5	174.3	11.02	4.85	5.88
121	−18	25	1	183.3	8.91	4.65	5.69
122	−18	25	2	179.8	9.65	4.69	5.73
123	−18	25	3	177.1	10.22	4.74	5.77
124	−18	25	4	175.4	10.63	4.80	5.84
125	−18	25	5	174.5	10.86	4.88	5.92

Each indicator was transformed into its variation in per unit basis through Equation (1).

$$V_{ij} = \frac{X_{ij} - X_{jref}}{X_{jref}} \tag{1}$$

where X_{jref} is the value corresponding to criterion *j* in the case without pre-injection.

As indicated previously, an important step in a MCDM approach is to determine the criteria weights, and this issue will be treated in Sections 3.1–3.5. Once determined, the adequacy index for each i-th case, AI_i, was calculated through the SAW (simple additive weighting) method, Equation (2). Obviously, the most appropriate solution corresponds to the minimum value of *AI*.

$$AI_i = \sum_{j=1}^{n} w_j V_{ij} \tag{2}$$

where w_j is the weight of the j-th criterion and n the number of criteria, i.e., four (SFC, NO_x, CO, and HC).

3.1. Subjective Weighting Method

In this method, the levels of importance were set by experts in the field. According to these experts, the same importance was given to consumption (50%) and emissions (50%). Regarding emissions, the importance was also distributed equally between NO_x (33.3%), CO (33.3%), and HC (33.3%). According to this, the weights of SFC, NO_x, CO, and HC result in $w_1 = 0.5$, $w_2 = 0.16$, $w_3 = 0.16$, and $w_4 = 0.16$, respectively.

3.2. Entropy Weighting Method

This method measures the uncertainty in the information, and the criteria weights are given by Equation (3).

$$w_j = \frac{1 - E_j}{\sum\limits_{j=1}^{n} (1 - E_j)} \tag{3}$$

In the equation above, 1—*Ej* represents the degree of diversity of the information related to the j-th criterion and E_j is the entropy value of the j-th criterion, given by Equation (4). In this equation, p_{ij} are the normalized data, Equation (5).

$$E_j = -\frac{\sum\limits_{i=1}^{m} p_{ij} \ln(p_{ij})}{\ln(m)} \tag{4}$$

$$p_{ij} = \frac{V_{ij}}{\sum\limits_{i=1}^{m} V_{ij}} \tag{5}$$

According to the equations above, the range of the entropy value is 0–1. A low entropy value indicates that the degree of disorder corresponding to criterion j is low and thus leads to a high weight.

3.3. CRITIC Weighting Method

In this method, the criteria weights are obtained by

$$w_j = \frac{C_j}{\sum\limits_{j=1}^{n} C_j} \tag{6}$$

where C_j, Equation (7), represents a measure of the conflict created by criterion j with respect to the decision situation defined by the rest of the criteria. As the scores of the alternatives in criteria i and j become more discordant, the value of l_{ij} is lowered. The higher the value C_j, the larger the amount of information transmitted by the corresponding criterion and the higher the relative importance for the decision-making process. The objective weights are derived by normalizing these values to unity, as indicated above through Equation (6).

$$C_j = \sigma_j \sum_{i=1}^{m} (1 - l_{ij}) \tag{7}$$

where σ_j is the standard deviation of the *j*-th criterion and l_{ij} is the correlation coefficient, Equation (8). These correlation coefficients represent linear correlation coefficients between the criteria values in the matrix.

$$l_{ij} = \frac{\sum\limits_{k=1}^{m} (V_{ki} - \overline{V}_i)(V_{kj} - \overline{V}_j)}{\sqrt{\sum\limits_{k=1}^{m} (V_{ki} - \overline{V}_i)^2}\sqrt{\sum\limits_{k=1}^{m} (V_{kj} - \overline{V}_j)^2}} \tag{8}$$

3.4. Variance Weighting Method

The variance procedure method determines the criteria weights in terms of their statistical variances, $\sigma_j{}^2$, through the following equation:

$$w_j = \frac{\sigma_j^2}{\sum\limits_{j=1}^{n} \sigma_j^2} \tag{9}$$

3.5. Standard Deviation Weighting Method

The standard deviation method determines the criteria weights in terms of their standard deviations through the following equation:

$$w_j = \frac{\sigma_j}{\sum\limits_{j=1}^{n} \sigma_j} \tag{10}$$

To summarize, Table 4 summarizes the criteria weights obtained using these weighting methods. As can be seen, the entropy method assigns an important weight to SFC, while the CRITIC, variance, and standard deviation methods assign NO_x as the most relevant criterion. Both variance and standard deviation procedures measure the spread, i.e., the degree to which each sample is different from the mean. As can be seen in Table 3 shown above, CO and HC emissions remain practically constant and thus lead to low values of both variance and standard deviation. On the other hand, NO_x and, to a lesser extent, SFC present more spread and thus higher values of variance and standard deviation. For this reason, the variance and standard deviation methods provide a significant weight to NO_x and low weights to CO and HC. Since the standard deviation is the square root of the variance, the variance method assigns a higher weight to NO_x than the standard deviation method. The CRITIC method assigns high values of weights to those criteria with high standard deviation and low correlation with other responses. According to this, the results obtained through the CRITIC method are very similar to those obtained through the standard deviation method, but with less differences between the criteria weights. The entropy method also takes into account the uncertainty in the information and thus assigns low weights to CO and HC. Besides uncertainty, the entropy is based on the degree of disorder and thus assigns an important weight to SFC.

Table 4. Criteria weights according to the subjective, entropy, CRITIC, variance, and standard deviation weighting methods.

Weighting Method	Criteria Weights, w_j			
	SFC ($j=1$)	NO_x ($j=2$)	CO ($j=3$)	HC ($j=4$)
Subjective	0.50	0.16	0.16	0.16
Entropy	0.50	0.19	0.16	0.14
CRITIC	0.27	0.41	0.16	0.16
Variance	0.09	0.87	0.02	0.02
Standard deviation	0.20	0.62	0.09	0.09

Table 5 outlines the results of the 125 cases analyzed using these procedures. As can be seen, the subjective weighting method provides case 91, with an adequacy index of $AI_{91} = -0.122$, as the most appropriate injection pattern. This case corresponds to the $-19°$ CA ATDC pre-injection starting instant, 20% pre-injection rate, and $1°$ CA pre-injection duration. Since the subjective method assigns an important weight to NO_x, this 91st solution provides a significant NO_x reduction with a low increment of SFC, CO, and HC. This solution provides an important pre-injection rate, 20%, due to its importance on NO_x reduction. Retarding the pre-injection instant also reduces NO_x noticeably but at expenses of important increments on consumption. This reason leads to the CRITIC, variance, and standard deviation weighting methods to provide case 105, corresponding to the $-18°$ CA ATDC pre-injection starting instant, 5% pre-injection rate, and $5°$ CA pre-injection duration, as the most appropriate injection pattern, mainly due to the important weight of NO_x over the other criteria and lower weight of SFC in comparison with the subjective weighting method. A value of $-18°$ CA ATDC leads to important NO_x reduction with a noticeable SFC penalty. Basically, the NO_x reduction achieved with a high pre-injection rate or by a late pre-injection rate is reached through a reduction in the combustion temperature, since the high combustion temperatures reached in the combustion chamber are responsible for most NO_x emitted to the atmosphere [41,42]. On the other hand, the entropy method provides case 25 as the most appropriate injection pattern, with a $-22°$ CA ATDC pre-injection starting instant, 25% pre-injection rate, and $5°$ CA pre-injection duration. Since the entropy method assigns more weight to SFC and, to a lesser extent, to NO_x, it provides an earlier pre-injection starting instant, which leads to a reduction in SFC and a higher pre-injection rate, which leads to a reduction in NO_x.

Table 5. Adequacy index according to subjective, entropy, CRITIC, variance, and standard deviation weighting methods.

Case (i)	S (°CA ATDC)	R (%)	D (°CA)	AI_i				
				Subjective	Entropy	CRITIC	Variance	Standard Deviation
1	−22	5	1	−0.007	−0.020	−0.140	−0.376	−0.247
2	−22	5	2	−0.006	−0.017	−0.128	−0.347	−0.227
3	−22	5	3	−0.004	−0.015	−0.118	−0.325	−0.211
4	−22	5	4	−0.001	−0.012	−0.110	−0.309	−0.200
5	−22	5	5	0.002	−0.008	−0.103	−0.299	−0.192
6	−22	10	1	−0.004	−0.020	−0.170	−0.463	−0.302
7	−22	10	2	−0.003	−0.017	−0.155	−0.427	−0.278
8	−22	10	3	−0.001	−0.014	−0.143	−0.399	−0.259
9	−22	10	4	0.003	−0.011	−0.132	−0.380	−0.244
10	−22	10	5	0.007	−0.006	−0.124	−0.368	−0.235
11	−22	15	1	−0.002	−0.019	−0.190	−0.522	−0.340
12	−22	15	2	−0.001	−0.017	−0.173	−0.482	−0.313
13	−22	15	3	0.002	−0.014	−0.159	−0.451	−0.291
14	−22	15	4	0.006	−0.009	−0.148	−0.428	−0.274
15	−22	15	5	0.010	−0.005	−0.138	−0.415	−0.264
16	−22	20	1	0.000	−0.019	−0.206	−0.569	−0.369

Table 5. *Cont.*

Case (i)	S (°CA ATDC)	R (%)	D (°CA)	AI_i				
				Subjective	Entropy	CRITIC	Variance	Standard Deviation
17	−22	20	2	0.002	−0.016	−0.188	−0.525	−0.340
18	−22	20	3	0.004	−0.013	−0.172	−0.491	−0.316
19	−22	20	4	0.008	−0.008	−0.159	−0.467	−0.298
20	−22	20	5	0.014	−0.003	−0.149	−0.452	−0.286
21	−22	25	1	0.002	−0.018	−0.219	−0.608	−0.394
22	−22	25	2	0.004	−0.016	−0.199	−0.561	−0.362
23	−22	25	3	0.006	−0.012	−0.183	−0.525	−0.337
24	−22	25	4	0.011	−0.007	−0.169	−0.499	−0.318
25	−22	25	5	0.016	−0.001	−0.158	−0.483	−0.305
26	−21	5	1	−0.010	−0.020	−0.112	−0.296	−0.195
27	−21	5	2	−0.008	−0.017	−0.100	−0.267	−0.176
28	−21	5	3	−0.006	−0.015	−0.090	−0.245	−0.160
29	−21	5	4	−0.004	−0.012	−0.082	−0.229	−0.148
30	−21	5	5	0.000	−0.008	−0.075	−0.219	−0.141
31	−21	10	1	−0.009	−0.021	−0.136	−0.364	−0.239
32	−21	10	2	−0.007	−0.018	−0.121	−0.329	−0.215
33	−21	10	3	−0.005	−0.015	−0.109	−0.301	−0.196
34	−21	10	4	−0.002	−0.012	−0.098	−0.281	−0.181
35	−21	10	5	0.002	−0.008	−0.090	−0.270	−0.172
36	−21	15	1	−0.008	−0.021	−0.152	−0.411	−0.269
37	−21	15	2	−0.006	−0.019	−0.136	−0.371	−0.242
38	−21	15	3	−0.004	−0.016	−0.121	−0.339	−0.220
39	−21	15	4	0.000	−0.011	−0.110	−0.317	−0.204
40	−21	15	5	0.005	−0.007	−0.101	−0.304	−0.193
41	−21	20	1	−0.007	−0.022	−0.165	−0.448	−0.293
42	−21	20	2	−0.005	−0.019	−0.147	−0.404	−0.263
43	−21	20	3	−0.002	−0.015	−0.131	−0.370	−0.239
44	−21	20	4	0.001	−0.011	−0.118	−0.346	−0.221
45	−21	20	5	0.007	−0.006	−0.109	−0.331	−0.210
46	−21	25	1	−0.006	−0.022	−0.175	−0.479	−0.312
47	−21	25	2	−0.004	−0.019	−0.156	−0.432	−0.280
48	−21	25	3	−0.001	−0.015	−0.139	−0.395	−0.255
49	−21	25	4	0.003	−0.011	−0.126	−0.369	−0.236
50	−21	25	5	0.008	−0.005	−0.115	−0.354	−0.224
51	−20	5	1	−0.011	−0.019	−0.091	−0.236	−0.156
52	−20	5	2	−0.010	−0.017	−0.079	−0.207	−0.137
53	−20	5	3	−0.008	−0.014	−0.068	−0.184	−0.121
54	−20	5	4	−0.005	−0.011	−0.060	−0.168	−0.109
55	−20	5	5	−0.002	−0.008	−0.054	−0.159	−0.102
56	−20	10	1	−0.011	−0.021	−0.110	−0.290	−0.191
57	−20	10	2	−0.010	−0.018	−0.095	−0.254	−0.167
58	−20	10	3	−0.007	−0.015	−0.083	−0.226	−0.148
59	−20	10	4	−0.004	−0.012	−0.072	−0.207	−0.134
60	−20	10	5	0.000	−0.008	−0.064	−0.195	−0.124
61	−20	15	1	−0.011	−0.022	−0.123	−0.327	−0.215
62	−20	15	2	−0.010	−0.019	−0.107	−0.287	−0.188
63	−20	15	3	−0.007	−0.016	−0.092	−0.256	−0.166
64	−20	15	4	−0.003	−0.012	−0.081	−0.233	−0.150
65	−20	15	5	0.001	−0.007	−0.072	−0.220	−0.139
66	−20	20	1	−0.011	−0.023	−0.134	−0.356	−0.234
67	−20	20	2	−0.009	−0.020	−0.115	−0.312	−0.205
68	−20	20	3	−0.007	−0.016	−0.100	−0.278	−0.181
69	−20	20	4	−0.003	−0.012	−0.087	−0.254	−0.163
70	−20	20	5	0.003	−0.007	−0.077	−0.240	−0.151
71	−20	25	1	−0.010	−0.023	−0.142	−0.381	−0.250
72	−20	25	2	−0.009	−0.021	−0.123	−0.334	−0.218
73	−20	25	3	−0.006	−0.017	−0.106	−0.298	−0.193
74	−20	25	4	−0.002	−0.012	−0.093	−0.272	−0.174
75	−20	25	5	0.004	−0.006	−0.082	−0.256	−0.161
76	−19	5	1	−0.011	−0.018	−0.076	−0.195	−0.130
77	−19	5	2	−0.010	−0.015	−0.064	−0.166	−0.110

Table 5. *Cont.*

Case (i)	S (°CA ATDC)	R (%)	D (°CA)	AI_i				
				Subjective	Entropy	CRITIC	Variance	Standard Deviation
78	−19	5	3	−0.008	−0.013	−0.054	−0.143	−0.094
79	−19	5	4	−0.005	−0.010	−0.045	−0.127	−0.083
80	−19	5	5	−0.002	−0.007	−0.039	−0.118	−0.075
81	−19	10	1	−0.012	−0.020	−0.092	−0.239	−0.159
82	−19	10	2	−0.010	−0.017	−0.077	−0.204	−0.135
83	−19	10	3	−0.008	−0.014	−0.065	−0.176	−0.115
84	−19	10	4	−0.005	−0.011	−0.054	−0.156	−0.101
85	−19	10	5	−0.001	−0.006	−0.046	−0.145	−0.091
86	−19	15	1	−0.012	−0.021	−0.103	−0.270	−0.179
87	−19	15	2	−0.011	−0.019	−0.086	−0.230	−0.152
88	−19	15	3	−0.008	−0.015	−0.072	−0.199	−0.130
89	−19	15	4	−0.004	−0.011	−0.061	−0.176	−0.114
90	−19	15	5	0.000	−0.006	−0.052	−0.163	−0.103
91	−19	20	1	−0.012	−0.022	−0.112	−0.294	−0.194
92	−19	20	2	−0.011	−0.019	−0.094	−0.251	−0.165
93	−19	20	3	−0.008	−0.016	−0.078	−0.216	−0.141
94	−19	20	4	−0.004	−0.011	−0.065	−0.192	−0.123
95	−19	20	5	0.001	−0.006	−0.056	−0.178	−0.112
96	−19	25	1	−0.012	−0.023	−0.119	−0.315	−0.207
97	−19	25	2	−0.011	−0.020	−0.100	−0.268	−0.176
98	−19	25	3	−0.008	−0.016	−0.083	−0.231	−0.151
99	−19	25	4	−0.004	−0.012	−0.069	−0.205	−0.132
100	−19	25	5	0.002	−0.006	−0.059	−0.190	−0.119
101	−18	5	1	−0.010	−0.016	−0.068	−0.173	−0.116
102	−18	5	2	−0.009	−0.014	−0.056	−0.144	−0.096
103	−18	5	3	−0.007	−0.011	−0.045	−0.122	−0.080
104	−18	5	4	−0.004	−0.008	−0.037	−0.106	−0.068
105	−18	5	5	−0.001	−0.005	−0.031	−0.096	−0.061
106	−18	10	1	−0.011	−0.018	−0.082	−0.213	−0.141
107	−18	10	2	−0.009	−0.015	−0.067	−0.177	−0.117
108	−18	10	3	−0.007	−0.012	−0.055	−0.150	−0.098
109	−18	10	4	−0.004	−0.009	−0.044	−0.130	−0.084
110	−18	10	5	0.001	−0.004	−0.036	−0.118	−0.074
111	−18	15	1	−0.011	−0.019	−0.092	−0.241	−0.159
112	−18	15	2	−0.009	−0.016	−0.075	−0.200	−0.132
113	−18	15	3	−0.007	−0.013	−0.061	−0.169	−0.110
114	−18	15	4	−0.003	−0.009	−0.049	−0.147	−0.094
115	−18	15	5	0.002	−0.004	−0.040	−0.134	−0.083
116	−18	20	1	−0.011	−0.020	−0.100	−0.262	−0.173
117	−18	20	2	−0.009	−0.017	−0.081	−0.218	−0.143
118	−18	20	3	−0.007	−0.013	−0.066	−0.184	−0.120
119	−18	20	4	−0.003	−0.009	−0.053	−0.160	−0.102
120	−18	20	5	0.003	−0.004	−0.043	−0.146	−0.090
121	−18	25	1	−0.011	−0.020	−0.106	−0.280	−0.185
122	−18	25	2	−0.009	−0.017	−0.087	−0.233	−0.153
123	−18	25	3	−0.007	−0.014	−0.070	−0.197	−0.128
124	−18	25	4	−0.002	−0.009	−0.057	−0.171	−0.109
125	−18	25	5	0.003	−0.003	−0.046	−0.156	−0.096

4. Conclusions

The main goal of the present paper is to characterize the most appropriate pre-injection pattern in a marine diesel engine, the Wärtsilä 6L 46, in order to optimize the pilot injection process. A CFD model previously validated with experimental results was employed to obtain data corresponding to a set of 125 injection patterns using pilot injection. The pre-injection rate, duration, and starting instant were varied in the ranges of 5% to 25%, 1° to 5° CA, and −22° to −18° CA ATDC, respectively. Since the manipulation of these parameters has conflicting results on consumption and emissions of NO_x, CO, and HC, a MCDM approach was employed to select the most appropriate injection pattern. Due to the importance of criteria weights on the overall result, several criteria weighting methods were compared. In particular, a subjective weighting method was compared with four objective weighting methods: entropy, CRITIC, variance, and standard deviation. The CRITIC, variance, and standard deviation methods led to the same injection pattern: $-19°$ CA pre-injection starting angle, 20% pre-injection rate, and $5°$ CA pre-injection duration. Nevertheless, the entropy method provided a $-22°$ CA pre-injection starting angle, 25% pre-injection rate, and $5°$ CA pre-injection duration as the most appropriate injection pattern, and the subjective method determined this as a $-19°$ CA pre-injection starting angle, 20% pre-injection rate, and $1°$ CA pre-injection duration. The main contribution of the present work consists in emphasizing the differences between the results obtained using various methods for the determination of the criteria weights, showing the advantage of subjectivism over objectivism. Based on the overall results, the subjective method is recommended since the criteria weights are defined by experts in the field. In fact, in practical applications, subjective methods are more frequently employed than objective ones. Objective methods are only recommended when the objectivity of the research is too important or when there is no agreement between the weights proposed by the experts.

Author Contributions: Conceptualization, M.I.L. and C.G.R.; methodology, M.I.L. and C.G.R.; software, M.I.L. and C.G.R.; validation, M.I.L. and C.G.R.; formal analysis, M.I.L. and L.C.-S.; investigation, M.I.L., C.G.R., and L.C.-S.; resources, M.I.L. and C.G.R.; writing—original draft preparation, M.I.L. and L.C.-S.; writing—review and editing, M.I.L. and L.C.-S. All authors have read and agreed to the published version of the manuscript.

Funding: This research received no external funding.

Acknowledgments: The authors would like to express their gratitude to Norplan Engineering S.L. and recommend the courses "CFD with OpenFOAM" and "C ++ applied to OpenFOAM" available at www.technicalcourses.net.

Conflicts of Interest: The authors declare no conflict of interest.

Nomenclature

AI	Adequacy index
C	Measure of the conflict
E	Entropy value
i	Case
j	Criterion
m	Number of cases analyzed
n	Number of criteria
p	Normalized data
σ	Standard deviation
σ^2	Statistical variance
V	Value in per unit basis
w	Criterion weight
X	Value

Abbreviations

ATDC	After top dead center
CRITIC	Criteria importance through intercriteria correlation
CA	Crank angle
CFD	Computational fluid dynamics
CO	Carbon monoxide
CO_2	Carbon dioxide
D	Pre-injection duration
HC	Hydrocarbons
IMO	International Maritime Organization

References

1. Sinay, J.; Puskar, M.; Kopas, M. Reduction of the NOx emissions in vehicle diesel engine in order to fulfill future rules concerning emissions released into air. *Sci. Total Environ.* **2018**, *624*, 1421–1428. [CrossRef] [PubMed]
2. Sui, S.; de Vos, P.; Stapersma, D.; Visser, K.; Ding, Y. Fuel consumption and emissions of ocean-going cargo ship with hybrid propulsion and different fuels over voyage. *J. Mar. Sci. Eng.* **2020**, *8*, 588. [CrossRef]
3. Perez, J.R.; Reusser, C.A. Optimization of the emissions profile of a marine propulsion system using a shaft generator with optimum tracking-based control scheme. *J. Mar. Sci. Eng.* **2020**, *8*, 221. [CrossRef]
4. Shen, H.; Zhang, J.; Yang, B.; Jia, B. Development of a marine two-stroke diesel engine MVEM with in-cylinder pressure trace predictive capability and a novel compressor model. *J. Mar. Sci. Eng.* **2020**, *8*, 204. [CrossRef]
5. Sencic, T.; Mrzljak, V.; Blecich, P.; Bonefacic, I. 2D CFD simulation of water injection strategies in a large marine engine. *J. Mar. Sci. Eng.* **2019**, *7*, 296. [CrossRef]
6. Puskar, M.; Kopas, M.; Sabadka, D.; Kliment, M.; Marieta Soltesova, M. Reduction of the gaseous emissions in the marine diesel engine using biodiesel mixtures. *J. Mar. Sci. Eng.* **2020**, *8*, 330. [CrossRef]
7. Pistek, V.; Kucera, P.; Fomin, O.; Alyona, A. Effective mistuning identification method of integrated bladed discs of marine engine turbochargers. *J. Mar. Sci. Eng.* **2020**, *8*, 379. [CrossRef]
8. Witkowski, K. Research of the effectiveness of selected methods of reducing toxic exhaust emissions of marine diesel engines. *J. Mar. Sci. Eng.* **2020**, *8*, 452. [CrossRef]
9. Seddiek, I.S.; El Gohary, M.M.; Ammar, N.R. The hydrogen-fuelled internal combustion engines for marine applications with a case study. *Brodogradnja* **2015**, *66*, 23–38.
10. El Gohary, M.M.; Ammar, N.R.; Seddiek, I.S. Steam and SOFC based reforming options of PEM fuel cells for marine applications. *Brodogradnja* **2015**, *66*, 61–76.
11. International Maritime Organization (IMO). *Third IMO GHG Study 2014. Executive Summary and Final Report, MEPC 67/6/INF.3, June 2014*; International Maritime Organization: London, UK, 2014.
12. Winnes, H.; Fridell, E.; Moldanova, J. Effects of marine exhaust gas scrubbers on gas and particle emissions. *J. Mar. Sci. Eng.* **2020**, *8*, 299. [CrossRef]
13. Nehmer, D.A.; Reitz, R.D. *Measurement of Effect of Injection Rate and Split Injections on Diesel Engine, Soot and NOx Emissions*; SAE Technical Paper 940668; SAE International: Warrendale, PA, USA, 1998. [CrossRef]
14. Han, Z.; Uludogan, A.; Hampson, G.; Reitz, R. *Mechanism of Soot and NOx Emission Reduction Using Multiple-Injection in a Diesel Engine*; SAE Technical Paper 960633; SAE International: Warrendale, PA, USA, 1996. [CrossRef]
15. Ikegami, M.; Nakatani, K.; Tanaka, S.; Yamane, K. *Fuel Injection Rate Shaping and Its Effect on Exhaust Emissions in A Direct-Injection Diesel Engine Using a Spool Acceleration Type Injection System*; SAE Techical Paper 970347; SAE International: Warrendale, PA, USA, 1997. [CrossRef]
16. Carlucci, A.P.; Ficarella, A.; Laforgia, D. Control of the combustion behaviour in a diesel engine using early injection and gas addition. *Appl. Therm. Eng.* **2006**, *26*, 2279–2286. [CrossRef]
17. Mohan, B.; Yang, W.; Chou, S.K. Fuel injection strategies for performance improvement and emissions reduction in compression ignition engines. A review. *Renew. Sustain. Energy Rev.* **2013**, *28*, 664–676. [CrossRef]
18. Chen, S.K. *Simultaneous Reduction of NOx and Particulate Emissions by Using Multiple Injections in A Small Diesel Engine*; SAE Technical Paper 2000-01-3084; SAE International: Warrendale, PA, USA, 2000. [CrossRef]
19. Fang, T.; Coverdill, R.; Lee, C.F.; White, R.A. Effects of injection angles on combustion process using multiple injection strategies in an HSDI diesel engine. *Fuel* **2008**, *87*, 3232–3239. [CrossRef]
20. Chen, G.; Wang, K.; Yang, J.; Zhang, W.; Xu, J.; Luo, J. Effects of fuel injection strategy with EGR on diesel engine combustion process and CDPF regeneration performance. *Chin. Intern. Combust. Engine Eng.* **2017**, *38*, 131–141. [CrossRef]
21. Cha, J.; Yang, S.H.; Naser, N.; Ichim, A.I.; Chung, S.H. *High Pressure and Split Injection Strategies for Fuel Efficiency and Emissions in Diesel Engine*; SAE Technical Paper 127084; SAE International: Warrendale, PA, USA, 2015. [CrossRef]
22. Shi, J.; Wang, T.; Zhao, Z.; Yang, T.; Zhang, Z. Experimental study of injection parameters on the performance of a diesel engine with Fischer–Tropsch fuel synthesized from coal. *Energies* **2018**, *11*, 3280. [CrossRef]

23. Mardani, A.; Jusoh, A.; Zavadskas, E.K.; Cavallaro, F.; Khalifah, Z. Sustainable and renewable energy–an overview of the application of multiple criteria decision-making techniques and approaches. *Sustainability* **2015**, *7*, 13947–13984. [CrossRef]

24. Vinogradova, I.; Podvezko, V.; Zavadskas, E.K. The recalculation of the weights of criteria in MCDM methods using the Bayes approach. *Symmetry* **2018**, *10*, 205. [CrossRef]

25. Deng, H.; Yeh, C.H.; Willis, R.J. Inter-company comparison using modified TOPSIS with objective weights. *Comput. Oper Res.* **2000**, *27*, 963–973. [CrossRef]

26. Diakoulaki, D.; Mavrotas, G.; Papayannakis, L. Determining objective weights in multiple criteria problems: The critic method. *Comput. Oper. Res.* **1995**, *22*, 763–770. [CrossRef]

27. Jahan, A.; Mustapha, F.; Sapuan, S.M.; Ismail, M.Y.; Bahraminasab, M. A framework for weighting of criteria in ranking stage of material selection process. *Int. J. Adv. Manuf. Tech.* **2012**, *58*, 411–420. [CrossRef]

28. Lamas, M.I.; Rodríguez, C.G.; Rebollido, J.M. Numerical model to study the valve overlap period in the Wärtsilä 6L46 four-stroke marine engine. *Pol. Marit. Res.* **2012**, *1*, 31–37. [CrossRef]

29. Lamas, M.I.; Rodríguez, C.G. Numerical model to study the combustion process and emissions in the Wärtsilä 6L 46 four-stroke marine engine. *Pol. Marit. Res.* **2013**, *20*, 61–66. [CrossRef]

30. Lamas, M.I.; Rodriguez, C.G.; Rodriguez, J.D.; Telmo, J. Internal modifications to reduce pollutant emissions from marine engines. A numerical approach. *Int. J. Nav. Archit. Mar. Eng.* **2013**, *5*, 493–501. [CrossRef]

31. Lamas, M.I.; Rodriguez, C.G.; Telmo, J.; Rodriguez, J.D. Numerical analysis emissions from marine engines using alternative fuels. *Pol. Marit. Res.* **2015**, *22*, 48–52. [CrossRef]

32. Lamas, M.I.; Rodríguez, J.d.; Castro-Santos, L.; Carral, L.M. Effect of multiple injection strategies on emissions and performance in the Wärtsilä 6L 46 marine engine. A numerical approach. *J. Clean. Prod.* **2019**, *206*, 1–10. [CrossRef]

33. Galdo, M.I.L.; Castro-Santos, L.; Carlos, G.R.V. Selection of an appropriate pre-injection pattern in a marine diesel engine through a multiple-criteria decision making approach. *Appl. Sci.* **2020**, *10*, 2482. [CrossRef]

34. Galdo, M.I.L.; Castro-Santos, L.; Vidal, C.G.R. Numerical analysis of NOx reduction using ammonia injection and comparison with water injection. *J. Mar. Sci. Eng.* **2020**, *8*, 109. [CrossRef]

35. Ra, Y.; Reitz, R. A reduced chemical kinetic model for IC engine combustion simulations with primary reference fuels. *Combust. Flame* **2008**, *155*, 713–738. [CrossRef]

36. Yang, H.; Krishnan, S.R.; Srinivasan, K.K.; Midkiff, K.C. Modeling of NOx emissions using a superextended Zeldovich mechanism. In Proceedings of the ICEF03 2003 Fall Technical Conference of the ASME Internal Combustion Engine Division, Erie, PA, USA, 7–10 September 2003.

37. Miller, J.A.; Glarborg, P. Modeling the formation of N2O and NO2 in the thermal DeNOx process. *Springer Ser. Chem. Phys.* **1996**, *61*, 318–333.

38. Dukowicz, J.K. A particle-fluid numerical model for liquid sprays. *J. Comput. Phys.* **1980**, *35*, 229–253. [CrossRef]

39. Ricart, L.M.; Xin, J.; Bower, G.R.; Reitz, R.D. *In-Cylinder Measurement and Modeling of Liquid Fuel Spray Penetration in A Heavy-Duty Diesel Engine*; SAE Technical Paper 971591; SAE International: Warrendale, PA, USA, 1997. [CrossRef]

40. Gao, J.; Kuo, T.W. Toward the accurate prediction of soot in engine applications. *Int. J. Engine Res.* **2018**, *20*, 706–717. [CrossRef]

41. Lamas, M.I.; Rodriguez, C.G. NOx reduction in diesel-hydrogen engines using different strategies of ammonia injection. *Energies* **2019**, *12*, 1255. [CrossRef]

42. Lamas, M.I.; Rodriguez, C.G. Numerical model to analyze NOx reduction by ammonia injection in diesel-hydrogen engines. *Int. J. Hydrogen Energy* **2017**, *42*, 26132–26141. [CrossRef]

Publisher's Note: MDPI stays neutral with regard to jurisdictional claims in published maps and institutional affiliations.

MDPI
St. Alban-Anlage 66
4052 Basel
Switzerland
Tel. +41 61 683 77 34
Fax +41 61 302 89 18
www.mdpi.com

Journal of Marine Science and Engineering Editorial Office
E-mail: jmse@mdpi.com
www.mdpi.com/journal/jmse